Stearoyl-CoA Desaturase Genes in Lipid Metabolism

James M. Ntambi
Editor

Stearoyl-CoA Desaturase Genes in Lipid Metabolism

Springer

Editor
James M. Ntambi, Ph.D.
Departments of Biochemistry
and Nutritional Sciences
University of Wisconsin-Madison
Madison, WI, USA

ISBN 978-1-4614-7968-0 ISBN 978-1-4614-7969-7 (eBook)
DOI 10.1007/978-1-4614-7969-7
Springer New York Heidelberg Dordrecht London

Library of Congress Control Number: 2013943586

Printed on acid-free paper

Springer is part of Springer Science+Business Media (www.springer.com)

Preface

This is the first edition of this book that provides the reader with current information on the stearoyl-CoA desaturase gene family that encode stearoyl-CoA desaturase, a previously unknown enzyme and its role in metabolism in different organisms. Whereas today many genes and gene families are defined and analyzed bioinformatically, early events using the murine 3T3-L1 system provided the foothold that have enabled our current understanding of what has become one of the most interesting and important loci in cellular metabolism. In addition, the reader will learn about the important and widely diversified field of fatty acids and their health implications in humans. However since the precise role of fatty acids in the etiology of the various metabolic disorders has yet to be delineated it is not the intention of the book to present a unified view on the healthy implications of monounsaturated fatty acids or provide guidelines for fatty acid consumption.

Work over the last 20+ years has focused on identifying and analyzing numerous SCD isoforms in a variety of species that are in turn controlled by multiple signaling systems. Given the diverse role of monounsaturated fatty acids in metabolic processes it is not surprising that multiple genes exist and that multiple regulatory elements control isoform expression. In addition, tissue-specific dietary control of isoform expression occurs via a series of complex signal transduction schemes making SCD one of the most highly studied gene families. Conditional alleles and corresponding tissue-specific knockout models for many of the murine SCD genes have provided a wealth of information on not only tissue-specific fatty acid metabolism but also the key transcription factors that regulate SCD expression under a variety of metabolic and genetic backgrounds. Our studies using mice with whole-body and tissue-specific deletions of one of the mouse SCD isoforms (SCD1), together with several human studies have all agreed that the expression of the SCD1 gene isoform represents a key step in partitioning of lipids between storage and oxidation. High SCD1 expression (high oleate levels) favors fat storage leading to obesity while reduced SCD1 (low oleate levels) expression favors fat burning and leanness. However, as we understand more about the detrimental roles of cellular saturated fatty acids, it is becoming clear that unqualified complete inhibition of SCD1 activity, whether systemic or restricted to specific organ systems, may not come without a plethora of undesirable side effects, including an increased propensity to inflammatory disease. A proper ratio of saturated/monounsaturated fat may be required for normal health and physiological function. Nevertheless, novel insights gained into the multifaceted roles of these intrinsic loci in cellular metabolism may yet lead to new models for pharmacological interventions for the treatment of metabolic disease.

I am very grateful to many people who helped me make possible the creation of the first edition on the role of SCD genes in metabolism. First of all I would like to thank my early research mentors, Dr. Paul Englund and M. Daniel Lane both at the Johns Hopkins University School of Medicine who introduced me to research in biochemistry and molecular biology. David Bernlohr presently at the University of Minnesota handed me the PAL122 cDNA clone that I used as a probe to isolate the first murine SCD gene (mSCD1). I would like to express my sincere appreciation to all the contributors of the diverse chapters on SCD genes for their cooperation and excellent work. Without their participation, this initial project would not have been possible. I would like to thank Anna Marsicek who has edited the various chapters of the book. I would also like to thank Springer for their assistance and support during the course of this project. They saw the potential for this project and helped me in many ways to craft this book. Finally I wish to thank my family for their encouragement, support, and patience over the past many years.

Madison, WI, USA James M. Ntambi, Ph.D.

Contents

Contributors

Silvia Graciela Altabe Facultad de Ciencias Bioquímicas y Farmacéuticas, Departamento de Microbiología, Instituto de Biología Molecular y Celular de Rosario (IBR), Consejo Nacional de Investigaciones Científicas y Técnicas, Universidad Nacional de Rosario, Rosario, Argentina

Pappas Apostolos Skin Research Center, Johnson & Johnson, CPPW, Skillman, NJ, USA

L. Bernard INRA, UMR1213 Herbivores, Saint-Genès-Champanelle, France

Clermont Université, VetAgro Sup, UMR Herbivores, Clermont-Ferrand, France

David A. Bernlohr Department of Biochemistry, Molecular Biology and Biophysics, University of Minnesota, Minneapolis, MN, USA

Laura M. Bond Department of Biochemistry, University of Wisconsin-Madison, Madison, WI, USA

Jennifer L. Cantley Departments of Internal Medicine and Cellular and Molecular Physiology, Yale School of Medicine, New Haven, CT, USA

Y. Chilliard INRA, UMR1213 Herbivores, Saint-Genès-Champanelle, France

Clermont Université, VetAgro Sup, UMR Herbivores, Clermont-Ferrand, France

Paul Cohen Cardiovascular Division, Brigham and Women's Hospital, Boston, MA, USA

Michael P. Czech Program in Molecular Medicine, University of Massachusetts Medical School, Worcester, MA, USA

Agnieszka Dobrzyn Laboratory of Cell Signaling and Metabolic Disorders, Nencki Institute of Experimental Biology, Warsaw, Poland

Pawel Dobrzyn Laboratory of Molecular and Medical Biochemistry, Nencki Institute of Experimental Biology, Warsaw, Poland

Matthew W. Hulver Department of Human Nutrition, Foods, and Exercise, Virginia Tech, Blacksburg, VA, USA

Thilak Jayalath Department of Medicine, Faculty of Medicine, University of Peradeniya, Peradeniya, Sri Lanka

Nishan Sudheera Kalupahana Department of Physiology, Faculty of Medicine, University of Peradeniya, Peradeniya, Sri Lanka

M. Daniel Lane Department of Biological Chemistry, Johns Hopkins University School of Medicine, Baltimore, MD, USA

C. Leroux INRA, UMR1213 Herbivores, Saint-Genès-Champanelle, France

Clermont Université, VetAgro Sup, UMR Herbivores, Clermont-Ferrand, France

María Cecilia Mansilla Facultad de Ciencias Bioquímicas y Farmacéuticas, Departamento de Microbiología, Instituto de Biología Molecular y Celular de Rosario (IBR), Consejo Nacional de Investigaciones Científicas y Técnicas, Universidad Nacional de Rosario, Rosario, Argentina

Diego de Mendoza Facultad de Ciencias Bioquímicas y Farmacéuticas, Departamento de Microbiología, Instituto de Biología Molecular y Celular de Rosario (IBR), Consejo Nacional de Investigaciones Científicas y Técnicas, Universidad Nacional de Rosario, Rosario, Argentina

Makoto Miyazaki Division of Renal Diseases and Hypertension, University of Colorado Denver, Aurora, CO, USA

Naima Moustaid-Moussa Nutritional Sciences, Texas Tech University, Lubbock, TX, USA

James M. Ntambi Departments of Biochemistry and Nutritional Sciences, University of Wisconsin-Madison, Madison, WI, USA

Lucas M. O'Neill Department of Biochemistry, University of Wisconsin-Madison, Madison, WI, USA

Satish Parimoo Johnson & Johnson Skin Biology TRC, Skillman, NJ, USA

Shaikh M. Rahman Nutritional Sciences, Texas Tech University, Lubbock, TX, USA

Stephen B. Smith Department of Animal Science, Texas A&M University, College Station, TX, USA

Joseph Stevens Department of Human Nutrition, Foods, and Exercise, Virginia Tech, Blacksburg, VA, USA

Tabitha C. Ting Division of Renal Diseases and Hypertension, University of Colorado Denver, Aurora, CO, USA

Shu Wang Nutritional Sciences, Texas Tech University, Lubbock, TX, USA

Jennifer L. Watts School of Molecular Biosciences, Washington State University, Pullman, WA, USA

Chapter 1
Early Studies on Role of Stearoyl-CoA Desaturase During Preadipocyte Differentiation

David A. Bernlohr and M. Daniel Lane

Background

Having extensively investigated the catalytic mechanism of action of acetyl-CoA carboxylase (ACC) (Polakis et al. 1972, 1974; Guchhait et al. 1971, 1974a, b; Moss and Lane 1971) and its allosteric regulation by citrate (Moss and Lane 1972; Gregolin et al. 1966; Ryder et al. 1967; Beaty and Lane 1983a, b), in the late 1970s and early 1980s, our lab decided to determine how expression of ACC is controlled. To conduct such studies we needed a lipogenic cell line that could be cultured for extended periods of time (i.e., greater than a week), during which gene expression and protein translation could be assessed. About this time Howard Green and colleagues, who had recently moved from the New York University Medical School to the Massachusetts Institute of Technology (MIT), had established several preadipocyte cell lines—notably the 3T3-L1 and 3T3F442 cell lines (Green and Kehinde 1974, 1975, 1976)—for the study of adipocyte biology. These lines could be induced to differentiate into cells with the phenotype of adipocytes including high rates of lipogenesis (Mackall et al. 1976; Mackall and Lane 1977; Coleman et al. 1978; Student et al. 1980) and responsiveness to insulin (Reed et al. 1977, 1981; Reed and Lane 1980). Since then the 3T3-L1 preadipocyte line has become the "gold standard" for studies on adipocyte differentiation. We (Student et al. 1980) and others (Lai et al. 1982) developed protocols to induce differentiation and demonstrated that expression of fatty acid synthase (FAS), ACC, and numerous other lipogenic proteins paralleled the acquisition of adipocyte morphology. These early observations

D.A. Bernlohr
Department of Biochemistry, Molecular Biology and Biophysics,
University of Minnesota, Minneapolis, MN 55455, USA

M.D. Lane (✉)
Department of Biological Chemistry, Johns Hopkins University School of Medicine,
725 N. Wolfe St. 512 WBSB, Baltimore, MD 21205, USA
e-mail: dlane@jhmi.edu

J.M. Ntambi (ed.), *Stearoyl-CoA Desaturase Genes in Lipid Metabolism*,
DOI 10.1007/978-1-4614-7969-7_1, © Springer Science+Business Media New York 2013

verified the system and established the primary parameters linked to the utilization of the 3T3-L1 model including (1) an accumulation of cytoplasmic fat as revealed by Oil Red O staining (Mackall et al. 1976), (2) a dramatic increase in the rate of fatty acid synthesis (Mackall et al. 1976; Student et al. 1980), (3) a coordinate increase in the enzymatic activities of ACC (Mackall et al. 1976), FAS, and complex lipids synthesis (triacylglycerol and phospholipids) (Coleman et al. 1978), and (4) hormonally regulated lipid metabolism including insulin-stimulated glucose uptake and lipogenesis and catecholamine-stimulated lipolysis (Reed et al. 1977, 1981; Reed and Lane 1980).

The most widely used procedure for induction of adipocyte differentiation (Student et al. 1980) involves: allowing 3T3-L1 preadipocytes to proliferate until growth arrest at confluence (G_1 stage of the cell cycle) and after two additional days in G_1, treating the cells with an induction "cocktail" consisting of cAMP (or methylisobutylxanthine), dexamethasone, and a high level of insulin (or a low level of IGF1). Following withdrawal of cAMP and dexamethasone the induced cells synchronously reenter the cell cycle, undergo ~2 rounds of division (mitotic clonal expansion) (Tang et al. 2003), and return to the state of G_1 growth arrest. At this point expression of the lipogenic proteins/enzymes of the adipocyte differentiation program occurs synchronously (Tang et al. 2003; Tang and Lane 2012). The use of such established preadipocyte lines have been indispensable for the identification and characterization of the key steps in the differentiation program. A large body of evidence (Tang and Lane 2012) shows that these preadipocyte models faithfully recapitulate differentiation of mouse embryonic fibroblasts (MEFs) in cell culture (Tang et al. 2003; Zhang et al. 2003). Likewise, the authenticity of model systems has been validated in vivo by demonstrating that these preadipocyte lines give rise to normal adipose tissue when implanted subcutaneously into athymic mice without exogenous inducers (Mandrup et al. 1997).

Gene Expression During Adipocyte Differentiation

The utilization of the 3T3-L1 cell line in culture represents a useful model system for investigating the mechanisms of cellular differentiation (Tang and Lane 2012). Under appropriate conditions, 3T3-L1 preadipocytes, which initially resemble fibroblast, differentiate readily into adipocytes in cell culture (Green and Kehinde 1975; Mackall et al. 1976). The differentiation process is accompanied by a dramatic increase in the activities of the enzymes associated with de novo lipogenesis (Mackall et al. 1976; Mackall and Lane 1977; Coleman et al. 1978; Student et al. 1980) and triglyceride mobilization (Hirsch and Rosen 1984), as well as an increased sensitivity to lipogenic and lipolytic hormones (Kohanski et al. 1986; Hirsch and Rosen 1984; Watkins et al. 1982). The cells accumulate triglyceride (Green and Kehinde 1975; Mackall et al. 1976) and acquire the morphological characteristics of adipocytes (Novikoff et al. 1980) isolated from normal adipose tissue. Strong evidence indicated that for some of the lipogenic enzymes (e.g., FAS), the increases in activity during differentiation arise from increased rates of enzyme synthesis (Student et al. 1980) suggesting increased mRNA abundance. Moreover, it was subsequently shown that changes in mRNA expression closely correlated with the increased activity and abundance of the proteins that were differentially expressed (Mandrup and Lane 1997; MacDougald and Lane 1995; Bernlohr et al. 1985a, b; Angus et al. 1981). Therefore, it became evident that to understand the genetic mechanisms responsible for these changes in phenotype, it would be necessary to identify genes whose expression is activated during differentiation of 3T3-L1 preadipocytes into adipocytes. This ultimately led to the discovery that one of the key regulatory genes in this process was stearoyl-CoA desaturase (Ntambi et al. 1988).

Identification of mRNAs Whose Expression Is Induced During Differentiation of 3T3-L1 Preadipocytes into Adipocytes

An initial approach towards determining the mechanism of increased expression of key lipogenic and lipolytic enzymes was in vitro mRNA translation using ^{35}S-methionine followed by immunoprecipitation (Bernlohr et al. 1985a). In the early 1980s antibodies to several lipogenic enzymes were available (ACC, FAS, malic enzyme, pyruvate carboxylase, etc.) and by using in vitro translation of mRNA isolated from either 3T3-L1 preadipocytes or adipocytes followed by immunoprecipitation, one could assess the relative expression of certain mRNAs. William Angus, a PhD student at the time, utilized such an approach and demonstrated increased mRNA levels for pyruvate carboxylase (Angus et al. 1981) and acetyl-CoA carboxylase (MacDougald and Lane 1995). These findings suggested that differentiation of 3T3-L1 preadipocytes in culture likely involved the activation of the entire adipogenic gene expression program. While technically simple, the method was limited by the availability of mono-specific antibodies and was time-consuming and relatively inefficient. This consideration drove the race towards cloning differentially expressed mRNAs (MacDougald and Lane 1995; Cornelius et al. 1994).

A second approach towards analysis of gene expression during differentiation involved the isolation and characterization of cDNAs corresponding to genes that are activated during differentiation. Indeed, during the 1980s the identification of cDNA clones corresponding to differentially expressed genes occupied the focus of a number of laboratories. Toward that end, we constructed and characterized a cDNA library from mRNAs expressed by differentiated 3T3-L1 adipocytes (Bernlohr et al. 1984). A similar approach was taken by Spiegelman and colleagues (Ross et al. 1990) with 3T3-F442A cells, a related mouse preadipocyte cell line. Once a cDNA library was developed a dual-label analysis method was utilized to identify such clones that corresponded to mRNAs that were adipose-specific (Bernlohr et al. 1984). In this method, ^{3}H-labeled cDNA was synthesized using mRNA from preadipocytes and mixed with ^{32}P-labled cDNA synthesized from differentiated adipocytes. The mixed cDNA population was used in colony hybridization studies to identify cDNAs that contained differentially expressed mRNA

(Bernlohr et al. 1984). Using such a dual-label screening method we identified several cDNA clones that correspond to mRNAs that increase in abundance upon differentiation. Many of these clones (e.g., pAL122, pAL421, pAL422) represent mRNAs that are detectable primarily in differentiated cells. From such a screen came the identification of a cDNA, pAL122, corresponding to stearoyl-CoA desaturase (Ntambi et al. 1988).

Identification of Stearoyl-CoA Desaturase as a Differentiation-Induced Enzyme During Adipocyte Differentiation

Differential screening led to the identification of a family of adipocyte mRNAs (Ntambi et al. 1988) that included: (1) The pAL422 mRNA (later identified as encoding FABP4, the adipocyte fatty acid-binding protein also known as aP2) of which numerous clones were obtained owing to its abundant expression in differentiated 3T3-L1 adipocytes (Bernlohr et al. 1984, 1985b) and (2) The pAL122 mRNA was later identified as encoding the stearoyl-CoA desaturase one mRNA (Ntambi et al. 1988).

It should be noted that at the time of cloning these cDNAs their identities were unknown. The utilization of databases and searchable sequence repositories was in its infancy and even DNA sequencing was time-consuming and laborious. Indeed, since the original cDNA cloning was accomplished using the GC-tailing method, sequencing of such cloned inserts often proved problematic and required many sequencing runs and overlapping replicates to develop reproducible trustworthy results. The pAL422 cDNA was identified from a protein sequence data base as a putative fatty acid-binding protein, aP2, that exhibited ~60 % amino acid sequence identity to mP2, a myelin fatty acid-binding protein (Bernlohr et al. 1985b). James Ntambi undertook the identification of pAL122. It should be noted that Ntambi had little to go on as the pAL122 cDNA nucleotide sequence contained largely 3′-untranslated sequence and only a small amount of coding information upstream of the translational stop site (Ntambi et al. 1988). Northern blotting had revealed that the pAL122 mRNA sequence was ~5 kb, suggesting that the pAL 122 cDNA clone corresponded to less than 30 % of the complete cDNA.

In an act of heroic proportion James Ntambi began a manual "nucleotide-by-nucleotide" sequence comparison of the pAL122 mRNA with any new sequences that appeared in the literature. Remarkably, in 1986 Ntambi found a match with the nearly identical nucleotide sequence of the rat liver SCD1 from Philip Strittmater's laboratory that appeared in the Journal of Biological Chemistry in 1986 (Thiede et al. 1986). Thus, James' persistence paid off with the identification of pAL122 as murine stearoyl-CoA desaturase1 (Ntambi et al. 1988). Armed with the knowledge that pAL122 corresponded to SCD1, Ntambi utilized the pAL122 1.4-kb cDNA to isolate a 4.9-kb full-length cDNA clone and determined the complete mouse SCD1 sequence (Ntambi et al. 1988).

Isolation and Structure of the SCD1 Gene

Stearoyl-CoA desaturase is the key enzyme in the synthesis of unsaturated fatty acids and is under complex transcriptional and posttranscriptional regulation. To develop tools for the molecular analysis of SCD1, Ntambi, along with Susan Buhrow, Klaus Kaestner, Eric Sibley, and Robert Christy in our lab, was able to isolate genomic clones corresponding to murine SCD1 and to identify binding sites in the 5′-flanking regions of the gene that are involved in the coordinate differentiation-induced expression in 3T3-L1 preadipocytes (Ntambi et al. 1988; Christy et al. 1989). From these studies emerged the structural organization of the murine SCD1 gene and comparison to the murine 3T3-L1 SCD1 cDNA (Fig. 1.1). At the same time, genomic clones corresponding to pAL422 (Kaestner et al. 1989) had been obtained and armed with sequence information in the 5′ upstream region from both genes, control elements that may provide clues towards coordinate regulation during differentiation were identified.

Identification and Cloning of a Second Stearoyl-CoA Desaturase, SCD2

Southern blotting of mouse DNA with the SCD1 cDNA suggested that there was a second similar sequence in the genome, and therefore, potentially a second SCD gene. Armed with the SCD1 cDNA, Klaus Kaestner and James Ntambi screened the 3T3-L1 cDNA library and identified a cDNA clone that encoded a second stearoyl-CoA desaturase, SCD2, from 3T3-Ll adipocytes. This second SCD gene was

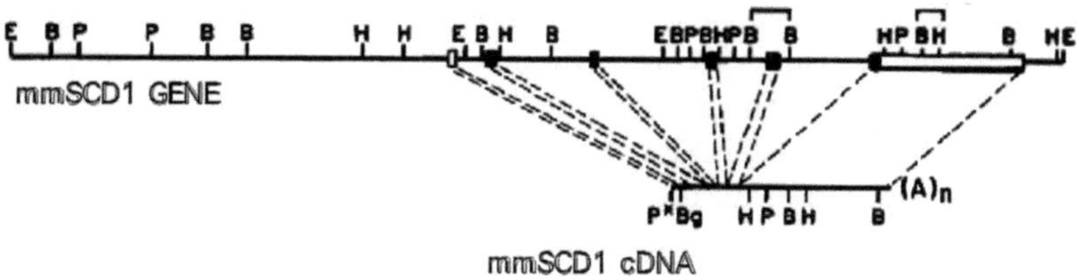

Fig. 1.1 Structural organization of the Mus musculus SCD1 gene and cDNA. The *top scale* indicates nucleotides in kilobases. The *middle scale* indicates the organization of the murine SCD1 gene where *boxes* indicate the exonic regions with open regions representing noncoding and filled regions coding sequence. The *lower scale* indicates the organization of the SCD1 cDNA isolated from 3T3-L1 adipocytes. *Single letters* represent restriction sites; *B* BamH1, *E* EcoR1, *H* HindIII, *P* Pst1, *B* BglII, *kb* kilobase pair. Adapted from (Ntambi et al. 1988)

expressed at tenfold higher levels in 3T3-L1 adipocytes compared to the preadipocytes. The full-length SCD2 cDNA predicted a protein of 358 amino acids with 87 % amino acid sequence identity to murine SCDl (Christy et al. 1989). Utilizing the SCD2 cDNA the murine SCD2 gene was cloned and analyzed to assess its regulation and expression patterns (Kaestner et al. 1989).

SCD2 mRNA is expressed constitutively at a high level in brain and its expression in kidney, adipose, and lung tissue is increased greatly by shifting mice from a diet containing unsaturated fatty acids to a diet devoid of fat (Landschulz et al. 1994). Presumably, the major function of SCD2 in brain is to supply unsaturated fatty acids constitutively for the synthesis of membrane phospholipids, particularly for myelination. In high carbohydrate, fat-free diets, fatty acids are synthesized de novo and are desaturated for membrane phospholipid synthesis and stored as triacylglycerol in the monounsaturated form. Strikingly, SCD2 mRNA is absent from liver regardless of the dietary treatment (Christy et al. 1991). The difference in tissue-specific expression and dietary control of SCD1 and SCD2 expression suggests distinct control mechanisms for the expression of each of the two genes. This in turn led to an analysis of the structure and regulation of SCD2.

The SCD2 gene spans approximately 15 kb pairs and consists of six exons and five introns, with intron–exon junctions similar to those of SCDl (Kaestner et al. 1989; Christy et al. 1989). As determined by primer extension analysis the start site of transcription maps 300 nucleotides upstream of the initiator methionine codon. Unlike the SCDl gene, SCD2 lacks a typical "TATA" box in the 5′ flanking region but has two "CCAAT" boxes at positions −90 and −135 relative to the transcription initiation site. The SCD2 promoter contains a 140-base pair sequence (located between nucleotides −54 and −201) that possesses 77 % sequence identity to a region (located between nucleotides −472 and −325) in the SCD1 promoter. There is a GC-rich sequence in the SCD2 promoter (at nucleotide −175) similar to the binding site for the nuclear transcription factor Sp1 as well as an element with homology to the core consensus sequence of the glucocorticoid receptor positioned a −500 and a potential CCAAT box/enhancer binding protein sequence at position −540. Although the promoter of the two genes differs markedly, there is one region with high sequence identity. A nucleotide sequence comparison of the 5′ flanking regions of the SCD2 and SCDl genes revealed a stretch of 146 bp of striking sequence homology. The sequence between −201 and −54 bp in the SCD2 gene is 77 % identical to the sequence between −472 and −325 in the SCDl gene, including conservation of the two CCAAT boxes. This marked sequence homology suggests that duplication occurred during evolution of the SCD gene family. Presumably after gene duplication, either a sequence insertion or deletion occurred to yield the different promoters for SCDl and SCD2. The SCD2 promoter lacks the potential cAMP core regulatory sequence present in the SCDl promoter, but contains a potential binding site for the CCAAT/enhancer binding protein (Kaestner et al. 1989).

C/EBPα as Transcriptional Inducer of Gene Expression During Preadipocyte Differentiation

In view of the coordinate nature of the increase in expression of the 422/aP2 and SCD1 genes (Bernlohr et al. 1985a; Kaestner et al. 1989), the possibility was considered that a common transcription factor might activate expression of this group of adipocyte genes. Thus, we employed both DNase I footprinting and gel retardation analysis to show that nuclear factors present in differentiated—but not undifferentiated—3T3-L1 adipocytes bind to regions within the 5′-flanking regions of the promoters of the 422/aP2 and SCD1 genes (Christy et al. 1989). Subsequently we searched for and found that a factor present in nuclear extracts of differentiated—but not undifferentiated—3T3-L1 cells "foot-printed" the similar sites in the 5′-flanking regions of several differentiate-induced genes, notably the 422/aP2 and SCD1 genes (Christy et al. 1991). It was also shown that the differentiation-induced nuclear factor was CCAAT/enhancer binding protein alpha (C/EBPα) (Christy et al. 1991), a DNA-binding protein first isolated from rat liver (Landschulz et al. 1988). Bacterially expressed recombinant C/EBP binds to the same site at which the differentiation-specific nuclear factor interacts within the promoter of each gene. Northern analysis with RNA from 3T3-L1 cells shows that C/EBPα mRNA abundance increases markedly during differentiation. Co-transfection of 422(aP2)-chloramphenicol acetyl transferase or SCD1-chloramphenicol acetyl transferase constructs with a C/EBPα expression vector into 3T3-L1 preadipocytes revealed that C/EBPα trans-activated both the 422(aP2) and SCDl gene promoters, thus implicating C/EBPα in differentiation-specific gene expression. Finally, transient co-transfection studies with a C/EBPα expression vector showed that C/EBPα could serve as trans-activator of both the 422(aP2) and SCDl gene promoters (Fig. 1.2). Later we found that C/EBPα also served as a transcriptional activator of the leptin (Hwang et al. 1996, 1997; MacDougald et al. 1995) and GLUT4 genes (Kaestner et al. 1990) during the differentiation process (Fig. 1.2).

As C/EBPα appeared to serve as a pleiotropic activator of gene expression during preadipocyte differentiation (Kaestner et al. 1989, 1990; Hwang et al. 1996, 1997; MacDougald et al. 1995; Lin and Lane 1992), it was of interest to determine how the C/EBPα gene itself is controlled during differentiation. Consistent with its suspected role as a pleiotropic activator, we found that the increase in the C/EBPα message level during differentiation was due to an increase in the rate of transcription of the gene (Christy et al. 1991). We then set out to determine the structure of the C/EBPα gene and the nucleotide sequence of its 5′ flanking region. DNase I footprinting of the 5′ flanking region by nuclear factors from differentiated and undifferentiated 3T3-L1 cells revealed two sites of differential binding (Christy et al. 1991). One site in the gene promoter between nucleotides –252 and –239 binds a nuclear factor(s) present in preadipocytes that is lost or modified upon

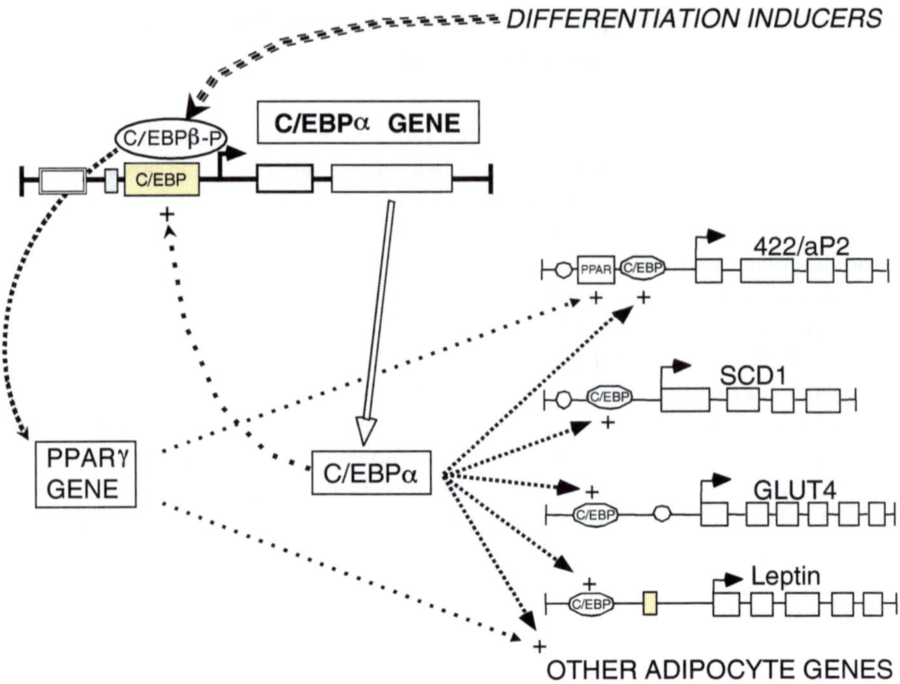

Fig. 1.2 Roles of C/EBPα, C/EBPβ, and PPARγ in the transcriptional activation of adipocyte genes during differentiation of 3T3-L1 preadipocytes. Following induction of differentiation, C/EBPβ is expressed that in turn activates the expression of C/EBPα and PPARγ. C/EBPα and PPARγ act as pleiotropic activators of many adipocyte genes including those encoding 422/aP2 (aP2 or FABP4), a fatty acid-binding protein; *SCD1* stearoyl-CoA desaturase-1, GLUT4 the insulin-responsive glucose transporter; and the leptin gene. C/EBPα also activates its own gene thereby maintaining the differentiated state. C/EBP refers to CCAAT/enhancer binding protein and PPAR to peroxisome proliferator-activated receptor

differentiation. The second site, located between nucleotides −203 and −176, exhibited different—but overlapping—footprints by nuclear factors present in differentiated and undifferentiated cells. Gel shift analysis with oligonucleotides corresponding to these sites revealed protein-oligonucleotide complexes containing these differentially expressed nuclear factors. The factor present in differentiated cells that binds at this site was identified as C/EBP presumably C/EBPα (Kaestner et al. 1989; Christy et al. 1991). These findings suggested that C/EBPα may regulate expression of its own gene (Fig. 1.2). It appears that the continued expression of C/EBPα allowed the maintenance of the differentiated state through interaction of the gene product (e.g., C/EBPα) with the promoter of its own gene. It was later found that C/EBPβ, a related C/EBP family member, also binds at this site and is responsible for activating expression of the C/EBPα gene following induction much earlier in the differentiation program (Fig. 1.2) (Tang and Lane 2012).

Conclusions

From the simple beginning using the differentiating 3T3-L1 model cell system has come a broad and comprehensive analysis of the SCD gene family in many different organisms. Work over the last 20+ years has focused on identifying and analyzing numerous SCD isoforms in a variety of species that are in turn controlled by multiple signaling systems. The genes for SCD have been cloned from human, mice, rats, sheep, hamster, *Drosophila*, and *C. elegans* (reviewed by Paton and Ntambi 2009). In the mouse, four isoforms (SCD1, 2, 3 and 4) have been identified while in the rat only two isoforms have been characterized. In humans, two genes have been identified; a ubiquitously expressed SCD1 and a brain, pancreas form termed SCD5. Given the diverse role of monounsaturated fatty acids in metabolic processes such as the synthesis of phospholipids, triglycerides, cholesteryl esters, wax esters, and alkyldiacylglycerols, it is not surprising that multiple genes exist and that multiple regulatory elements control isoform expression. In addition, tissue-specific dietary control of isoform expression occurs via a series of complex signal transduction schemes making SCD one of the most highly studied gene families. Conditional alleles and corresponding tissue-specific knockout models for many of the murine SCD genes have provided a wealth of information on not only tissue-specific fatty acid metabolism, but also the key transcription factors that regulate SCD expression under a variety of metabolic and genetic backgrounds. Whereas today many genes and gene families are defined and analyzed bioinformatically, early events using the 3T3-L1 system provided the foothold that enabled our current understanding of one of the most interesting and important loci in cellular metabolism.

Acknowledgments We would like to thank the members of the Lane laboratory who contributed to the development of the 3T3-L1 model cell system and whose work laid the foundation for studies on SCD1.

References

Angus CW, Lane MD, Rosenfeld PJ, Kelly TJ (1981) Increase in translatable mRNA for mitochondrial pyruvate carboxylase during differentiation of 3T3-L1 preadipocytes. Biochem Biophys Res Commun 103:1216–1222

Beaty NB, Lane MD (1983a) Kinetics of activation of acetyl-CoA carboxylase by citrate: relationship to the rate of polymerization of the enzyme. J Biol Chem 258:13043–13050

Beaty NB, Lane MD (1983b) The polymerization of acetyl-CoA carboxylase. J Biol Chem 258:13051–13055

Bernlohr DA, Angus CW, Lane MD, Bolanowski MA, Kelly TJ (1984) Expression of specific mRNAs during adipose differentiation: identification of an mRNA encoding a homologue of myelin P2 protein. Proc Natl Acad Sci U S A 81:5468–5472

Bernlohr DA, Bolanowski MA, Kelly TJ, Lane MD (1985a) Evidence for an increase in transcription of specific mRNAs during differentiation of 3T3-L1 preadipocytes. J Biol Chem 260:5563–5567

Bernlohr DA, Doering TL, Kelly TJ, Lane MD (1985b) Tissue specific expression of p422 protein, a putative lipid carrier, in mouse adipocytes. Biochem Biophys Res Commun 132:850–855

Christy RJ, Yang VW, Ntambi JM, Geiman DE, Landschulz WH, Friedman AD, Nakabeppu Y, Kelly TJ, Lane MD (1989) Differentiation-induced gene expression in 3T3-L1 preadipocytes: CCAAT/enhancer binding protein interacts with and activates the promoters of two adipocyte-specific genes. Genes Dev 3:1323–1335

Christy RJ, Kaestner KH, Geiman DE, Lane MD (1991) The CCAAT/enhancer binding protein gene promoter: binding of nuclear factors during differentiation of 3T3-L1 preadipocytes. Proc Natl Acad Sci U S A 88:2593–2597

Coleman RA, Reed BC, Mackall JC, Student AK, Lane MD, Bell RM (1978) Selective changes in microsomal enzymes of triacylglycerol, phosphatidylcholine and phosphatidylethanolamine biosynthesis during differentiation of 3T3-L1 preadipocytes. J Biol Chem 253:7256–7261

Cornelius P, MacDougald OA, Lane MD (1994) Regulation of adipocyte development. Annu Rev Nutr 14:99–129

Green H, Kehinde O (1974) Sublines of mouse 3T3 cells that accumulate lipid. Cell 1:113–116

Green H, Kehinde O (1975) An established preadipose cell line and its differentiation in culture. II. Factors affecting the adipose conversion. Cell 5:19–27

Green H, Kehinde O (1976) Spontaneous heritable changes leading to increased adipose conversion in 3T3 cells. Cell 7:105–113

Gregolin C, Ryder E, Warner RC, Kleinschmidt AK, Lane MD (1966) Liver acetyl-CoA carboxylase: the dissociation-reassociation process and its relation to catalytic activity. Proc Natl Acad Sci U S A 56:1751–1758

Guchhait RB, Moss J, Sokolski W, Lane MD (1971) The carboxyl transferase component of acetyl-CoA carboxylase: structural evidence for intersubunit translocation of the biotin prosthetic group. Proc Natl Acad Sci U S A 68:653–657

Guchhait RB, Polakis SE, Dimroth P, Stoll E, Moss J, Lane MD (1974a) Acetyl-CoA carboxylase system of *E. coli.* purification and properties of the biotin carboxylase, carboxyltransferase, and carboxyl carrier protein components. J Biol Chem 249:6633–6645

Guchhait RB, Polakis SE, Hollis D, Fenselau C, Lane MD (1974b) Acetyl-CoA carboxylase system of *E. coli.* Site of carboxylation of biotin and enzymatic reactivity of 1′-N-(ureido)-carboxybiotin derivatives. J Biol Chem 249:6646–6656

Hirsch AM, Rosen OM (1984) Lipolytic stimulation modulates the subcellular distribution of hormone-sensitive lipase in 3T3-L1 cells. J Lipid Res 25:665–677

Hwang CS, Mandrup S, MacDougald OA, Geiman DE, Lane MD (1996) Transcriptional activation of the obese gene by CCAAT/enhancer binding proteina. Proc Natl Acad Sci U S A 93:873–877

Hwang CS, Loftus TM, Mandrup S, Lane MD (1997) Adipocyte differentiation and leptin expression. Annu Rev Cell Dev Biol 13:231–259

Kaestner KH, Ntambi JM, Kelly TJ, Lane MD (1989) Differentiation-induced gene expression in 3T3-L1 preadipocytes: a second differentially-expressed gene encoding stearoyl-CoA desaturase. J Biol Chem 264:14755–14761

Kaestner KH, Christy RJ, Lane MD (1990) The mouse insulin-responsive glucose transporter gene: characterization of the gene and transactivation by the CCAAT/enhancer binding protein. Proc Natl Acad Sci U S A 87:251–255

Kohanski RA, Frost SC, Lane MD (1986) Insulin-dependent phosphorylation of the insulin receptor-protein kinase and activation of glucose transport in 3T3-L1 adipocytes. J Biol Chem 261:12272–12281

Lai E, Rosen OM, Rubin CS (1982) Dexamethasone regulates the beta-adrenergic receptor subtype expressed by 3T3 L1 preadipocytes and adipocytes. J Biol Chem 257:6691–6696

Landschulz WH, Johnson PF, Adashi EY, Graves BJ, McKnight SL (1988) Isolation of a recombinant copy of the gene encoding C/EBP. Genes Dev 2:786–800

Lin FT, Lane MD (1992) Antisense CCAAT/enhancer binding protein α RNA suppresses coordinate gene expression and triglyceride accumulation during differentiation of 3T3-L1 preadipocytes. Genes Dev 6:533–544

MacDougald OA, Lane MD (1995) Transcriptional regulation of gene expression during adipocyte differentiation. Annu Rev Biochem 64:345–373

MacDougald OA, Hwang CS, Fan H, Lane MD (1995) Regulated expression of the obese gene product (leptin) in white adipose tissue and 3T3-L1 adipocytes. Proc Natl Acad Sci U S A 92:9034–9037

Mackall JC, Lane MD (1977) Role of pyruvate carboxylase in fatty acid synthesis: alterations during preadipocyte differentiation. Biochem Biophys Res Commun 79:720–725

Mackall JC, Student AK, Polakis SE, Lane MD (1976) Induction of lipogenesis during differentiation in a "preadipocyte" cell line. J Biol Chem 251:6462–6464

Mandrup S, Lane MD (1997) Regulating adipogenesis. J Biol Chem 272:5367–5370

Mandrup S, Loftus TM, MacDougald OA, Kuhajda F, Lane MD (1997) Leptin is expressed at in vivo levels by fat pads derived from subcutaneously implanted 3T3-F442A preadipocytes. Proc Natl Acad Sci U S A 94:4300–4305

Moss J, Lane MD (1971) The biotin-dependent enzymes. Adv Enzymol 35:321–442

Moss J, Lane MD (1972) Acetyl-CoA carboxylase. III. Further studies on the relation of catalytic activity to polymeric state. J Biol Chem 247:4944–4951

Novikoff AB, Novikoff PM, Rosen OM, Rubin CS (1980) Organelle relationships in cultured 3T3-L1 preadipocytes. J Cell Biol 87:180–196

Ntambi JM, Buhrow S, Kaestner KH, Christy RJ, Sibley E, Kelly TJ, Lane MD (1988) Differentiation-induced gene expression in 3T3-L1 preadipocytes: characterization of a differentially-expressed gene encoding stearoyl-CoA desaturase. J Biol Chem 263:17291–17300

Paton CM, Ntambi JM (2009) Biochemical and physiological functions of stearoyl-CoA desaturase. Am J Physiol Endocrinol Metab 297:E28–E37

Polakis SE, Guchhait RB, Lane MD (1972) On the possible involvement of a carbonyl phosphate intermediate in the ATP-dependent carboxylation of biotin. J Biol Chem 247:1335–1337

Polakis SE, Guchhait RB, Zwergel EE, Lane MD (1974) Acetyl-CoA carboxylase system of *E. coli.* Studies of the mechanisms of the biotin carboxylase- and carboxyltransferase-catalyzed reactions. J Biol Chem 249:6657–6667

Reed BC, Lane MD (1980) Insulin receptor synthesis and turnover in differentiating 3T3-L1 preadipocytes. Proc Natl Acad Sci U S A 77:285–289

Reed BC, Kaufmann SH, Mackall JC, Student AK, Lane MD (1977) Alterations in insulin binding accompanying differentiation of 3T3-L1 preadipocytes. Proc Natl Acad Sci U S A 74:4876–4880

Reed BC, Ronnett GV, Clements PR, Lane MD (1981) Regulation of insulin receptor metabolism: differentiation-induced alteration of receptor synthesis and degradation. J Biol Chem 256:3917–3925

Ross SR, Graves RA, Greenstein A, Platt KA, Shyu HL, Mellovitz B, Spiegelman BM (1990) A fat-specific enhancer is the primary determinant of gene expression for adipocyte P2 in vivo. Proc Natl Acad Sci U S A 87:9590–9594

Ryder E, Gregolin C, Chang HC, Lane MD (1967) Liver acetyl-CoA carboxylase: insight into the mechanism of activation by tricarboxylic acids and acetyl-CoA. Proc Natl Acad Sci U S A 57:1455–1462

Student AK, Hsu RY, Lane MD (1980) Induction of fatty acid synthase synthesis in differentiating 3T3-Ll preadipocytes. J Biol Chem 255:4745–4750

Tang QQ, Lane MD (2012) Adipogenesis: from stem cell to adipocyte. Annu Rev Biochem 81:715–736

Tang QQ, Otto TC, Lane MD (2003) Mitotic clonal expansion: a synchronous process required for adipogenesis. Proc Natl Acad Sci U S A 100:44–49

Thiede MA, Ozols J, Strittmatter P (1986) Construction and sequence of the cDNA for rat liver stearyl coenzyme A desaturase. J Biol Chem 261:13230–13235

Watkins PA, Moss J, Pekala PH, Lane MD (1982) Effect of differentiation on the adenylate cyclase system of 3T3-L1 preadipocytes: determination of choleragen substrates in differentiating 3T3-L1 and nondifferentiating 3T3-C2 cells. J Biol Chem 257:14719–14722

Zhang JW, Klemm DJ, Vinson C, Lane MD (2003) Transcriptional regulation of CCAAT/enhancer binding protein β gene during adipogenesis: role of CREB. J Biol Chem 279:4471–4478

Chapter 2
Skin Stearoyl-CoA Desaturase Genes

Satish Parimoo and Pappas Apostolos

Introduction

The hair follicle and its associated sebaceous gland constitute the skin pilosebaceous unit, the biogenesis of which occurs during embryogenesis. In certain skin areas such as eyelids, nipples, prepuce, and labia minora, sebaceous glands may form and function in the absence of hair follicles, but all hair follicles form and function normally in association with sebaceous glands. Although significant advances have been made in our understanding of hair biology during the last decade, cellular and molecular differences of sebaceous glands between various species and skin sites within an organism are poorly understood. Furthermore, it is currently unknown whether differences in sebaceous glands are related to the heterogeneity of hair follicles. It has been hypothesized that during evolution sebaceous glands arose first in the animal kingdom before hair follicles came into existence (Stenn et al. 2008). If this hypothesis is true, it can be logically concluded that sebaceous glands must have played some crucial role in the evolution of hair follicles. In fact, in vitro and in vivo studies suggest that sebaceous glands are important for normal formation or functioning of hair follicle (Williams and Stenn 1994; Allen et al. 2003). Given the relationship between sebaceous glands and hair follicles, an obvious interesting question that arises is what happens to skin and or hair follicles if sebaceous glands have defects in differentiation and as a result do not function properly. Such questions may be addressed with appropriate animal models (Porter et al. 2002; Lunny et al. 2005; Yager et al. 2012). A spontaneous mutant mouse, asebia (*ab*), is known to exist that serves as an excellent model for sebaceous gland deficiency.

S. Parimoo
Johnson & Johnson Skin Biology TRC, Grandview Road, Skillman, NJ 08558, USA
e-mail: sparimo@msn.com

P. Apostolos (✉)
Skin Research Center, Johnson & Johnson, CPPW, Skillman, NJ 08558, USA
e-mail: apappas@its.jnj.com

J.M. Ntambi (ed.), *Stearoyl-CoA Desaturase Genes in Lipid Metabolism*,
DOI 10.1007/978-1-4614-7969-7_2, © Springer Science+Business Media New York 2013

The recessive asebia mutation originated spontaneously in a colony of BALB/cCrglGa mice (Gates and Karasek 1965). The name asebia (*ab*) was initially given to mutant mice as it was believed that they lacked sebaceous glands; however, subsequent closer examination revealed that in fact sebaceous glands were not absent but hypoplastic, and other related glands (meibomian, perianal glands, and ceruminous glands) are affected to various extents in *ab* mice (Josefowicz and Hardy 1978; Pennycuik et al. 1986; Compton et al. 1989; Sundberg et al. 2000). Studies of dermal-epidermal recombination fetal grafts suggested that the functional defect in mutant mice was in the epidermis and not the dermis (Pennycuik et al. 1986). Other prominent features of *ab* mutant mice are their scant subcutaneous fat and corneal opacities.

Identification of *Scd1* Gene Defect in the *Asebia* Mouse by Positional Cloning

An allelic spontaneous mutation, ab^{J}, was identified at the Jackson Laboratories and was mapped to mouse Chromosome 19 (Sweet and Lane 1977). Later another allelic mutation, ab^{2J}, arose spontaneously at the Jackson Laboratories on the inbred mouse strain DBA/1LacJ and its characterization was reported by Sundberg et al. 2000. Differences in appearance of mutant and normal mice and their skin histology are shown in Fig. 2.1. Whereas well-differentiated sebaceous glands are obvious in normal mouse skin section, they are not apparent in ab^{2J} skin section. The gene

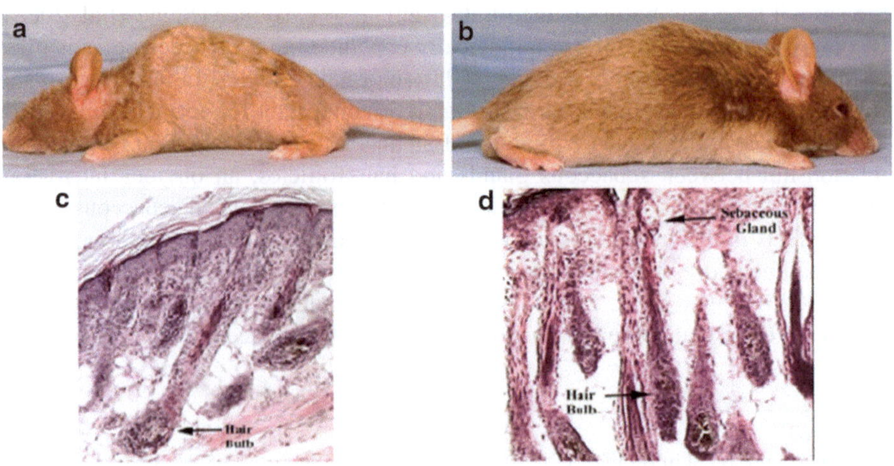

Fig. 2.1 Scaly skin with alopecia in an adult ab^{2J} mutant mouse (**a**) and normal fur of nonmutant littermate (**b**). Histology (H&E) skin sections of ab^{2J} and nonmutant mouse are shown in (**c**) and (**d**), respectively. While well-developed sebaceous glands, as indicated, are visible in (**d**), they are hypoplastic and not obvious in (**c**)

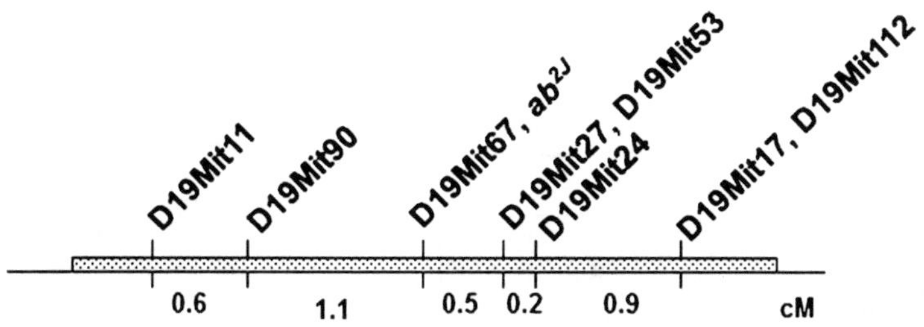

Fig. 2.2 Genetic linkage map of the mouse chromosome 19 region surrounding *ab*^{2J} locus and the associated microsatellite polymorphic markers that were used to obtain the map from backcross progeny as described by Eilertsen et al. (2000). The marker D19Mit67 is tightly linked to *ab*^{2J} as indicated. Inter-marker genetic distances are shown in centimorgans (cM). The *ab*^{2J} mutation is within the 1.6 cM genetic interval defined by D19Mit90 and 19Mit27/D19Mit53

defect in asebia mouse was identified by positional cloning involving two major steps. The first step was obtaining a high resolution genetic and physical map of the locus as reported by Eilertsen et al. (2000), followed by detailed identification and characterization of the genetic lesion in the mutant mouse using a variety of molecular techniques including nucleic acid sequencing (Zheng et al. 1999). These studies mapped the *ab*^{2J} mutation to mouse chromosome 19 that co-segregated with the polymorphic microsatellite marker D19Mit67 when DNA from 644 backcross mouse progeny were examined as reported by Eilertsen et al. 2000. The critical genetic interval in *ab*^{2J} mouse was identified to a fairly narrow region of DNA (1.6 cM) as shown in Fig. 2.2. Further detailed molecular analysis ultimately identified underlying genetic defects in both *ab*^{J} and *ab*^{2J}mice.

After probing skin RNA blots from *ab*^{J} and *ab*^{2J} mutant and normal mice with DNA probes derived from genes and expressed sequence tags (ESTs) from mouse chromosome 19 critical region, one gene probe derived from *Scd1* showed a lack of *Scd1* RNA expression on skin RNA blots from *ab*^{J} and *ab*^{2J} mice in contrast to normal mice (Zheng et al. 1999). Although this was good suggestive evidence that the *Scd1* was most likely the gene that was mutated in *ab* mice, more in-depth analysis of Southern and Northern blots, RT-PCR and DNA sequence analysis from mutant and normal mice, subsequently revealed that the *Scd1* gene expression was indeed disrupted due to mutations in *Scd1* gene in both *ab*^{J} and *ab*^{2J} mice although the nature of genetic lesions was different. Whereas a large deletion of DNA (exon 1–4) is responsible for gene inactivity in *ab*^{J} mice, a more subtle deletion of 18 nucleotides at exon 2 and intron 2 boundary was identified in *ab*^{2J} mice (Zheng et al. 1999). The gel electrophoresis of RT-PCR products from skin and liver RNA samples indicated the expected band of 787 base-pairs derived from *Scd1* RNA of the normal mouse, but mutant samples had either multiple bands due to aberrant RNA splicing in *ab*^{2J} or no bands at all due to larger genomic deletions in *ab*^{J} (Fig. 2.3a). DNA sequencing of RT-PCR products established that all three aberrantly spliced transcripts had in-frame stop codons in *ab*^{2J} (Fig. 2.3b). Deletion of 18 nucleotides in *ab*^{2J} causes loss of the 5′ splice site (GT) at the exon 2–intron 2 boundary used for *Scd1* RNA splicing. As a consequence of this 5′ splice site deletion, genomic DNA

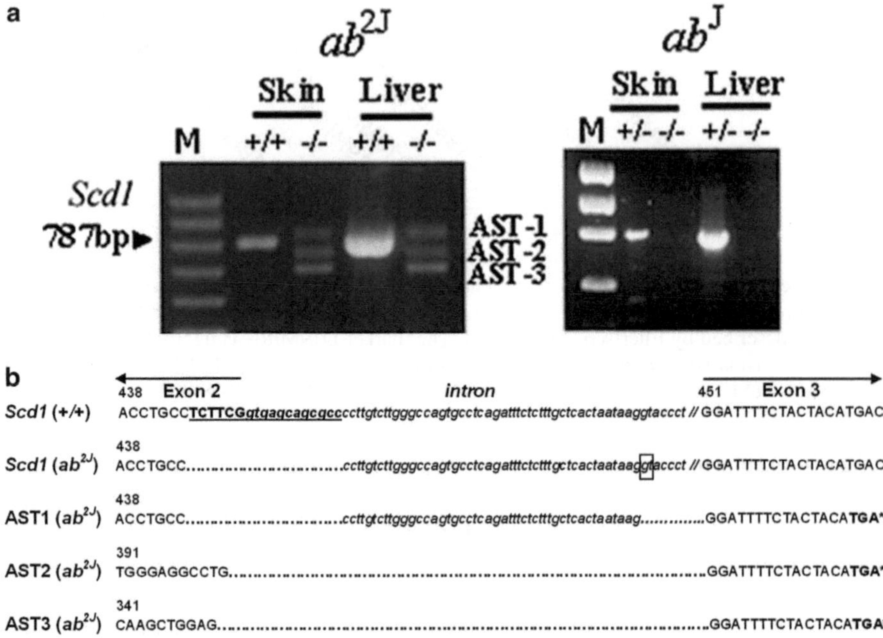

Fig. 2.3 (**a**) Characterization of RT-PCR products from mutants *ab*^{J}, *ab*^{2},^{J} and nonmutant tissues on agarose gel. *M* marker ladder. AST-1, AST-2, and AST-3 are three aberrantly spliced transcripts from *ab*^{2J} tissues. *Arrowheads* indicate *scd-1*-specific transcript. +/+ = nonmutant; −/− = mutant; +/− = heterozygote that is phenotypically normal. Reproduced from Zheng et al. (1999) *Nat Genet* 23:268–270. (**b**) Comparison of DNA sequences from genomic DNA of nonmutant (+/+), *ab*^{2J} mouse and cDNA derived from aberrantly spliced transcripts (AST-1, AST-2, and AST-3) from *ab*^{2J} mouse. *Asterisk* indicates in-frame stop codons. The GT dinucleotide used as cryptic 5′ splice for the generation of AST-1 is shown as boxed within the intron 2 of genomic sequence of *Scd1* (*ab*^{2J}). The exon sequences are in the *upper case* and intronic sequences are in the *italics* and *lower case*. The eighteen-nucleotides, encompassing exon 2 and intron 2 junction, that are deleted in *ab*^{2J} mutant are shown in *bold* and *underlined* in *Scd1* genomic sequence (+/+). *Dots* reflect deletions and // indicates intronic sequence present in the genomic DNA but not shown in the figure. The sequence numbering is same as that of the published cDNA (Ntambi et al. 1988. *J Biol Chem* 263:17291–17300)

codes for the aberrantly spliced transcript (AST-1) in *ab*^{2J} by utilizing a 5′ cryptic splice site (GT) within the intron as shown in Fig. 2.3b. The AST-1 from *ab*^{2J} has an insertion of 47 nucleotides from intron 2 as reflected by its larger sized RT-PCR product in comparison to that derived from the nonmutant mouse samples and confirmed by DNA sequencing (Fig. 2.3). Smaller sized RT-PCR products derived from aberrantly spliced transcripts, AST-2 and AST-3, apparently use other upstream 5′ cryptic sites and as a result generate smaller sized RT-PCR products (Fig. 2.3). None of the aberrantly spliced transcripts in *ab*^{2J} can code for Scd protein since all of them have in-frame stop codons (Fig. 2.3b). The fact that the *Scd1* gene mutation is responsible for the asebia phenotype was also supported by *Scd1* gene knockout studies as reported by Miyazaki et al. (2001). Details of how the *Scd1* gene defect in asebia impacts sebum synthesis or controls sebaceous gland morphogenesis and function are still. However, the role of various members of fatty acid desaturase gene family in skin has been reviewed recently (Sampath and Ntambi 2011).

Impact of *Scd1* Gene Defect in Mouse

Scd1 mutation causes several distinct changes in mouse when compared to heterozygote or normal mice. Adult homozygous asebia mice develop generalized alopecia, scaly skin, and have a hunched back. The hair shafts formation occurs normally, but they are sparse and matted (Josefowicz and Hardy 1978; Compton et al. 1989; Sundberg 1994). In adult asebia mice, hair follicles protrude deep in subcutis and hair cycles (anagen, catagen, and telogen) last longer than those in the controls (Josefowicz and Hardy 1978; Pennycuik et al. 1986). A subtle abnormality in mutant mice hair shaft includes retention of inner root sheath (IRS) above sebaceous gland that causes the plugging of the hair canal (pilary canal). The abnormally long hair follicles deep in dermis in mutant mice are believed to exist as a consequence of the retention of this cornified cell plug in the pilary canal that impedes the normal exit of hair shaft (Sundberg et al. 2000). The protrusion of hair shaft deep within the skin causes foreign body reaction with the resultant dermal scarring and eventual destruction of the hair follicle itself resulting in progressive alopecia with age in mutant mice. These observations underscore the importance of the sebaceous gland in controlling normal hair shaft growth and hair shaft-sheath interactions. Other studies also suggest that the separation of the hair shaft from the sheath is dependent on an intact mid-follicle/sebaceous gland (Williams and Stenn 1994; Gemmell and Chapman 1971; Williams et al. 1996). Given the similarity between hair follicle destruction observed in ab^{2J} mouse and human scarring alopecia, the asebia mouse has been postulated to be a good model for the pathogenesis of some forms of human scarring alopecia (Sundberg et al. 2000). Besides hair follicle abnormalities, asebia mice have thickened epidermis and dermis which is marked by increased vascularity and cellular infiltrate of mast cells and macrophages (Brown and Hardy 1988, 1989). Dermal inflammation and epidermal hyperplasia in mutant mice are believed to be due to macrophages containing lipid crystals (Brown and Hardy 1988). Skin surface lipid analysis of the asebia mutant mouse showed a marked reduction in sterol esters and cholesterol with apparently total loss of diol diesters compared to littermate controls (Wilkinson and Karasek 1966; Sundberg et al. 2000).

Although adult mutant mice are easy to distinguish visibly from normal mice, they are very difficult to distinguish phenotypically from littermate controls at the time of birth until approximately 1 week of age when hair fibers emerge from the follicles. Histological examination of skin reveals the presence of very small sebaceous glands in mutant skin as compared to controls (Sundberg et al. 2000). Ultrastructural studies of the glands indicate fewer lipid droplets, intracellular abnormalities such as impaired mitochondrial development and dilated smooth endoplasmic reticulum (Josefowicz and Hardy 1978).

The fact that disruption of *Scd1* expression in asebia mice causes dramatic phenotypic changes highlights the importance of this enzyme for normal skin function including sebaceous gland differentiation. How exactly the enzyme or its product is able to control the fate of sebaceous gland morphogenesis is unknown at present. Stearoyl-CoA desaturase (Scd) is a key iron-containing membrane-bound enzyme that catalyzes the rate-limiting step of oxidative conversion of saturated fatty acids to monounsaturated fatty acids, involving the insertion of a cis-double bond between the C9 and C10 positions in the acyl-CoA derivatives of saturated fatty acids such as palmitic and stearic acids to produce palmitoleoyl and oleoyl-CoA, respectively (Tocher et al. 1998; Paton and Ntambi 2009). This process is dependent upon electron flow from NAD(P)H through NADH-cytochrome b5 reductase to cytochrome b5, to Scd and finally to molecular oxygen that is reduced to water. Unsaturated fatty acids are not only involved in energy metabolism, but also in lipid-activated signal transduction (Shinomura et al. 1991; Kasai et al. 1998). Hence, one can conclude that monounsaturated fatty acids or their metabolite products, through an unknown mechanism, affect some pathways that are crucial for the normal differentiation of sebaceous gland and other related glands.

Cloning of *Scd3*, a Gene with Restricted Expression in Skin

Until late 1990s, only two members of *Scd* gene family were known to express in mouse tissues. A novel member of the *Scd* gene family was subsequently discovered while work on DNA sequencing of cDNA clones derived from asebia and normal skin was in progress. This novel member was named *Scd3* and characterization of this gene was reported by Zheng et al. 2001. Nucleotide sequence analysis of cDNA clones indicated that the *Scd3* transcript has 91 and 88 % identity with *Scd1* and *Scd2* coding sequences, respectively, with very limited homology in noncoding regions, suggesting independent control of gene expression in these genes. The amino acid identity of the Scd3-deduced protein sequence, compared to Scd1 and Scd2, is 88.7 % and 84.9 %, respectively. The Scd3 protein sequence of 359 amino acids, like the other two family members (Sdc1, Scd2), possesses four transmembrane domains and all the conserved histidine residues typically found in the desaturases and hydroxylases of several species. *Scd3* gene exists in close proximity to *Scd1* and *Scd2* genes on chromosome 19 as revealed by in situ hybridization and PCR analysis of BAC clones (Zheng et al. 2001).

Expression of Mouse *Scd* Gene Family in Sebaceous Glands and During Hair Cycle

RNA in situ hybridization of frozen mouse skin sections using a *Scd-1* specific probe demonstrated that *Scd-1* expression could be observed in the sebaceous gland but not in the hair follicle or any other cutaneous structure of wild-type mouse skin

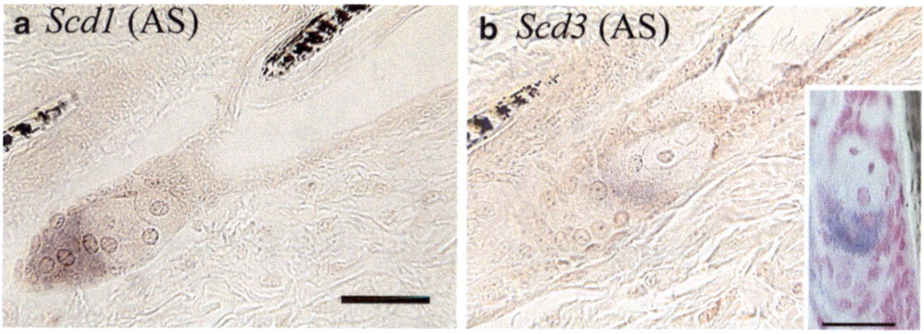

Fig. 2.4 RNA in situ hybridization of mouse skin sections comparing expression of *Scd1* and *Scd3* in the sebaceous gland. The same sebaceous gland is shown on both (**a**) and (**b**) as pictures were derived from two consecutive sections of mouse skin hybridized separately with the *Scd1* and *Scd3* probes. The inset shows an independent section where the slide was stained with hematoxylin and eosin (H&E) after obtaining *Scd3*-specific signal. *AS* antisense probe. The bar is 30 μM. Reproduced from Zheng et al. (2001) *Genomics* 71:182–91 with permission from the publisher (Elsevier)

(Zheng et al. 1999, 2001). *Scd-1* RNA expression was observed in sebaceous glands of skin irrespective of hair cycle stage as reported by Zheng et al. 1999. Similar to *Scd1* expression, *Scd3* expression by in situ RNA hybridization was observed to be restricted to sebaceous glands only (Fig. 2.4). However, the interesting feature of expression of the two genes (*Scd1* and *Scd3*) was that they did not co-express in the same region of sebaceous gland as observed from their expression in two consecutive tissue sections by in situ hybridization. This difference in gene expression may be attributed to the sebocytes of different stages of differentiation and also highlight the independent roles of *Scd1* and *Scd3* in different sebocytes within the sebaceous gland.

It is a well-known fact that hair follicle undergoes cyclic changes of growth (anagen), regression (catagen), and quiescence (telogen). The duration of each stage of the cycle is characteristic of both the hair type and species. Given the close association of sebaceous gland and hair follicle as mentioned earlier, it is of interest to know if hair cycle-induced changes in skin also impact expression of the *Scd* gene family. Such changes can be observed in natural or induced (plucked hair) hair cycles. As shown in Fig. 2.5a, steady-state levels of *Scd-1* transcript change over the natural hair cycle. The level of *Scd1* transcript was very low in skin at day 1 post-birth, but its expression increases considerably in the skin derived from mouse in anagen of the first hair cycle (day 8) and then decreases in regression/resting phases of hair cycle followed by increase in the next growth phase of the second hair cycle (days 28 and 35). *Scd-1* expression is much lower in the telogen hair cycle (days 21 and 42). In contrast to *Scd1*, *Scd2* did not show hair cycle-dependent changes in expression in the skin. This phenomenon of enhanced *Scd1*gene expression in skin with hair follicles in the anagen phase followed by decreased expression in skin containing telogen hair follicles was also observed in induced hair cycles (Fig. 2.5b). These results as reported by Zheng et al. (1999, 2001) do not answer the question if

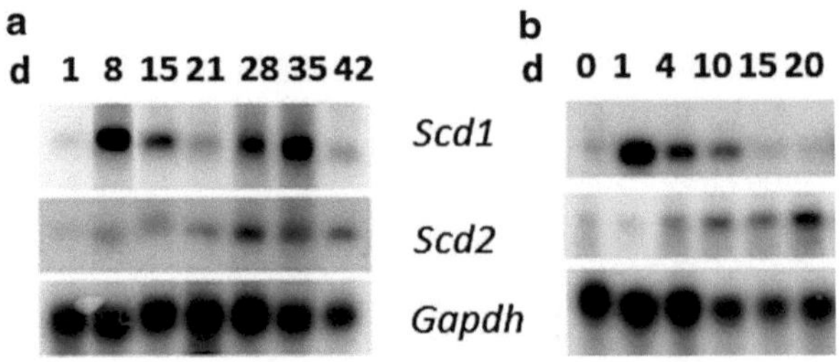

Fig. 2.5 Changes in *Scd-1* expression during spontaneous and induced hair cycles. (**a**) RNA blots of the samples from spontaneous hair cycle obtained from dorsal skin RNA of mice normal mice (DBA/1LacJ) at the age of mice in days (d) as indicated were probed with *Scd1*, *Scd2,* or Gapdh (control)-specific probes. (**b**) RNA blots of the samples from induced hair cycle14 procured from dorsal skin of normal mice (C57BL/6) probed with the same probes as in (**a**). Day 0 (d 0) and day 1 (d 1) represent telogen (quiescent) and early anagen (growth), respectively. The anagen phase ends at about day 18. Adapted from Zheng et al. (1999) *Nat Genet* 23:268–270

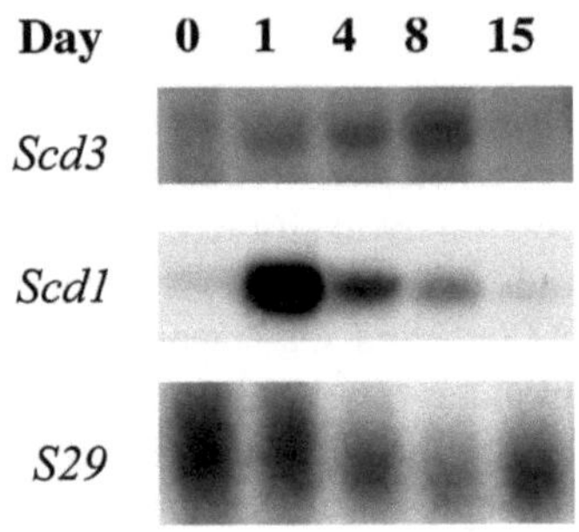

Fig. 2.6 Mouse hair cycle-dependent expression of *Scd3* in comparison to *Scd1*. RNA samples from the skin of 8-week-old mice with hair follicles in quiescent phase (telogen—day 0) and at days 1, 4, 8, and 15 after induction of follicular growth (anagen) by hair depilation were analyzed by RNA blots and probed with *Scd1*- and *Scd3*-derived probes. S29 is a ribosomal protein control probe. Reproduced from Zheng et al. (2001) *Genomics* 71:182–91 with permission from the publisher (Elsevier)

the changes in expression of skin *Scd-1* over the hair cycle are because of changes in *Scd1* expression in the sebaceous gland alone, or whether other tissues such as subcutaneous fat cells also contribute to these changes. Interestingly, in comparison to *Scd1*, *Scd3* transcript level increases gradually during induced hair cycle reaching the highest level at day 8 of anagen (late anagen), and then decreases at catagen (regression phase), which is day 15 of the induced hair cycle (Fig. 2.6). *Scd1* expression is very high in early anagen but relatively low in late anagen. These data also highlight the fact that *Scd1* gene expression is much higher compared to *Scd3* in mouse skin, particularly in early anagen phase (day1) of the induced hair cycle.

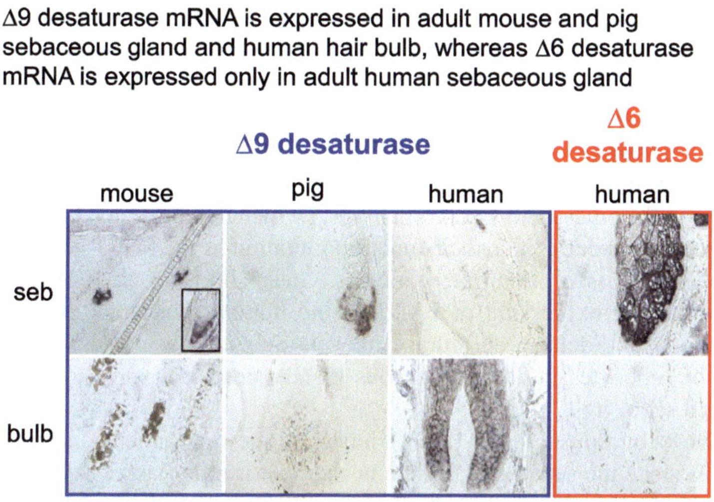

Fig. 2.7 Mouse SCD-1 expression is restricted to the sebaceous gland. Human SCD is weakly expressed in the sebaceous gland, but is highly expressed in the hair follicle bulb and lower hair shaft

Human Skin Desaturases

The human SCD1 gene was cloned by screening a human keratinocyte cDNA library and analysis of 3 h-RACE (rapid amplification of cDNA ends) products from various tissues yielded a 5.2 kb cDNA encoding a 359 amino acid protein with a calculated molecular mass of 41.5 kDa. Analysis of 3 h-RACE products suggested that alternative usage of polyadenylation sites generates two transcripts of 3.9 and 5.2 kb, a result consistent with Northern analysis. Southern analysis demonstrated the existence of two *SCD* loci in the human genome. Chromosomal mapping localized one locus to chromosome 10, and the second locus to chromosome 17. The locus on chromosome turned out to be transcriptionally inactive, fully processed pseudogene. Although the primary sequence and intron–exon structure of *SCD* is phylogenetically conserved, divergence between rodent and human is seen in the number of *SCD* genes and in the generation of alternative transcripts, suggesting a species-specific component of *SCD* regulation of unsaturated fatty acid metabolism.

In skin, mouse SCD-1 is restricted to and abundantly expressed in mouse sebaceous gland, whereas human SCD is weakly expressed in human sebaceous gland (Fig. 2.7). In skin, human SCD is highly expressed in the hair follicle bulb, lower hair shaft, and in the eccrine sweat gland, suggesting a role in the production of hair and sweat. Pig SCD is similar to mouse with sebaceous gland expression, and similar to human with sweat gland (apocrine) expression. However, pig differs from human in that the hair follicle lacks SCD expression. The differences between species in pilosebaceous expression of SCD do not seem to correlate with the number of SCD genes. The difference in gene number, polyadenylation, and expression pattern indicate that the SCD genes underwent species-specific selection during evolution, and that species-specific function of the pilosebaceous unit may contribute to this diversification. Among human, pig, mouse, and sheep, humans are the only species to express the delta 6 desaturase in sebaceous glands and if we consider that they are the only species that develop acne, that makes it a valid target for that human-specific disease. As described above, the essential role of the SG in hair follicle growth comes from the asebia mouse, which demonstrated hypoplastic SGs and alopecia, and due to a loss-of-function mutation in the Scd-1 gene; highly and specifically expressed in murine sebocytes (Gates and Karasek 1965; Zheng et al. 1999). From the expression of FAD2 in the human sebaceous gland we could conclude that the delta 6 desaturase may participate towards the differentiation process of sebocytes in adult sebaceous glands since its expression is mainly in suprabasal sebocytes.

Human sebaceous glands (SG's) have unique anatomical and biochemical properties since the sebaceous follicle is characterized by large, multi-lobular SGs that are attached to a hair follicle bearing a small hair shaft (Kligman 1974). But the most striking exceptionality is that the human sebum contains an abundant fatty acid that is unique among the animal kingdom. This fatty acid is named sapienic acid, thus "sapien" is the root of its name, and is a 16 carbon monounsaturated fatty bearing the double bond on the *n*10 or delta 6 position (Nicolaides 1974; Stewart et al. 1986). This fatty acid is implicated in the pathogenesis of acne, a disease that is mainly manifested in humans (Downing et al. 1986; Stewart et al. 1986). The levels of sapienic acid are multiple folds higher than any of its derivatives or isomers or other monounsaturated fatty acids found in sebum, which combined with the lack of SCD expression in human SG suggested that other members of the desaturase superfamily, particularly Δ-6 desaturase (also known as *f*atty *a*cid *des*aturase-2, FADS2), may be expressed in human SG (Ge et al. 2003; Pappas et al. 2002; Pappas 2009).

Concluding Remarks

Based on the dramatic cutaneous pathology observed in asebia mouse, it has become clear that *Scd-1* gene product plays an important role for maintaining normal skin and eye phenotype. Mechanistic understanding of how the *Scd-1* gene product contributes to normal differentiation and function of sebaceous glands or Meibomian glands is unclear at present. Histological studies of normal and asebia mice suggest that the sebaceous gland also plays an important role in dissociating the hair follicle sheath from the outgrowing shaft during hair growth cycle, but molecular basis for this process is again unclear. It is possible that monounsaturated fatty acids or their derivatives may play an important role in sebocyte membrane synthesis, in sebum production, or in sebocyte signaling pathways (Shinomura et al. 1991; Kasai, et al. 1998). Monounsaturated fatty acids have been implicated in maintaining retinol homeostasis based on the observation that skin retinol/retinoic acid and the retinoic acid-induced genes were elevated in *Scd1* gene knockout mice (Flowers et al. 2011).

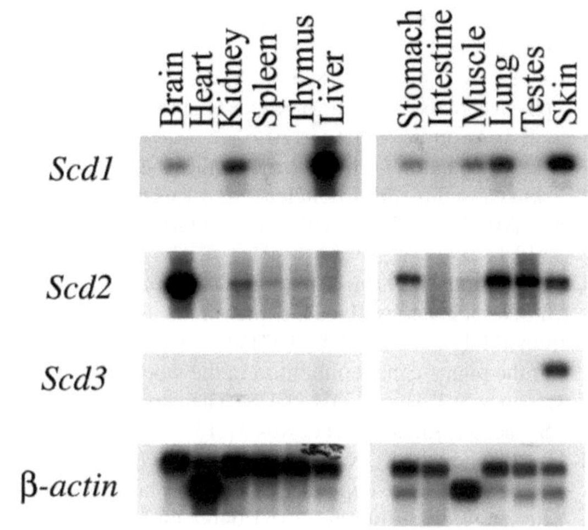

Fig. 2.8 Tissue-specific expression pattern of *Scd3* from RNA blots. The membrane blots were probed with cDNA probes from the noncoding regions specific for each *Scd* gene. The β-actin probe was used as a control. The exposure times for *Scd3* blot was adjusted more than fivefold in order to achieve comparable signal intensity to that of *Scd1* probed blot. Reproduced from Zheng et al. (2001) *Genomics* 71:182–91 with permission from the publisher (Elsevier)

Expression of multiple *Scd* genes has been reported in rodent skin (Zheng et al. 1999, 2001). Although the predicted Scd1 and Scd3 protein sequences have high sequence homologies, it is unknown if there are any differences in the activity of the two gene products. However, differences in the gene promoter sequences suggest different gene control mechanisms that may be responsible for observed tissue specificity with respect to their expression (Fig. 2.8). It is noteworthy that the lack of expression of *Scd1* in asebia mouse does not alter the expression of *Scd3* gene, at least in young mice (Zheng et al. 2001). Decrease of *Scd3* expression with advanced age in mutant mice is most likely related to the destruction of follicles and sebaceous glands due to the scarring of the pilosebaceous units (Sundberg et al. 2000). However, it is obvious that in asebia mutant mice with rudimentary sebocytes, *Scd3* gene expression in young mice cannot compensate for the lack of *Scd1* expression. The spatial expression of *Scd1* and *Scd3* in the mouse sebaceous gland underscores their roles in distinct cellular populations of sebaceous gland. Interestingly *Scd3* gene expression is slightly higher in male skin than in female skin (Zheng et al. 2001). Whether this differential regulation is achieved by androgen-mediated up-regulation in male or estrogen-mediated down-regulation in female mouse is unknown, although an estrogen receptor recognition site was observed in the *Scd3* distal promoter (Zheng et al. 2001).

References

Allen M, Grachtchouk M, Sheng H, Grachtchouk V, Wang A, Wei L et al (2003) Hedgehog signaling regulates sebaceous gland development. Am J Pathol 163:2173–2178

Brown WR, Hardy MH (1988) A hypothesis on the cause of chronic epidermal hyperproliferation in asebia mice. Clin Exp Dermatol 13:74–77

Brown WR, Hardy MH (1989) Mast cells in asebia mouse skin. J Invest Dermatol 93:708

Compton JG, Dunstan RW, Sundberg JP (1989) Asebia, a mutation affecting sebaceous glands, mouse. In: Jones TC, Mohr U, Hunt RD (eds) Monographs on pathology of laboratory animals. Integument and mammary glands. Springer, Heidelberg, pp 218–222

Downing DT, Stewart ME, Wertz PW, Strauss JS (1986) Essential fatty acids and acne. J Am Acad Dermatol 14(2 Pt 1):221–225

Eilertsen KJ, Tran T, Sundberg JP, Stenn KS, Parimoo S (2000) High resolution genetic and physical mapping of the mouse asebia locus: a key gene locus for sebaceous gland differentiation. J Exp Anim Sci 40:165–170

Flowers MT, Paton CM, O'Byrne SM, Schiesser K, Dawson JA et al (2011) Metabolic changes in skin caused by *Scd1* deficiency: a focus on retinol metabolism. PLoS One 6(5):e19734

Gates AH, Karasek M (1965) Hereditary absence of sebaceous glands in the mouse. Science 148:1471–1473

Ge L, Gordon JS, Hsuan C, Stenn K, Prouty SM (2003) Identification of the delta-6 desaturase of human sebaceous glands: expression and enzyme activity. J Invest Dermatol 120(5):707–714

Gemmell RT, Chapman RE (1971) Formation and breakdown of the inner root sheath and features of the pilary-canal epithelium in the wool follicle. J Ultrastruct Res 36:355–366

Josefowicz WJ, Hardy MH (1978) The expression of the gene asebia in the laboratory mouse. 3. Sebaceous glands. Genet Res 31:157–166

Kasai T, Ohguchi K, Nakashima S, Ito Y, Naganawa T, Kondo N, Nozawa Y (1998) Increased activity of oleate-dependent type phospholipase D during actinomycin D-induced apoptosis in Jurkat T cells. J Immunol 161:6469–6474

Kligman AM (1974) An overview of acne. J Invest Dermatol 62:268–287

Lunny DP, Weed E, Nolan PM et al (2005) Mutations in gasdermin 3 cause aberrant differentiation of the hair follicle and sebaceous gland. J Invest Dermatol 124:615–621

Miyazaki M, Man WC, Ntambi JM (2001) Targeted disruption of stearoyl-CoA desaturase1 gene in mice causes atrophy of sebaceous and meibomian glands and depletion of wax esters in the eyelid. J Nutr 131:2260–2268

Nicolaides N (1974) Skin lipids: their biochemical uniqueness. Science 186:19–26

Pappas A (2009) Epidermal surface lipids. Dermatoendocrinol 1(2):72–76

Pappas A, Anthonavage M, Gordon JS (2002) Metabolic fate and selective utilization of major fatty acids in human sebaceous gland. J Invest Dermatol 118:164–171

Paton CM, Ntambi JM (2009) Biochemical and physiological function of stearoyl-CoA desaturase. Am J Physiol Endocrinol Metab 297:E28–E37

Pennycuik PR, Raphael KA, Chapman RE, Hardy MH (1986) The site of action of the asebia locus (ab) in the skin of the mouse. Genet Res 48:179–185

Porter RM, Jahoda CAB, Lunny DP, Henderson G, Ross J, McLean WH et al (2002) Defolliculated: a dominant mutation leading to poor sebaceous gland differentiation and total elimination of pelage follicles. J Invest Dermatol 119:32–37

Sampath H, Ntambi JM (2011) The role of fatty acid desaturases in epidermal metabolism. Dermatoendocrinol 3:62–64

Shinomura T, Asaoka Y, Oka M, Yoshida K, Nishizuka Y (1991) Synergistic action of diacylglycerol and unsaturated fatty acid for protein kinase C activation: its possible implications. Proc Natl Acad Sci U S A 88:5149–5153

Stenn KS, Zheng Y, Parimoo S (2008) Phylogeny of the hair follicle: the sebogenic hypothesis. J Invest Dermatol 128:1576–1578

Stewart ME, Grahek MO, Cambier LS, Wertz PW, Downing DT (1986) Dilutional effect of increased sebaceous gland activity on the proportion of linoleic acid in sebaceous wax esters and in epidermal acylceramides. J Invest Dermatol 87:733–736

Sundberg JP (ed) (1994) The asebia (*ab*, *ab*^{J}) mutations, chromosome 19. Handbook of mouse mutations with skin and hair abnormalities. Animal models and biomedical tools. CRC, Boca Raton, FL, pp 171–177

Sundberg JP, Bogess D, Sundberg BA, Eilertsen K, Parimoo S, Filippi M et al (2000) Asebia-2J (Scd1ab2J): a new allele and a model for scarring Alopecia. Am J Pathol 156:2067–2075

Sweet HO, Lane PW (1977) Asebia-J is now known to be very close to ru on Chr 19. Mouse News Lett 57:20

Tocher DR, Leaver MJ, Hodgson PA (1998) Recent advances in the biochemistry and molecular biology of fatty acyl desaturases. Prog Lipid Res 37:73–117

Wilkinson DI, Karasek MA (1966) Skin lipids of a normal and a mutant (asebic) mouse strain. J Invest Dermatol 47:449–455

Williams D, Stenn KS (1994) Transection level dictates the pattern of hair follicle sheath growth in vitro. Dev Biol 165:469–479

Williams D, Siock P, Stenn KS (1996) 13-cis-retinoic acid affects sheath shaft interactions of equine hair follicles in vitro. J Invest Dermatol 106:356–361

Yager JA, Gross TL, Shearer D, Rothstein E, Power H, Sinke JD, Kraus H, Gram D, Cowper E, Foster A, Welle M (2012) Abnormal sebaceous gland differentiation in 10 kittens ('sebaceous gland dysplasia') associated with generalized hypotrichosis and scaling. Vet Dermatol 23:136–144

Zheng Y, Eilersten KJ, Ge L, Zhang L, Sundberg JP, Prouty S, Stenn KS, Parimoo S (1999) Scd1 is expressed in sebaceous glands and is disrupted in the asebia mouse. Nat Genet 23:268–270

Zheng Y, Prouty SM, Harmon A, Sundberg JP, Stenn KS, Parimoo S (2001) Scd3-a novel gene of the stearoyl-CoA desaturase family with restricted expression in skin. Genomics 71:182–191

Chapter 3
Stearoyl-CoA Desaturase-1 Is a Biological Regulator of Energy Homeostasis

James M. Ntambi

Role of Hepatic SCD1 in Regulation of Lipogenesis and Fatty Acid Oxidation

Stearoyl-CoA desaturase-1 (SCD1) has since been found to be very important for the synthesis of oleate despite the fact that mammalian diets supply abundant dietary oleate. If this is true, then why is SCD1 a highly regulated enzyme? Using the asebia mouse strains (ab^{j} and ab^{2j}) that have a naturally occurring mutation in SCD1 isoform (Zheng et al. 1999) as well as a mouse model with a targeted disruption (SCD1−/−) (Miyazaki et al. 2001a, b), our laboratory showed that SCD1−/− mice are deficient in triglycerides, cholesterol esters, wax esters, and alkyldiacylglycerols (Miyazaki et al. 2001a, b; Attie et al. 2002). The levels of palmitoleate (16:1n7) and oleate (18:1n9) are reduced in the plasma and tissue lipid fractions of SCD1−/− mice while palmitate and stearate are increased. These changes are correlated with a decrease in desaturation index (18:1/18:0 or 16:1/16:0 ratio) in liver tissue and plasma (Attie et al. 2002).

Normally, a high carbohydrate diet fed to mice or rats induces the expression of the hepatic SCD1 gene and other lipogenic genes through insulin-mediated sterol regulatory element binding protein-1 (SREBP-1)-dependent mechanism. This mechanism results in an increase of monounsaturated fatty acids (MUFAs) and hepatic triglycerides (Shimomura et al. 1999a, b). One of our observations is that SCD1−/− mice on a high carbohydrate diet increase plasma insulin levels but they fail to accumulate hepatic triglycerides and cholesterol esters (Miyazaki et al. 2001a, b). Supplementation of the high carbohydrate diet with high levels of triolein or tripalmitolein can normalize cholesterol ester levels but the triglyceride levels cannot be returned to the levels found in the wild type mouse (Miyazaki et al. 2001a, b). In

J.M. Ntambi, Ph.D. (✉)
Departments of Biochemistry and Nutritional Sciences,
University of Wisconsin-Madison, Madison, WI 53706, USA
e-mail: ntambi@biochem.wisc.edu

J.M. Ntambi (ed.), *Stearoyl-CoA Desaturase Genes in Lipid Metabolism*,
DOI 10.1007/978-1-4614-7969-7_3,

addition, the SCD1−/− mice have very low levels of triglycerides in the very low-density lipoprotein (VLDL) and low-density lipoprotein (LDL) fractions compared to their wild type counterparts. Furthermore, the rate of VLDL-triglyceride secretion, as measured by inhibition of VLDL clearance using Triton-1339, was dramatically reduced in the SCD1−/− mice (Cohen et al. 2002). Transient transfections of an SCD1 expression vector into Chinese hamster ovary (CHO) cells result in increased SCD1 activity and esterification of cholesterol to cholesterol esters (Miyazaki et al. 2000). These observations reveal that endogenously synthesized MUFAs by SCD most likely serve as the main substrates for the synthesis of hepatic triglycerides and cholesterol esters. The enzymes involved in the de novo synthesis of triglycerides and cholesterol ester including SCD, acyl-CoA:cholesterol acyl transferase (ACAT), diacylglycerol acyl transferase (DGAT), and microsomal glycerol phosphate acyl transferase (GPAT) are located in the endoplasmic reticulum (ER) membrane. A possible physiological explanation for the requirement of SCD expression in the synthesis of the triglycerides and cholesterol esters is the production of more easily accessible MUFAs within the vicinity of ACAT, DGAT, and microsomal GPAT. These enzymes probably exist in a complex promoting substrate channeling and fuel partitioning into various metabolic pathways. It was demonstrated previously that SCD1 and DGAT-2 colocalize in ER of liver (Man et al. 2006).

It became interesting to determine whether the impairment in acyl-CoA desaturation in the SCD1−/− mice would alter the whole-animal energy homeostasis, or whether dietary MUFAs would ameliorate the deficiency. Although the growth curves of male SCD1−/− mice were similar to that of the wild type sibs on chow diet, a high fat diet revealed large differences in weight gain in both males and females. On average, the SCD1−/− mice consumed 10–15 % more food than wild type mice even though they were leaner and accumulated less fat in their adipose tissue. The epididymal fat pad mass was markedly reduced in male SCD1−/− vs. wild type mice on a chow diet and a high fat diet. Decreases in fat pads' weights were observed in subcutaneous, mesenteric, and retroperitoneal both on chow and high fat diets. The livers of the wild type and SCD1−/− mice were grossly normal and of similar mass on a chow diet. However, on a high fat diet the livers of the wild type were lighter in color than knockout mice suggesting steatosis. Masses of white adipose depots in SCD1−/− mice were universally decreased compared with wild type, regardless of diet. Thus, SCD1−/− mice were resistant to diet-induced weight gain and fat accumulation, despite increased food intake (Ntambi et al. 2002).

What then is the fate of the dietary fat in excess? Indirect calorimetry was carried out to investigate whether the resistance to weight gain was due to increased energy expenditure. The SCD1−/− mice exhibited consistently higher rates of oxygen consumption (higher metabolic rates) than their wild type littermates throughout day and night. It was hypothesized that the increased energy expenditure in the SCD1−/− mice was due to increased lipid catabolism. Although ketone bodies were undetectable in plasma from either strain during postprandial conditions, β-hydroxybutyrate levels were consistently much higher in knockout mice following a 4-h fast indicating a higher rate of β-oxidation in SCD1−/− mice.

DNA micro arrays were employed to identify genes whose expression was altered in the liver of SCD1−/− mice. Two hundred mRNAs that were significantly different between the livers of SCD1−/− and wild type mice were identified. The most striking pattern was the genes involved in lipogenesis and fatty acid β-oxidation. Lipid oxidation genes such as acyl-CoA oxidase (ACO), very long chain acyl-CoA dehydrogenase (VLCAD), carnitine palmitoyltransferase-1 (CPT-1), and fasting-induced adipocyte factor (FIAF) were up regulated while lipid synthesis genes such as SREBP-1, FAS, and mitochondrial glycerol phosphate acyl-CoA transferase (GPAT) were down regulated in the SCD1−/− mice (Ntambi et al. 2002). SREBP-1c is the main SREBP-1 isoform expressed in liver and regulates the expression of lipogenic genes (Shimomura et al. 1999a, b). Insulin levels, dietary carbohydrate, fatty acids, and cholesterol regulate the SREBP-1 gene expression and protein maturation. Thus, the down regulation of SREBP-1 gene expression in the SCD1−/− could have numerous effects on various metabolic pathways regulated by SREBP-1. For instance, the induction of SREBP-1 by insulin and cholesterol greatly enhances the synthesis and secretion of triglycerides by the liver (Shimomura et al. 1999a, b). However, in the SCD1−/− mice, as well as in the liver-specific knockout of SCD1 mice, carbohydrate feeding fails to induce SREBP-1 and lipogenic gene expression to the same level found in the wild type mice (Miyazaki et al. 2001a, b; Miyazaki et al. 2007).

CPT, ACO, VLCAD, and FIAF are known targets of peroxisome proliferator-activated receptor-α (PPARα) and contain PPARα response regions in their promoters (Kersten et al. 2000). Since PPARα mRNA level is unchanged (Miyazaki et al. 2004), the up regulation of enzymes of fatty acid β-oxidation in the SCD1−/− mice must be downstream of PPARα transcription. Thus, the characteristics exhibited by the SCD1−/− mice are consistent with presence of a PPARα activator with reduced activity in wild type mice but this was not found to be the case (Miyazaki et al. 2004). The SCD1−/− mice exhibit increase in the contents of saturated fatty acids (C16:0 and C18:0) while the contents of the polyunsaturated fatty acids (PUFAs) of the n-6 and n-3 are not changed. Changes in the levels of saturated fats may indirectly stimulate the burning of fat by mitochondria.

If SCD1 is blocked, cells may no longer generate normal levels of monounsaturated fats, which are required for normal synthesis of VLDL and triglycerides. If that happens, saturated fatty acids might build up. Scientists reported in the 1970s that fatty acids inhibit acetyl-CoA carboxylase-1 and 2 (ACC1 and ACC2) which produce malonyl-CoA. Such a buildup of saturated fatty acids might act to decrease malonyl-CoA levels and stimulate fatty acid import into mitochondria and in turn the burning of fat. Malonyl-CoA is known to inhibit the catabolism of fats and one possibility is that somehow SCD1 controls malonyl-CoA levels. In this mechanism the saturated fatty acyl-CoAs would allosterically inhibit ACC, thus reducing cellular levels of malonyl-CoA (Lunzer et al. 1977). Malonyl-CoA is required for fatty acid biosynthesis and also inhibits the mitochondrial carnityl palmitoyl transferase shuttle system, the rate-limiting step in the import and oxidation of fatty acids in mitochondria (McGarry et al. 1978). Thus, reduced levels of SCD1 would lead to a decrease in the cellular levels of malonyl-CoA and derepress fatty acid oxidation. The findings in the SCD1−/− mice are therefore similar to those observed in mice lacking

ACC2, which also have increased fatty acid oxidation in skeletal muscle and possess a lean phenotype (Abu-Elheiga et al. 2001).

Another mechanism that could account for the increased energy expenditure and protection against diet-induced adiposity and insulin resistance in the SCD1-deficient mice is associated with increased activity of AMPK (Dobrzyn et al. 2004; Hulver et al. 2005), an enzyme that has been shown to stimulate fatty acid oxidation following leptin administration. SCD1 deficiency could also be associated with direct or indirect effects on uncoupling proteins that are involved in thermogenesis (Lee et al. 2004).

Finally the fur and lipid abnormalities associated with defects in SCD1-deficient mice could be associated with increased energy dissipation. The SCD1−/− mice as well as the asebia mutant mice show cutaneous abnormalities with atrophic sebaceous glands and narrow eye fissure with atrophic meibomian glands, suggesting an important role of MUFAs in skin homeostasis (Zheng et al. 1999; Sundberg et al. 2000). Studies on the role of lipids in cutaneous biology have mainly focused on PUFAs (alpha-linoleic acid) and their role in skin barrier function (Marcelo et al. 1992). The role of MUFA in skin homeostasis is unknown. However, it is possible that these fatty acids, which are the products of SCD, play an important role in differentiation of the sebaceous gland and a parallel can be drawn to the activation of SCD and production of MUFA during 3T3-L1 preadipocyte differentiation (Casimir and Ntambi 1996; Christianson et al. 2008). Recent studies have attributed the failure of SCD1−/− mice to regulate adaptive thermogenesis and to protect against diet-induced obesity was caused by defective skin lipid barrier due to deficiency in ceramides and transepidermal water loss (TEWL) (Binczek et al. 2007). However, earlier studies had showed that the SCD1-deficient mice have a normal skin barrier but had reduced hydration due to deficiency in glycerol. In addition, the defective skin lipid barrier is due to deficiency of SCD2 isoform in the epidermis of neonates (Miyazaki et al. 2005). Further, there are studies showing that obese humans have increased TEWL (Loffler et al. 2002) supporting our contention that the hypermetabolism in SCD1-deficient mice may not be due to TEWL and an impaired epidermal lipid barrier.

We have recently found that mice with a specific deletion of SCD1 in the skin (SKO) recapitulate some of the phenotypes observed in GKO mice including increased whole-body energy expenditure, protection against dietary-induced adiposity, hepatic steatosis, and glucose intolerance (Sampath et al. 2009). The increased energy expenditure in SKO mice does not result simply from heat loss through the skin. Rather, thermogenesis appears to be constitutively activated in these mice regardless of changes in the ambient temperature (Flowers et al. 2011). Genes of fat oxidation, lipolysis, and thermogenesis, including uncoupling proteins and PPARγ co-activator-1α, are up regulated in peripheral tissues of SKO mice. However, unlike mice globally deficient in SCD1, SKO mice have an intact hepatic lipogenic response to acute high carbohydrate feeding. The up regulation of thermogenic processes for temperature maintenance at the expense of fuel economy illustrates the cross talk between the skin and peripheral tissues in maintaining energy homeostasis.

The SKO mice exhibit a dramatic decrease in the levels of 18:1n9 and 16:1n7 MUFAs in neutral lipid fractions of the sebaceous glands despite sufficient levels of these fatty acids in the diets. Furthermore, the levels of retinol and the expression of retinoic acid (RA)-target genes including the gene for the secreted adipokine, lipocalin-2, are increased in the skin of SKO mice. Severe alterations in retinol availability or retinoic acid (RA) levels can affect the regulation of adaptive thermogenesis and adiposity, two processes which are fundamental to energy balance (Bonet et al. 2003). For example, retinaldehyde inhibits adipogenesis and blocking retinaldehyde metabolism to RA reduces diet-induced obesity (Ziouzenkova et al. 2007). The net effect of retinoids on adipogenesis and other physiological processes results from a complex balance between RA metabolism and relative amounts of retinoid receptor subtypes (RXR and RAR) (Villarroya et al. 1999). RA is a potent signaling molecule that is essential for many biological processes, and its levels are tightly regulated by mechanisms that are only partially understood.

The synthesis of RA from its precursor retinol, or vitamin A, is a major mode in the regulation of RA levels in cells and tissues (Napoli 1999). To generate RA, retinol is oxidized in two sequential reactions, catalyzed by retinol and retinal dehydrogenases (Napoli 1999), whose activities regulate RA homeostasis. The balance between retinol and retinyl esters may also regulate the availability of retinol for these reactions. Retinol esterification is carried out by two distinct enzymatic activities. One is mediated by LRAT, which catalyzes the covalent joining of a fatty acyl moiety from lecithin (phosphatidylcholine) to retinol that is bound to cellular retinol-binding protein-1 (CRBP-1) (MacDonald and Ong 1988). Unbound retinol is esterified by acyl-CoA:retinol acyltransferase (ARAT) activity that has been very difficult to purify. However, DGAT-1 has ARAT activity and it has recently been shown that DGAT-1 functions as the major ARAT in murine skin (Yen et al. 2008) where it acts to maintain retinoid homeostasis and prevent retinoid toxicity. When dietary retinol is abundant, DGAT-1 deficiency results in elevated levels of RA in skin and cyclical hair loss; both are prevented by dietary retinol deprivation (Shih et al. 2009). Since the skin and metabolic phenotypes of DGAT-1 (Smith et al. 2000) and SCD1 knockout mice are similar, it is very likely that the endogenous MUFAs produced by SCD1 are not available for esterification of excess unbound retinol to retinyl esters by DGAT-1 leading to accumulation of retinol. Both DGAT-1 and SCD1 are therefore important for handling free retinol that exceeds the capacity of LRAT. Some of the excess retinol is subsequently converted to retinoic acid, which initiates the signaling that leads to the phenotype observed in both the SCD1 and DGAT-1 knockout mice. Interestingly SCD1-gene expression is activated by retinol in liver (Miller et al. 1997) and muscle possibly to enhance esterification of excess retinol. In addition, we demonstrated previously that SCD1 and DGAT-2 colocalize in ER of liver (Man et al. 2006). It is very possible that SCD1 and DGAT-1 colocalize in ER of the sebaceous glands.

Plasma levels of lipocalin-2 are also elevated in the SKO mice. Lipocalin-2 is a known secretory protein produced by the human sebaceous glands and has been implicated in obesity and insulin resistance both in humans and rodents (Yan et al. 2007) but its in vivo role in metabolic regulation is largely unknown. It is hypothesized (Fig. 3.1) that the endogenous 18:1n9 or 16:1n7 made by SCD1 are required for esterification of excess retinol into retinyl esters for storage and detoxification in the sebaceous glands. In the absence of SCD1, the excess free retinol is subsequently

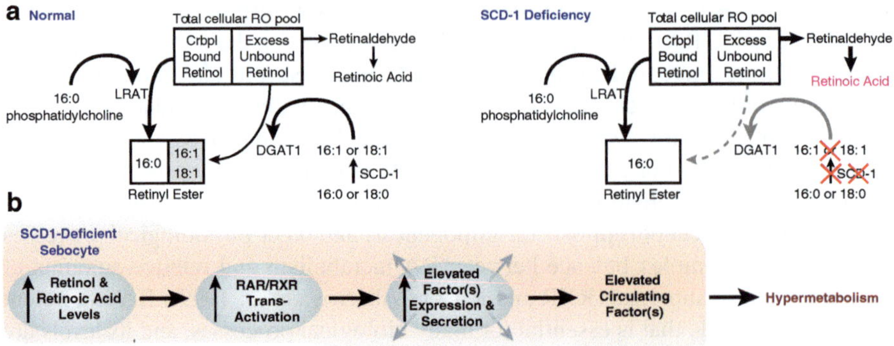

Fig. 3.1 The model proposed by Chad Paton showing how SCD1 deficiency modulates retinoic acid (RA) signaling in the sebaceous gland and leads to increased expression and secretion of a factor(s)

converted to RA, which initiates the RAR–RXR-mediated signaling in the sebaceous gland (Fig. 3.1b). The activation of RA signaling leads to secondary effects via elevations in sebaceous gland-derived lipocalin-2 or other circulating factor(s), that is, a sebokine or an adipokine released into the circulation that interact with peripheral tissues to affect whole-body energy expenditure (Fig. 3.1b). However, potentially TEWL could be the mechanism to hypermetabolism. The hypermetabolism/TEWL could still be due to sebocyte dysfunction but not dependent on a circulating factor. Retinol toxicity (or potentially cholesterol toxicity) may be causing sebocyte atrophy and a disturbed epidermal lipid barrier. This could be the reason for the hypermetabolism, and the effect on a circulating factor is just an associated gene expression change. This should be considered as an alternative hypothesis and should be pursued in case the circulating factor hypothesis does not fully explain the origin of the skin SCD1 deficiency phenotype.

Conclusion

The recent studies using the knockout mouse models have revealed the phenotypes generated as a result of SCD1 gene deficiency. SCD is critical in the biosynthesis of several classes of lipids including phospholipids, triglycerides, cholesterol esters, wax esters, and 1-alkyl-2,3-diacylglycerol. SCD1 deficiency either directly or indirectly induces a signal that partitions fatty acids towards oxidation rather than synthesis. High SCD1 expression favors fat storage while low SCD expression leads to leanness, increased metabolic rate, and insulin sensitivity. Current studies have demonstrated that liver SCD1 deficiency protects mice against high carbohydrate-induced obesity while skin SCD1 deficiency protects mice against high fat diet-induced obesity mediated by a circulating factor derived from the skin. More experiments will be required to provide physiologically relevant information on the role that skin SCD1 gene plays in energy homeostasis.

Acknowledgements This work has been supported by NIH grant NIDDK-RO162388, AHA Grant-in Aid, Wisconsin Affiliate, USDA (Hatch) grant # 142-4306.

References

Abu-Elheiga L, Matzuk MM, Abo-Hashema KA, Wakil SJ (2001) Continuous fatty acid oxidation and reduced fat storage in mice lacking acetyl-CoA carboxylase 2. Science 291:2613–2616

Attie AD, Krauss RM, Gray-Keller MP, Brownlie A, Miyazaki M, Kastelein JJ, Lusis AJ, Stalenhoef AF, Stoehr JP, Hayden MR, Ntambi JM (2002) Relationship between stearoyl-CoA desaturase activity and plasma triglycerides in human and mouse hypertriglyceridemia. J Lipid Res 43:1899–1907

Binczek E, Jenke B, Holz B, Gunter RH, Thevis M, Stoffel W (2007) Obesity resistance of stearoyl-CoA desaturase-deficient (SCD1-/-) mouse results from disruption of epidermal lipid barrier and adaptive thermoregulation. Biol Chem 388:405–418

Bonet ML, Ribot J, Felipe F, Palou A (2003) Vitamin A and the regulation of fat reserves. Cell Mol Life Sci 7:1311–1321

Casimir DA, Ntambi JM (1996) cAMP activates the expression of stearoyl-CoA desaturase gene 1 during early preadipocyte differentiation. J Biol Chem 271:29847–29853

Christianson JL, Nicoloro S, Straubhaar J, Czech MP (2008) Stearoyl-CoA desaturase 2 is required for peroxisome proliferator-activated receptor gamma expression and adipogenesis in cultured 3T3-L1 cells. J Biol Chem 283:2906–2916

Cohen P, Miyazaki M, Socci ND, Hagge-Greenberg A, Liedtke W, Soukas AA, Sharma R, Hudgins LC, Ntambi JM, Friedman JM (2002) Role for stearoyl-CoA desaturase-1 in leptin-mediated weight loss. Science 297:240–243

Dobrzyn P, Dobrzyn A, Miyazaki M, Cohen P, Asilmaz E, Hardie DG, Friedman JM, Ntambi JM (2004) Stearoyl-CoA desaturase-1 deficiency increases fatty acid oxidation by activating AMP-activated protein kinase in liver. Proc Natl Acad Sci USA 101:6409–6414

Flowers MT, Paton CM, O'Byrne SM, Schiesser K, Dawson JA, Blaner WS, Kendziorski C, Ntambi JM (2011) Metabolic changes in skin caused by Scd1 deficiency: a focus on retinol metabolism. PLoS One 5(5):e19734

Hulver MW, Berggren JR, Carper MJ, Miyazaki M, Ntambi JM, Hoffman EP, Thyfault JP, Stevens R, Dohm GL, Houmard JA, Muoio DM (2005) Elevated stearoyl-CoA desaturase-1 expression in skeletal muscle contributes to abnormal fatty acid partitioning in obese humans. Cell Metab 2:251–261

Kersten S, Mandard S, Tan NS, Escher P, Metzger D, Chambon P, Gonzalez FJ, Desvergne B, Wahli W (2000) Characterization of the fasting-induced adipose factor FIAF, a novel peroxisome proliferator-activated receptor target gene. J Biol Chem 275:28488–28493

Lee SH, Dobrzyn A, Dobrzyn P, Rahman SM, Miyazaki M, Ntambi JM (2004) Lack of stearoyl-CoA desaturase 1 upregulates basal thermogenesis but causes hypothermia in a cold environment. J Lipid Res 45:1674–1682

Loffler H, Aramaki JU, Effendy I (2002) The influence of body mass index on skin susceptibility to sodium lauryl sulphate. Skin Res Technol 8:19–22

Lunzer MA, Manning JA, Ockner RK (1977) Inhibition of rat liver acetyl coenzyme A carboxylase by long chain acyl coenzyme A and fatty acid. Modulation by fatty acid-binding protein. J Biol Chem 252:5483–5487

MacDonald PN, Ong DE (1988) Acyl-CoA-independent esterification of retinol bound to cellular retinol-binding protein (type II) by microsomes from rat small intestine. J Biol Chem 263:12478–12482

Man WC, Miyazaki M, Chu K, Ntambi JM (2006) Co-localization of SCD1 and DGAT2: implying preference for endogenous monounsaturated fatty acids in triglyceride synthesis. J Lipid Res 47:1928–1939

Marcelo CL, Duell EA, Rhodes LM, Dunham WR (1992) In vitro model of essential fatty acid deficiency. J Invest Dermatol 99:703–708
McGarry JD, Leatherman GF, Foster DW (1978) Carnitine palmitoyltransferase I. The site of inhibition of hepatic fatty acid oxidation by malonyl-CoA. J Biol Chem 253:4128–4136
Miller CW, Waters KM, Ntambi JM (1997) Regulation of hepatic stearoyl-CoA desaturase gene 1 by vitamin A. Biochem Biophys Res Commun 231:206–210
Miyazaki M, Kim YC, Gray-Keller MP, Attie AD, Ntambi JM (2000) The biosynthesis of hepatic cholesterol esters and triglycerides is impaired in mice with a disruption of the gene for stearoyl-CoA desaturase 1. J Biol Chem 275:30132–30138
Miyazaki M, Kim YC, Ntambi JM (2001a) A lipogenic diet in mice with a disruption of the stearoyl-CoA desaturase 1 gene reveals a stringent requirement of endogenous monounsaturated fatty acids for triglyceride synthesis. J Lipid Res 42:1018–1024
Miyazaki M, Man WC, Ntambi JM (2001b) Targeted disruption of stearoyl-CoA desaturase1 gene in mice causes atrophy of sebaceous and meibomian glands and depletion of wax esters in the eyelid. J Nutr 131:2260–2268
Miyazaki M, Dobrzyn A, Sampath H, Lee SH, Man WC, Chu K, Peters JM, Gonzalez FJ, Ntambi JM (2004) Reduced adiposity and liver steatosis by stearoyl-CoA desaturase deficiency are independent of peroxisome proliferator-activated receptor-alpha. J Biol Chem 279:35017–35024
Miyazaki M, Dobrzyn A, Elias PM, Ntambi JM (2005) Stearoyl-CoA desaturase-2 gene expression is required for lipid synthesis during early skin and liver development. Proc Natl Acad Sci USA 102:12501–12506
Miyazaki M, Flowers MT, Sampath H, Chu K, Otzelberger C, Liu X, Ntambi JM (2007) Hepatic stearoyl-CoA desaturase-1 deficiency protects mice from carbohydrate-induced adiposity and hepatic steatosis. Cell Metab 6:484–496
Napoli JL (1999) Interactions of retinoid binding proteins and enzymes in retinoid metabolism. Biochim Biophys Acta 1440:139–162
Ntambi JM, Miyazaki M, Stoehr JP, Lan H, Kendziorski CM, Yandell BS, Song Y, Cohen P, Friedman JM, Attie AD (2002) Loss of stearoyl-CoA desaturase-1 function protects mice against adiposity. Proc Natl Acad Sci USA 99:11482–11486
Sampath H, Flowers MT, Liu X, Paton CM, Sullivan R, Chu K, Zhao M, Ntambi JM (2009) Skin-specific deletion of stearoyl-CoA desaturase-1 alters skin lipid composition and protects mice from high-fat diet-induced obesity. J Biol Chem 284:19961–19973
Shih MYS, Kane MA, Zhou P, Yen CL, Streeper RS, Napoli JL, Farese RV (2009) Retinol esterification by DGAT is essential for retinoid homeostasis in murine skin. J Biol Chem 284:4292–4299
Shimomura I, Bashmakov Y, Horton JD (1999a) Increased levels of nuclear SREBP-1c associated with fatty livers in two mouse models of diabetes mellitus. J Biol Chem 274:30028–30032
Shimomura I, Bashmakov Y, Ikemoto S, Horton JD, Brown MS, Goldstein JL (1999b) Insulin selectively increases SREBP-1c mRNA in the livers of rats with streptozotocin-induced diabetes. Proc Natl Acad Sci USA 96:13656–13661
Smith SJ, Cases S, Jensen DR, Sande E, Tow B, Sanan D, Raber J, Eckel RH, Farese RV Jr (2000) Obesity resistance and multiple mechanisms of triglyceride synthesis in mice lacking DGAT. Nat Genet 25:87–90
Sundberg JP, Boggess D, Sundberg BA, Eilertsen K, Parimoo S, Filippi M, Stenn K (2000) Asebia-2J (Scd1(ab2J)): a new allele and a model for scarring alopecia. Am J Pathol 156:2067–2075
Villarroya F, Giralt M, Iglesias R (1999) Retinoids and adipose tissues: metabolism, cell differentiation and gene expression. Int J Obes Relat Metab Disord 1:1–6
Yan QW, Yang Q, Mody N, Graham TE, Hsu CH, Xu Z, Houstis NE, Kahn BB, Rosen ED (2007) The adipokine lipocalin 2 is regulated by obesity and promotes insulin resistance. Diabetes 56:2533–2540

Yen CE, Stone SJ, Koliward S, Harris C, Farase RV (2008) DGAT enzymes and triacylglycerol biosynthesis. J Lipid Res 49:2283–2301

Zheng Y, Eilertsen KJ, Ge L, Zhang L, Sundberg JP, Prouty SM, Stenn KS, Parimoo S (1999) Scd1 is expressed in sebaceous glands and is disrupted in the asebia mouse. Nat Genet 23:268–270

Ziouzenkova O, Orasanu G, Sharlach M, Akiyama TE, Berger JP, Viereck J, Hamilton JA, Tang G, Dolnikowski GG, Vogel S et al (2007) Retinaldehyde represses adipogenesis and diet-induced obesity. Nat Med 6:695–702

Chapter 4
Role for Stearoyl-CoA Desaturase-1 in the Metabolic Effects of Leptin

Paul Cohen

Obesity and Metabolic Disease

Over the last few decades, there has been a worldwide increase in overweight and obesity. This has been associated with a dramatic rise in the incidence of diabetes and other metabolic diseases. The burden of these illnesses threatens to undermine many of the medical advances of the past generation. Obesity is also placing a strain on the healthcare system and economies of many nations, which could have broad societal implications.

In the United States, where obesity has had perhaps the most profound impact, more than one-third adults are obese and approximately two-thirds are overweight (Flegal et al. 2012). Body mass index (BMI), which is the most common metric of overweight and obesity, is associated with increased relative risk of diabetes, hypertension, and coronary heart disease (Kopelman 2000). Obesity is at the center of the metabolic syndrome, which has been defined based on increased visceral adiposity in conjunction with elevated blood pressure, elevated blood glucose, elevated triglycerides, and decreased HDL cholesterol. Obesity is also associated with an increased risk of dyslipidemia, nonalcoholic fatty liver disease, cancer, osteoarthritis, gout, gallstones, and sleep apnea. In addition to these comorbid conditions, obesity, and particularly visceral obesity, has been found in some studies to be linked to an increased risk of death (Pischon et al. 2008).

Obesity also carries a painful social stigma. Individuals who are overweight and obese tend to be viewed as lacking in willpower. Advances in basic science have started to dispel these misconceptions, with the increasing appreciation that body weight is under tight biological regulation (Friedman 2003). A deeper understanding of the mechanisms underlying body homeostasis will be crucial in helping to develop therapies for obesity and comorbid conditions.

P. Cohen, M.D., Ph.D. (✉)
Cardiovascular Division, Brigham and Women's Hospital, 75 Francis Street, Boston, MA 02115, USA
e-mail: pcohen3@partners.org

J.M. Ntambi (ed.), *Stearoyl-CoA Desaturase Genes in Lipid Metabolism*,
DOI 10.1007/978-1-4614-7969-7_4, © Springer Science+Business Media New York 2013

Leptin and the Regulation of Body Weight

The hormone leptin is the central regulator in an endocrine loop regulating appetite and energy homeostasis (Friedman 2009). The existence of such a pathway controlling homeostasis was suggested by parabiosis experiments done on the morbidly obese mouse mutants, *ob/ob* and *db/db*. These studies indicated the existence of a circulating factor regulating body weight, which *ob/ob* mice lack while *db/db* mice lack the ability to respond to it (Coleman 1978).

These predictions were confirmed with the positional cloning of the *ob* gene, which encodes the adipose-derived hormone leptin (Zhang et al. 1994). *Ob/ob* mice are massively obese and hyperphagic. They also develop severe insulin resistance and ectopic fat deposition in the liver and other peripheral tissues. Subsequent studies identified the leptin receptor as product of the gene mutated in *db/db* mice (Chen et al. 1996; Lee et al. 1996).

Leptin is a circulating hormone produced by adipocytes that acts via its receptor in the hypothalamus to coordinate food intake and whole body metabolism. The leptin receptor has five different isoforms (denoted ObRa-ObRe), which have also been detected on some peripheral tissues (Fei et al. 1997). In normal physiology, leptin is secreted from adipose tissue and signals the body's nutritional state to the hypothalamus, which coordinates food intake and energy expenditure to try and maintain energy balance. Treatment of wild-type or *ob/ob* mice with leptin results in voluntary reduction of food intake, increased energy expenditure, and weight loss (Halaas et al. 1995; Campfield et al. 1995; Pelleymounter et al. 1995). Leptin also acts to signal nutritional deprivation. Low leptin levels lead to an adaptive response to conserve energy, with increased food intake and suppression of the reproductive and other neuroendocrine axes (Ahima et al. 1996).

Central administration of leptin can recapitulate the effects of peripheral administration at doses that have no effect when delivered peripherally (Campfield et al. 1995). In addition, neuronal-specific deletion of the leptin receptor results in the same obesity phenotype as that seen in mice with whole body deletion of the receptor (Cohen et al. 2001). Leptin, via its receptor in the hypothalamus, activates anorexigenic pathways and inhibits orexigenic pathways (Saper et al. 2002). These pathways then signal to other brain centers to ultimately coordinate feeding behavior and energy expenditure.

Humans with mutations in leptin or the leptin receptor have also been described. They too have morbid obesity, hyperphagia, and insulin resistance (Strobel et al. 1998; Montague et al. 1997). Leptin replacement in humans with leptin mutations results in a dramatic correction of the obesity phenotype (Farooqi et al. 1999). Unfortunately, leptin mutations are very rare and a rather small percentage of obese humans are leptin-deficient. The great majority of overweight and obese individuals have increased plasma leptin levels, proportional to their fat mass (Maffei et al. 1995). Because most obesity is associated with leptin resistance, leptin, on its own, is not likely to be an effective weight loss drug. In order to develop therapies for obesity and metabolic diseases, it is important to understand the molecular basis for leptin's effects.

Metabolic Effects of Leptin

Several lines of evidence indicate that leptin produces unique metabolic effects distinct from those seen with food restriction alone. *Ob/ob* mice that are food restricted to the level that leptin-treated mice voluntarily consume (pair-fed) lose significantly less weight than leptin-treated counterparts (Levin et al. 1996). Additionally, leptin results in selective loss of fat mass, while food restriction leads to decreased fat and muscle mass. Food restriction is also associated with a compensatory decrease in energy expenditure, whereas leptin-mediated weight loss is associated with increased energy expenditure. Leptin deficiency is also characterized by ectopic lipid deposition in liver, muscle, and other peripheral tissues. This excess lipid is the hallmark of nonalcoholic fatty liver disease. It is also thought to play a pathophysiological role in the development of insulin resistance and diabetes. Leptin, much more so than pair-feeding, is able to potently deplete lipid from these tissues, and thereby results in rapid improvement in insulin resistance and other metabolic defects (Shimabukuro et al. 1997). Unlike with food restriction, leptin treatment is not associated with increased free fatty acids or ketones, supporting a unique mechanism of fatty acid oxidation. These findings indicate that leptin mediates weight loss, at least in part by activating a molecular program with unique effects on metabolism.

Identification of SCD-1 as a Leptin-Repressed Gene in Liver

Leptin-deficient *ob/ob* mice develop severely enlarged livers that are overloaded with lipid. Leptin treatment corrects the hepatic steatosis in these animals far more profoundly than food restriction. Following 12 days of leptin treatment, the liver of an *ob/ob* mouse is indistinguishable from that of a wild-type littermate, at both the gross and histological level. These dramatic effects led us to examine the molecular basis by which leptin depletes hepatic lipid stores. We postulated that this pathway would be important to liver metabolism and whole body energy homeostasis.

To determine the molecular mechanism by which leptin corrects hepatic steatosis we used oligonucleotide microarrays to identify leptin-regulated genes in the liver. *Ob/ob* mice were followed over a time course of either leptin treatment or vehicle treatment with pair-feeding for 2, 4, or 12 days. Free fed *ob/ob* mice treated with saline were also studied. Leptin-treated mice lost significantly more weight and showed a much greater correction of hepatic steatosis than their pair-fed counterparts. Liver RNA was isolated from mice from each group and hybridized to Affymetrix arrays. At the time this experiment was performed, the standard murine arrays contained 6,500 genes. In total, this experiment generated over 50,000 data points.

We developed computational approaches to identify leptin-regulated genes from this large data set (Soukas et al. 2000). Cluster analysis is a well described method for finding groups of genes with coordinate patterns of expression. In this data set, six clusters of genes were specifically regulated by leptin. However, this list still contained several hundred genes. To prioritize this further, we used a more focused approach to identify genes specifically repressed by leptin. We developed an algorithm to rank genes based on the following criteria: (1) increased expression in *ob/ob* relative to wild-type liver, (2) reduced expression with leptin treatment, and (3) maximally different expression between leptin and pair-feeding. This resulted in a prioritized list of genes with increased expression in *ob/ob* liver, which were specifically repressed by leptin. We hypothesized that downregulation of these genes might mediate some of the novel metabolic actions of leptin.

The gene encoding the enzyme stearoyl-CoA desaturase-1 (SCD-1) was the top-ranked gene with this approach (Cohen et al. 2002). SCD-1 is the rate-limiting enzyme in the biosynthesis of monounsaturated fats. The desaturation reaction requires NADPH, cytochrome b5, and cytochrome b5 reductase. The substrates of SCD-1 are palmitic (16:0) and stearic acid (18:0) which it desaturates to generate palmitoleic (16:1) and oleic acid (18:1). These products are among the most abundant fatty acids found in triglycerides, phospholipids, and cholesterol esters. SCD-1 is localized to the endoplasmic reticulum and undergoes rapid turnover in response to a number of nutritional and hormonal cues (Heinemann and Ozols 2003). SCD-1 is also under transcriptional regulation by SREBP-1, ChREBP, and other factors (Ntambi et al. 2004). SCD-1 is broadly expressed and is one of four characterized SCD genes in mice, which are in a cluster on chromosome 19. Humans have a single SCD gene with high homology to murine SCD-1. SCD homologs have also been described in yeast, flies, and worms.

To confirm that this algorithm properly identified SCD-1 as a leptin-regulated gene, an independent time course was done. Livers were isolated and SCD-1 RNA levels, enzymatic activity, and substrate/product levels were measured (Cohen et al. 2002). SCD-1 RNA levels were markedly elevated in *ob/ob* liver. Upon leptin treatment, SCD-1 RNA levels normalized by day 2, and within 4 days had fallen to levels below that in wild-type mice. Pair-fed mice showed a less robust and delayed decrease in SCD-1 expression. Measures of enzymatic activity reflected these trends in RNA levels. SCD enzymatic activity was more than sevenfold elevated in *ob/ob* livers. Leptin treatment completely normalized enzymatic activity, while pair-feeding had a more modest effect. The levels of 16:1 and 18:1, the products of SCD-1, were elevated in *ob/ob* liver and were preferentially normalized by leptin. Importantly, leptin also repressed SCD-1 in wild-type mice, suggesting that its effects on SCD-1 expression are not simply secondary to its correction of the obesity phenotype.

SCD-1 Repression Mediates the Metabolic Effects of Leptin

To determine whether repression of SCD-1 is important in mediating the metabolic effects of leptin, we studied mouse models of SCD-1 deficiency. Asebia mice (*ab^J/ab^J*) are a spontaneous mutant strain with a genomic deletion of the first four exons of

SCD-1. These mice have no functional SCD-1 mRNA or protein, though the expression of the related gene SCD-2 is unchanged (Zheng et al. 1999). SCD-1 knockout mice have also been generated by standard gene targeting ($SCD1^{-/-}$) (Ntambi et al. 2002). The phenotype of these two mutant strains is largely indistinguishable.

Asebia mice have severe ocular and cutaneous abnormalities and have received their name due to an absence of sebaceous glands. These mice also have decreased synthesis of palmitoleic and oleic acid and reduced levels of hepatic triglycerides, cholesterol esters, and VLDL production (Miyazaki et al. 2000). 16:1 and 18:1 are among the most abundant lipid species found in the diet. However, dietary supplementation is unable to correct the decrease in monounsaturated fats seen in mutant animals (Miyazaki et al. 2001). This indicates that there is a requirement for endogenous synthesis of these lipids.

Upon evaluating asebia mice, we appreciated that they were clearly leaner than their control littermates (Cohen et al. 2002). Despite normal body weight, asebia mice have reduced fat mass and plasma leptin levels. We next examined whether SCD-1 deficiency might replicate any of the effects of leptin. We predicted that *ob/ob* mice lacking SCD-1 would resemble leptin-treated *ob/ob* animals. Double mutant ab^J/ab^J;*ob/ob* mice were generated and found to have a dramatic reduction in body weight relative to *ob/ob* littermates. By 16 weeks of age, double mutants weighed 30 % less than *ob/ob* mice, a weight comparable to that seen in *ob/ob* mice after 12 days of leptin treatment. Body composition analysis confirmed a significant reduction in adiposity in double mutant mice, with an increase in lean body mass. Additionally, $SCD\text{-}1^{-/-}$ mice were protected from diet-induced obesity. The weight of every white adipose depot was significantly lower in knockout mice relative to controls (Ntambi et al. 2002).

To understand the mechanism whereby SCD-1-deficient mice were protected from obesity we measured food intake and energy expenditure (Cohen et al. 2002). Surprisingly, we found that mice lacking SCD-1 ate significantly more food than controls. Food intake in asebia mice was equivalent to that of *ob/ob* mice, and double mutant ab^J/ab^J;*ob/ob* mice consumed even more. We therefore predicted that mice lacking SCD-1 must have increased energy expenditure. We performed indirect calorimetry and found that asebia mice had significantly greater total and resting oxygen consumption than littermate controls. *Ob/ob* mice are known to have markedly reduced energy expenditure, since the absence of leptin signals a state of perceived starvation. However, *ob/ob* mice lacking SCD-1 had a complete correction of this hypometabolic state, with oxygen consumption equivalent to or greater than wild-type littermates.

SCD-1 and Fatty Liver Disease

Since leptin treatment potently depletes hepatic lipid while repressing SCD-1, we examined whether SCD-1 deficiency could protect *ob/ob* mice from hepatic steatosis. Livers from double mutant ab^J/ab^J;*ob/ob* mice were grossly and histologically indistinguishable from wild-type (Cohen et al. 2002). Triglyceride content was reduced more than threefold in double mutant relative to *ob/ob* livers, with levels comparable to those found in wild-type livers. Triglyceride levels in ab^J/ab^J mice were reduced even further than those of wild-type mice.

This protection from hepatic steatosis prompted us to evaluate whether SCD-1 deficiency could protect from lipid accumulation in other models. Lipodystrophy is defined by partial or complete absence of adipose tissue and is associated with leptin deficiency, insulin resistance, diabetes, and hepatic steatosis (Reitman et al. 2000). Human lipodystrophy can occur due to congenital mutations or due to acquired conditions such as with anti-retroviral therapy for HIV. Leptin replacement in a mouse model of lipodystrophy and in humans with the condition can markedly improve insulin resistance, diabetes, and fatty liver (Shimomura et al. 1999; Oral et al. 2002).

We examined whether hepatic SCD-1 is dysregulated in a mouse model of lipodystrophy. aP2-nSREBP1c mice have enlarged fatty livers with increased SCD-1 expression and enzymatic activity. Central administration of leptin (at doses that have no effect when delivered peripherally) completely normalized SCD activity in this lipodystrophic model (Asilmaz et al. 2004). This indicates that the effects of leptin on hepatic SCD-1 are centrally mediated. To assess the role of SCD-1 in the liver pathology in these mice, double mutant ab^J/ab^J*;aP2-nSREBP1c tg* mice were generated. Double mutant mice demonstrated a significant improvement in hepatic lipid overload, though it was not completely corrected.

Mechanism for the Metabolic Effects of SCD-1 Deficiency

Our data indicated that SCD-1 was likely to be an important metabolic control. Mice deficient in this enzyme were resistant to obesity and hepatic steatosis. This is due to markedly increased energy expenditure. To understand the mechanism underlying this phenomenon, we further examined hepatic metabolism in the setting of SCD-1 deficiency. Fatty acids in the liver are typically esterified to glycerol or cholesterol and can then (1) accumulate leading to hepatic steatosis, (2) be packaged into VLDL for export to other tissues, or (3) be oxidized (Cohen and Friedman 2004). Monounsaturated acids, which are the products of SCD-1, are required for triglyceride and cholesterol ester synthesis and for VLDL production. We predicted then that in the absence of SCD-1 there is a defect in lipid storage and VLDL synthesis, and that as a default, there is increased fatty acid oxidation. Histological analysis and triglyceride quantitation confirmed that lipid storage was impaired in SCD-1-deficient animals. We next measured VLDL production by treating mice with tyloxapol, an inhibitor of VLDL hydrolysis. Following injection, we measured plasma triglyceride levels over time. Since VLDL hydrolysis is blocked, the levels rise in a linear fashion in relation to the rate of VLDL production. We found that VLDL synthesis is significantly increased in *ob/ob* mice and reduced in ab^J/ab^J*;ob/ob* mice to levels similar to that in wild-type controls (Cohen et al. 2002).

We therefore assumed that SCD-1 ablation must result in increased fatty acid oxidation. This is consistent with the finding of decreased adiposity and increased energy expenditure, in the setting of increased food intake. Moreover, mice lacking SCD-1 have increased plasma ketones, a marker of fatty acid oxidation. Finally, gene expression profiling in SCD-$1^{-/-}$ mice demonstrated increased expression of genes involved in fatty acid oxidation (Ntambi et al. 2002).

We therefore considered a model to explain the mechanism by which SCD-1 deficiency might result in increased fatty acid oxidation. When SCD-1 is repressed, saturated fatty acyl CoAs cannot be desaturated and would therefore accumulate. Saturated fatty acyl CoAs potently allosterically inhibit acetyl CoA carboxylase (ACC), the enzyme that converts acetyl CoA into malonyl CoA, an intermediate in fatty acid biosynthesis (Lunzer et al. 1977). Malonyl CoA is known to inhibit carnityl palmitoyltransferase-1 (CPT-1), the rate-limiting enzyme in the transport of fatty acids from the cytosol, where they are synthesized, into the mitochondria, where they are oxidized (McGarry et al. 1977). When malonyl CoA levels fall, as would be expected in SCD-1 deficiency, CPT-1 is derepressed and fatty acids are shuttled into the mitochondria for oxidation. Decreased malonyl CoA would also result in reduced fatty acid biosynthesis. SCD-1 deficiency could also result in increased activity of AMP-activated protein kinase (AMPK), an intracellular control point for energy metabolism, that is activated by leptin (Minokoshi et al. 2002; Dobrzyn et al. 2004). In fact, unifying these two models, we found that SCD-1 deficiency is associated with increased AMPK activity along with reduced malonyl CoA levels, increased CPT activity, and increased fatty acid oxidation (Dobrzyn et al. 2004).

We also considered other mechanisms that could account for the increased energy expenditure in animals deficient in SCD-1. For example, inhibition of SCD-1 could increase the levels of a ligand for peroxisome proliferator-activated receptor α (PPARα) or another nuclear receptor. Interestingly, the PPARα target gene and rate-limiting enzyme in peroxisomal fatty acid oxidation, acyl CoA oxidase, is increased in livers of SCD-$1^{-/-}$ mice (Ntambi et al. 2002). SCD-1 inhibition could also modulate non-shivering thermogenesis, via actions on uncoupling proteins. Finally, SCD-1 null mice have defects in fur and skin lipids, which could result in increased energy dissipation.

In work done by other groups, it has become clear that inhibition of other genes in the same metabolic pathway as SCD-1 is also linked to increased energy expenditure. Mice with a knockout of ACC-2 demonstrate decreased malonyl CoA levels, increased fatty acid oxidation, and protection from obesity (Abu-Elheiga et al. 2001). The drug C75 causes weight loss by inhibiting the enzyme fatty acid synthase and by derepressing CPT-1 (Loftus et al. 2000). Additionally, deletion of acyl CoA:diacylglycerol acyltransferase (DGAT), which catalyzes the final step in fatty acid biosynthesis, results in some of the same cutaneous and metabolic findings as that seen in SCD-1 deficiency (Smith et al. 2000; Chen et al. 2002).

One of the roles of the triglyceride biosynthetic pathway in the liver is to generate products that can be packaged into VLDL for export to adipose and other tissues. Genetic mutations inhibiting VLDL delivery and hydrolysis are also associated with protection from obesity. Mice lacking the VLDL receptor are lean and protected from dietary and genetic obesity (Frykman et al. 1995; Weinstock et al. 1997). *Ob/ob* mice deficient in adipose tissue lipoprotein lipase, the enzyme required for triglyceride hydrolysis, also show reduced adiposity (Weinstock et al. 1997). Lastly, *ob/ob* mice overexpressing apolipoprotein C1, which inhibits VLDL hydrolysis, have a nearly complete correction of obesity (Jong et al. 2001).

SCD-1 and Other Components of the Metabolic Syndrome

Given that SCD-1 deficiency is associated with increased energy expenditure and resistance to obesity, we also evaluated whether it protects against other metabolic phenotypes, such as diabetes and hyperlipidemia. SCD-$1^{-/-}$ mice have improved glucose and insulin tolerance, suggesting that inhibition of SCD-1 might protect against diabetes (Ntambi et al. 2002). However, SCD-1 deficiency was not able to improve the severe diabetic phenotype seen in lipodystrophic mice (Asilmaz et al. 2004).

Mice lacking SCD-1 have significantly reduced synthesis of triglycerides, cholesterol esters, and VLDL (Miyazaki et al. 2000). Oleic acid, one of the products of SCD-1, is the preferred substrates for acyl CoA:cholesterol acyltransferase (ACAT), the rate-limiting enzyme in cholesterol esterification (Buhman et al. 2000). Mice deficient in SCD-1 have decreased cholesterol ester synthesis, despite normal ACAT activity. Additionally, in the *hyplip* murine model, hypertriglyceridemia is associated with SCD-1 activity (Attie et al. 2002). In a human cohort, SCD-1 activity was able to explain 44 % of the variance in triglyceride levels. These data suggest that inhibition of SCD-1 might protect against atherosclerosis.

SCD-1 as a Potential Therapeutic Target

Since our reports on the regulation of hepatic SCD-1 by leptin, additional studies have further clarified our original conclusions. Antisense oligonucleotides (ASOs) to SCD-1 reduced SCD-1 expression, decreased fatty acid synthesis, and increased fatty acid oxidation in primary hepatocytes. Administration of these ASOs to mice was able to prevent diet-induced obesity (Jiang et al. 2005). A separate study used ASOs to SCD-1 in mouse and rat models of insulin resistance. Insulin clamps showed that these ASOs were able to markedly improve hepatic insulin resistance (Gutierrez-Juarez et al. 2006).

Additional animal models have also further illuminated the role that SCD-1 in liver and other tissues contributes to body weight and metabolism. Mice with a liver-specific deletion of SCD-1 have been generated. These animals were protected from high carbohydrate, but not from high fat-induced obesity and hepatic steatosis (Miyazaki et al. 2007). This model suggests that SCD-1 activity in tissues other than liver plays a role in the phenotype seen with whole body deficiency. This was confirmed with the generation of mice with a skin-specific deletion of SCD-1. These mice were protected from high fat diet-induced obesity (Sampath et al. 2009).

These mice appear to have increased thermogenesis to compensate for defective insulating capacity by the skin.

In the aggregate, abundant evidence exists to support an important role for SCD-1 in whole body metabolism. Hepatic SCD-1 activity is markedly elevated in leptin-deficient animals. Leptin-mediated repression of hepatic SCD-1 appears to be an important component of the metabolic effects of leptin. Subsequent studies have shown that SCD-1 in the skin is important for maintaining insulation, and in its absence, animals are protected from obesity. Liver-specific small molecule inhibitors of SCD-1 have been generated (Oballa et al. 2011). These compounds may help further pinpoint the specific role of SCD-1 in obesity and other metabolic diseases. Further studies will be needed to determine the full spectrum of tissues where SCD-1 and other SCD isoforms impact upon energy homeostasis.

References

Abu-Elheiga L et al (2001) Continuous fatty acid oxidation and reduced fat storage in mice lacking acetyl-CoA carboxylase 2. Science 291(5513):2613–2616

Ahima RS et al (1996) Role of leptin in the neuroendocrine response to fasting. Nature 382(6588): 250–252

Asilmaz E et al (2004) Site and mechanism of leptin action in a rodent form of congenital lipodystrophy. J Clin Invest 113(3):414–424

Attie AD et al (2002) Relationship between stearoyl-CoA desaturase activity and plasma triglycerides in human and mouse hypertriglyceridemia. J Lipid Res 43(11):1899–1907

Buhman KF, Accad M, Farese RV (2000) Mammalian acyl-CoA: cholesterol acyltransferases. Biochim Biophys Acta 1529(1–3):142–154

Campfield LA et al (1995) Recombinant mouse OB protein: evidence for a peripheral signal linking adiposity and central neural networks. Science 269(5223):546–549

Chen H et al (1996) Evidence that the diabetes gene encodes the leptin receptor: identification of a mutation in the leptin receptor gene in db/db mice. Cell 84(3):491–495

Chen HC et al (2002) Leptin modulates the effects of acyl CoA:diacylglycerol acyltransferase deficiency on murine fur and sebaceous glands. J Clin Invest 109(2):175–181

Cohen P, Friedman JM (2004) Leptin and the control of metabolism: role for stearoyl-CoA desaturase-1 (SCD-1). J Nutr 134(9):2455S–2463S

Cohen P et al (2001) Selective deletion of leptin receptor in neurons leads to obesity. J Clin Invest 108(8):1113–1121

Cohen P et al (2002) Role for stearoyl-CoA desaturase-1 in leptin-mediated weight loss. Science 297(5579):240–243

Coleman DL (1978) Obese and diabetes: two mutant genes causing diabetes-obesity syndromes in mice. Diabetologia 14(3):141–148

Dobrzyn P et al (2004) Stearoyl-CoA desaturase 1 deficiency increases fatty acid oxidation by activating AMP-activated protein kinase in liver. Proc Natl Acad Sci U S A 101(17): 6409–6414

Farooqi IS et al (1999) Effects of recombinant leptin therapy in a child with congenital leptin deficiency. N Engl J Med 341(12):879–884

Fei H et al (1997) Anatomic localization of alternatively spliced leptin receptors (Ob-R) in mouse brain and other tissues. Proc Natl Acad Sci U S A 94(13):7001–7005

Flegal KM et al (2012) Prevalence of obesity and trends in the distribution of body mass index among US adults, 1999-2010. JAMA 307(5):491–497

Friedman JM (2003) A war on obesity, not the obese. Science 299(5608):856–858
Friedman JM (2009) Leptin at 14 y of age: an ongoing story. Am J Clin Nutr 89(3):973S–979S
Frykman PK et al (1995) Normal plasma lipoproteins and fertility in gene-targeted mice homozygous for a disruption in the gene encoding very low density lipoprotein receptor. Proc Natl Acad Sci U S A 92(18):8453–8457
Gutierrez-Juarez R et al (2006) Critical role of stearoyl-CoA desaturase-1 (SCD1) in the onset of diet-induced hepatic insulin resistance. J Clin Invest 116(6):1686–1695
Halaas JL et al (1995) Weight-reducing effects of the plasma protein encoded by the obese gene. Science 269(5223):543–546
Heinemann FS, Ozols J (2003) Stearoyl-CoA desaturase, a short-lived protein of endoplasmic reticulum with multiple control mechanisms. Prostaglandins Leukot Essent Fatty Acids 68(2): 123–133
Jiang G et al (2005) Prevention of obesity in mice by antisense oligonucleotide inhibitors of stearoyl-CoA desaturase-1. J Clin Invest 115(4):1030–1038
Jong MC et al (2001) Protection from obesity and insulin resistance in mice overexpressing human apolipoprotein C1. Diabetes 50(12):2779–2785
Kopelman PG (2000) Obesity as a medical problem. Nature 404(6778):635–643
Lee GH et al (1996) Abnormal splicing of the leptin receptor in diabetic mice. Nature 379(6566):632–635
Levin N et al (1996) Decreased food intake does not completely account for adiposity reduction after ob protein infusion. Proc Natl Acad Sci U S A 93(4):1726–1730
Loftus TM et al (2000) Reduced food intake and body weight in mice treated with fatty acid synthase inhibitors. Science 288(5475):2379–2381
Lunzer MA, Manning JA, Ockner RK (1977) Inhibition of rat liver acetyl coenzyme A carboxylase by long chain acyl coenzyme A and fatty acid. Modulation by fatty acid-binding protein. J Biol Chem 252(15):5483–5487
Maffei M et al (1995) Leptin levels in human and rodent: measurement of plasma leptin and ob RNA in obese and weight-reduced subjects. Nat Med 1(11):1155–1161
McGarry JD, Mannaerts GP, Foster DW (1977) A possible role for malonyl-CoA in the regulation of hepatic fatty acid oxidation and ketogenesis. J Clin Invest 60(1):265–270
Minokoshi Y et al (2002) Leptin stimulates fatty-acid oxidation by activating AMP-activated protein kinase. Nature 415(6869):339–343
Miyazaki M et al (2000) The biosynthesis of hepatic cholesterol esters and triglycerides is impaired in mice with a disruption of the gene for stearoyl-CoA desaturase 1. J Biol Chem 275(39): 30132–30138
Miyazaki M, Kim YC, Ntambi JM (2001) A lipogenic diet in mice with a disruption of the stearoyl-CoA desaturase 1 gene reveals a stringent requirement of endogenous monounsaturated fatty acids for triglyceride synthesis. J Lipid Res 42(7):1018–1024
Miyazaki M et al (2007) Hepatic stearoyl-CoA desaturase-1 deficiency protects mice from carbohydrate-induced adiposity and hepatic steatosis. Cell Metab 6(6):484–496
Montague CT et al (1997) Congenital leptin deficiency is associated with severe early-onset obesity in humans. Nature 387(6636):903–908
Ntambi JM et al (2002) Loss of stearoyl-CoA desaturase-1 function protects mice against adiposity. Proc Natl Acad Sci U S A 99(17):11482–11486
Ntambi JM, Miyazaki M, Dobrzyn A (2004) Regulation of stearoyl-CoA desaturase expression. Lipids 39(11):1061–1065
Oballa RM et al (2011) Development of a liver-targeted stearoyl-CoA desaturase (SCD) inhibitor (MK-8245) to establish a therapeutic window for the treatment of diabetes and dyslipidemia. J Med Chem 54(14):5082–5096
Oral EA et al (2002) Leptin-replacement therapy for lipodystrophy. N Engl J Med 346(8):570–578
Pelleymounter MA et al (1995) Effects of the obese gene product on body weight regulation in ob/ob mice. Science 269(5223):540–543
Pischon T et al (2008) General and abdominal adiposity and risk of death in Europe. N Engl J Med 359(20):2105–2120

Reitman ML et al (2000) Lipoatrophy revisited. Trends Endocrinol Metab 11(10):410–416
Sampath H et al (2009) Skin-specific deletion of stearoyl-CoA desaturase-1 alters skin lipid composition and protects mice from high fat diet-induced obesity. J Biol Chem 284(30):19961–19973
Saper CB, Chou TC, Elmquist JK (2002) The need to feed: homeostatic and hedonic control of eating. Neuron 36(2):199–211
Shimabukuro M et al (1997) Direct antidiabetic effect of leptin through triglyceride depletion of tissues. Proc Natl Acad Sci U S A 94(9):4637–4641
Shimomura I et al (1999) Leptin reverses insulin resistance and diabetes mellitus in mice with congenital lipodystrophy. Nature 401(6748):73–76
Smith SJ et al (2000) Obesity resistance and multiple mechanisms of triglyceride synthesis in mice lacking Dgat. Nat Genet 25(1):87–90
Soukas A et al (2000) Leptin-specific patterns of gene expression in white adipose tissue. Genes Dev 14(8):963–980
Strobel A et al (1998) A leptin missense mutation associated with hypogonadism and morbid obesity. Nat Genet 18(3):213–215
Weinstock PH et al (1997) Lipoprotein lipase controls fatty acid entry into adipose tissue, but fat mass is preserved by endogenous synthesis in mice deficient in adipose tissue lipoprotein lipase. Proc Natl Acad Sci U S A 94(19):10261–10266
Zhang Y et al (1994) Positional cloning of the mouse obese gene and its human homologue. Nature 372(6505):425–432
Zheng Y et al (1999) Scd1 is expressed in sebaceous glands and is disrupted in the asebia mouse. Nat Genet 23(3):268–270

Chapter 5
Regulation and Metabolic Functions of White Adipose Tissue Stearoyl-CoA Desaturase

Nishan Sudheera Kalupahana, Thilak Jayalath, Shu Wang, and Naima Moustaid-Moussa

Introduction

White adipose tissue (WAT) is the main site for the storage of excess energy, in the form of fat, in the body. In addition to its energy-storage function, adipose tissue also secretes numerous bioactive peptides collectively known as adipokines. These adipokines have important auto/paracrine as well as endocrine functions in regulation of energy balance, glucose homeostasis, and blood pressure. Therefore, adipose tissue is considered to be a dynamic endocrine organ.

As discussed in previous chapters, stearoyl-CoA desaturase (SCD) is a microsomal enzyme which catalyzes the rate limiting step of monounsaturated fatty acid (MUFA) synthesis from saturated fatty acid (SFA) precursors. Humans have two SCD isoforms, SCD1 and SCD5, of which WAT mainly expresses SCD1. Mice have four SCD isoforms, of which WAT expresses SCD1 and SCD2 encoded by *Scd1* and *Scd2,* respectively. The preferential substrates for SCD are palmitoyl-CoA and stearoyl-CoA, which are converted to palmitoleoyl-CoA and oleoyl-CoA, respectively. The latter acyl-CoAs are incorporated into triglycerides, with oleic acid being the predominant fatty acid in human adipose tissue triacylglycerols (Mauvoisin and Mounier 2011). Therefore SCD plays an important role in triglyceride storage in adipose tissue.

N.S. Kalupahana
Department of Physiology, Faculty of Medicine, University of Peradeniya, Peradeniya, Sri Lanka

T. Jayalath
Department of Medicine, Faculty of Medicine, University of Peradeniya, Peradeniya, Sri Lanka

S. Wang • N. Moustaid-Moussa (✉)
Nutritional Sciences, Texas Tech University, Lubbock, TX 79409, USA
e-mail: naima.moustaid-moussa@ttu.edu

J.M. Ntambi (ed.), *Stearoyl-CoA Desaturase Genes in Lipid Metabolism*,
DOI 10.1007/978-1-4614-7969-7_5, © Springer Science+Business Media New York 2013

In obese individuals, there is excessive triglyceride storage in adipocytes, which leads to adipocyte hypertrophy. This is also associated with development of a chronic low-grade inflammation in WAT, which leads to a dysregulation of adipokine secretory patterns. These changes lead to metabolic derangements and result in diseases such as type-2 diabetes and hypertension. Since SCD plays a key role in triglyceride storage in adipose tissue, it is likely that SCD plays an important role in modulating adipose tissue function as well as whole body metabolism. In this context, it is important to understand the factors which regulate SCD expression in WAT. This chapter provides an overview of adipose tissue structure and function, factors affecting WAT SCD expression and how adipose tissue SCD expression affects whole body metabolism.

Structure and Functions of WAT

Adipose tissue is considered as an endocrine organ, which consists of ~50 % fat cells (adipocytes) and ~50 % stromal vascular fraction that contains blood cells, endothelial cells, adipose tissue precursors, and macrophages in normal weight people (Wang and Nakayama 2010). These cells along with vascular and nervous tissue in the adipose tissue are supported by an extra-cellular matrix (Kalupahana et al. 2011). Adipose tissue macrophages (ATM) are classified according to the receptors they express as well as by their cytokine profile. Lean adipose tissue mainly contains M2 or alternatively activated macrophages, which secrete anti-inflammatory cytokines such as interleukin (IL)-4 and IL-10 (Kalupahana et al. 2012b). In contrast, adipose tissue from obese individuals contains predominantly M1 or classically activated macrophages, which secrete pro-inflammatory cytokines such as tumor necrosis factor-α (TNF-α), IL-1 and IL-6. In terms of lymphocytes, lean adipose tissue contains predominantly $FoxP3^+$ regulatory T cells (Tregs). This, however, changes with obesity so that the Tregs are reduced and other T cell subsets such as T helper 1 (Th1) and $CD8^+$ effector T cells are increased (Feuerer et al. 2009; Nishimura et al. 2009).

A major function of adipose tissue is to store excess energy in the form of triglycerides. Excess fatty acids from exogenous (dietary) and endogenous (mainly liver) sources are directed to the adipose tissue, while excess carbohydrates are converted to fatty acids (mainly SFA) by the liver (de novo lipogenesis) and then transported to the adipose tissue. De novo lipogenesis occurs to a certain extent in the adipose tissue as well. Fatty acids from both these sources are incorporated into triglycerides and stored in adipocytes. As described previously, SFA-induced lipogenesis is dependent upon SCD1. For example, in global SCD1 knockout mice, stearate does not induce lipogenesis (Sampath et al. 2007). Therefore, SCD is crucial in regulating triglyceride storage in adipocytes. Insulin is the main hormone which promotes lipid storage. In fasting stage, due to low insulin levels, lipolysis is stimulated and fatty acids are released from the adipose tissue to the blood. These fatty acids are taken up by tissues such as skeletal muscle, stored and/or oxidized to yield energy.

Increased lipid accumulation in skeletal muscle partially contributes to muscle insulin resistance. Defects in lipid storage by the adipose tissue have deleterious effects on body metabolism. For example, in insulin-resistant states and in conditions such as lipodystrophy, there is ectopic fatty acid deposition in organs such as liver, skeletal muscle, and pancreas, which leads to lipotoxicity and the development of diabetes (Hegele 2001). Therefore a normal lipid storage capacity of the adipose tissue is essential for homeostasis.

Adipokines, Obesity, and Adipose Tissue Inflammation

Adipose tissue secretes bioactive peptides termed adipokines. Some of the major adipokines secreted by the adipose tissue are leptin, adiponectin, resistin, angiotensinogen, and pro-inflammatory cytokines such as IL-6, monocyte chemotactic protein (MCP)-1, plasminogen activator inhibitor (PAI)-1, and TNF-α. Leptin functions as a major regulator of energy balance by reducing energy intake and increasing energy expenditure. Mice with genetic mutations of the leptin gene (Ob/Ob mice) or the leptin receptor (db/db mice) are severely obese, highlighting the critical role of this adipokine in energy homeostasis. Adiponectin is an insulin sensitizer, while resistin induces insulin resistance. Angiotensinogen (Agt), the precursor peptide of the hypertensive hormone angiotensin (Ang) II, is an important regulator of blood pressure and fluid balance. Recent evidence has identified Agt as an important regulator of glucose homeostasis as well (Kalupahana and Moustaid-Moussa 2012b). We and others have shown that Agt and Ang II derived from adipose tissue have important endocrine functions (Kalupahana et al. 2012a; Kalupahana and Moustaid-Moussa 2012a).

In states of chronic energy excess, adipose tissue expands by two mechanisms. First, new adipocytes are formed from adipocyte precursors such as preadipocytes and stem cells. This process is known as adipocyte hyperplasia. Second, excessive triglyceride storage within adipocytes leads to an increase in adipocyte size, which is known as adipocyte hypertrophy. Enlarged adipocytes change adipokine secretory patters, characterized by increased production of leptin, Agt, and pro-inflammatory cytokines such as MCP-1 and IL-6 and a reduced production of anti-inflammatory adipokines such as adiponectin (Kalupahana et al. 2011). Adipocyte hypertrophy also leads to adipocyte death. Endoplasmic reticulum stress and adipose tissue hypoxia are two mechanisms responsible for these changes (Boden et al. 2008; Pasarica et al. 2010). These changes in adipocytes are accompanied by changes in adipose tissue immune cell populations as described earlier, leading to a state of chronic low-grade inflammation in the adipose tissue. These changes in adipo/cytokine secretory patterns coupled with excessive lipolysis leading to lipotoxicity are considered to be causative factors for the generation of insulin resistance in the liver and skeletal muscle. Long-standing insulin resistance coupled with pancreatic dysfunction leads to the development of type-2 diabetes (Kalupahana et al. 2012b).

Regulation of Adipose Tissue SCD Expression

Hormonal Control

Human adipose tissue expresses SCD1, while murine adipose tissue expresses both SCD1 and SCD2 isoforms (Kaestner et al. 1989). In 3T3-L1 murine preadipocytes, SCD2 is expressed at higher levels than SCD1 (Kim et al. 2000). When differentiated by adding dexamethasone, methylisobutylxanthine, and insulin (DMI), SCD1 expression dramatically increases, while the increase in SCD2 is modest. Peroxisome proliferator-activated receptor gamma (PPARγ) is a transcription factor which is important in adipogenesis as well as maintaining normal systemic insulin sensitivity. PPARγ agonists have insulin-sensitizing effects. PPARγ agonist troglitazone increases 3T3-L1 preadipocyte differentiation. However, it suppresses SCD1 expression. Similar effects have been reported for pioglitazone, another PPARγ agonist (Kurebayashi et al. 1997). Expression of SCD2 is independent of both these pro-adipogenic mediators (DMI and troglitazone). In humans, WAT PPARγ expression is positively associated with SCD1 expression. Unlike the effect of troglitazone on SCD1 expression in murine adipocytes, patients treated with PPARγ agonist pioglitazone have a twofold increase in subcutaneous adipose tissue (SAT) SCD1 mRNA and protein expression (Yao-Borengasser et al. 2008). Further, pioglitazone increases SCD1 expression in cultured adipocytes (Yao-Borengasser et al. 2008). Therefore, while SCD expression increases with adipogenesis in humans and rodents, the effects of different PPARγ agonists on SCD expression is inconsistent. There appear to be response elements in the SCD promoter which respond to adipogenic agents (Ntambi et al. 1988).

WAT SCD expression is sensitive to both hormonal and nutritional signals, which are summarized in Fig. 5.1 and Table 5.1. Leptin is a major regulator of adipose tissue SCD expression. Similar to its action in the liver (Cohen et al. 2002), systemic administration of leptin decreases SCD1 expression in WAT (Macek Jilkova et al. 2010). Intracerebroventricular leptin administration also reduces WAT SCD1 expression, indicating that the central action of leptin on the brain has an effect on adipose tissue SCD expression (Gallardo et al. 2007; Lin et al. 2003). Since this effect of leptin is absent in the presence of a melanocortin antagonist, leptin-mediated suppression of WAT SCD1 appears to be mediated via melanocortin receptors (Lin et al. 2003). Conversely, low-leptinemic states such as caloric restriction are characterized by high adipose SCD1 levels (Macek Jilkova et al. 2010). While the anorexigenic hormone leptin decreases WAT SCD1 expression, central administration of the orexigenic hormone ghrelin increases WAT SCD1 expression (Theander-Carrillo et al. 2006). Considering the important function of SCD1 in mediating energy storage, this illustrates the opposite effects of leptin and ghrelin on energy balance—the former promoting energy expenditure and reducing energy storage, with the latter promoting energy storage.

Effects of other hormones involved in metabolism on WAT SCD expression are less clear. For example, WAT SCD1 is increased in mice lacking beta adrenergic

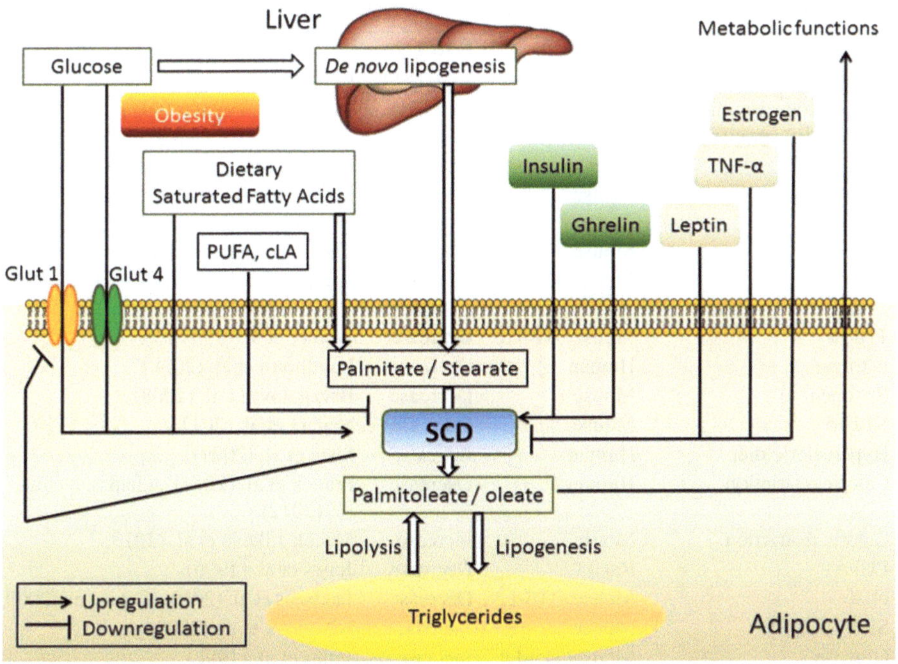

Fig. 5.1 Regulation and functions of white adipose tissue SCD. SCD catalyzes the desaturation of saturated fatty acids (palmitate and stearate) to from MUFAs (palmitoleate and oleate). Since these latter fatty acids are the preferred forms for triglyceride formation, SCD is crucial in regulating adipose tissue lipogenesis. Palmitoleate and oleate also have important regulatory functions in metabolism. In obesity, there is an increase in dietary intake of carbohydrates and saturated fats. Excess glucose from carbohydrates is converted into saturated fats by the liver (de novo lipogenesis). Both de novo-synthesized and dietary saturated fats upregulate SCD expression in adipose tissue, while PUFA and cLA downregulate its expression. Glucose enters adipocytes via Glut-4 and Glut-1 transporters and also upregulate SCD expression. SCD in turn has an inhibitory effect on Glut-1 expression. Hormonal control of SCD expression includes inhibitory effects from leptin, estrogen, and TNF-α and stimulatory effects from ghrelin and insulin. *PUFA* polyunsaturated fatty acids, *cLA* conjugated linoleic acid, *TNF* tumor necrosis factor

receptors. However, SCD1 expression is not suppressed by systemic administration of epinephrine (Mainieri et al. 2007). The effect of insulin on SCD1 expression is also inconsistent (Table 5.1). While we found that insulin did not have an independent effect on SCD1 gene transcription in 3T3-L1 adipocytes (Jones et al. 1998), others have reported that insulin increases SCD1 expression in the same adipocyte cell line (Weiner et al. 1991). TNF-α, a pro-inflammatory cytokine mediating insulin resistance in obese individuals, inhibits SCD1 expression in murine adipocytes in vitro (Weiner et al. 1991). Estrogen also appears to have an inhibitory effect on WAT SCD1 expression in humans (Lundholm et al. 2008) as well as mice (Bryzgalova et al. 2008). There are other hormones which influence SCD1 expression in a tissue-specific manner. For example, while growth hormone increases

Table 5.1 Factors which regulate white adipose tissue SCD1 expression

Factor	Species/cell line	Effect on expression	References
Troglitazone, pioglitazone (PPARγ agonists)	Mouse/3T3-L1	Decrease	Kurebayashi et al. (1997)
Pioglitazone	Human	Increase	Yao-Borengasser et al. (2008)
Leptin	Mouse	Decrease	Macek Jilkova et al. (2010), Gallardo et al. (2007), Lin et al. (2003)
Ghrelin	Mouse	Increase	Theander-Carrillo et al. (2006)
Insulin	Mouse/3T3-L1	Increase/no change	Weiner et al. (1991), Jones et al. (1998)
TNF-α	Mouse/3T3-L1	Decrease	Weiner et al. (1991)
Estrogen	Human	Decrease	Lundholm et al. (2008)
Estrogen	Mouse	Decrease	Bryzgalova et al. (2008)
KLF15	Mouse	Decrease	Nagare et al. (2011)
Hypercaloric diet	Human	Increase	Shea et al. (2009)
Caloric restriction	Human	Decrease	Franck et al. (2011), Johansson et al. (2012)
Caloric restriction	Mouse	Increase	Macek Jilkova et al. (2010)
PUFA	Rat	Decrease	Jones et al. (1996)
cLA	Mouse, 3T3-L1	Decrease	Jaudszus et al. (2010), Choi et al. (2000)
SFA	Mouse	Increase	Nikonova et al. (2008)
Glucose	Mouse/3T3-L1	Increase	Jones et al. (1998)
High-carbohydrate diet	Human	Increase/no change	Mangravite et al. (2007), Chong et al. (2008)

PPAR peroxisome proliferator-activated receptor, *TNF* tumor necrosis factor, *KLF* Kruppel-like factor, *PUFA* polyunsaturated fatty acid, *cLA* conjugated linoleic acid, *SFA* saturated fatty acid

SCD1 expression in the rodent liver, it has no effect on WAT SCD1 expression (Frick et al. 2002).

Certain transcription factors also regulate adipose SCD1 expression. For example, Kruppel-like factor 15 (KLF15) is a member of the Kruppel-like factor family of transcription factors, which is downregulated in WAT of obese animals. Transgenic overexpression of KLF15 downregulates adipose SCD1 expression (Nagare et al. 2011), possibly indicating that KLF15 is a negative regulator of SCD1 in WAT.

Nutritional Control

WAT SCD expression is regulated by both energy content as well as type of macronutrient in the diet. For example, consuming a 40 % hypercaloric diet for 7 days increases abdominal subcutaneous WAT SCD expression by fivefold in lean individuals (Shea et al. 2009). Conversely, caloric restriction leads to suppression of WAT SCD expression in humans (Franck et al. 2011; Johansson et al. 2012). WAT of both obese humans as well as animal models exhibit higher SCD expression, indicating a positive correlation of SCD expression with chronic energy excess.

We have shown that WAT SCD1 expression is increased in obese Zucker rats (Jones et al. 1996). Chronic food restriction/re-feeding also results in an increase in WAT SCD1 expression (Turyn et al. 2010).

WAT SCD1 exhibits a differential response to high-fat and high-carbohydrate diets. A high-polyunsaturated fatty acid (PUFA) diet can reduce the SCD1 mRNA expression by about 75 %, both in vivo as well as in vitro, suggesting that PUFA negatively regulates SCD1 expression in adipocytes (Jones et al. 1996). The *trans*-10, *cis*-12 isomer of conjugated linoleic acid (CLA) also suppresses SCD1 expression in rodents (Jaudszus et al. 2010; Choi et al. 2000). In contrast, SFA-rich diets increase SCD1 expression in murine WAT (Nikonova et al. 2008). Unlike in other tissues, the regulation of adipose SCD1 gene expression by PUFA appears to be at the posttranscriptional level. For example, arachidonic acid decreases the SCD1 mRNA half-life, while oleic and stearic acids do not have an effect on it (Sessler et al. 1996). Transcriptional or posttranslational regulation does not seem to be of major importance in the PUFA-mediated suppression of SCD1 expression in murine adipocytes. The reduction in SCD1 enzyme activity by PUFA can be accounted by the decrease in mRNA levels. The mRNA stability of SCD1 in adipocytes is possibly regulated via mRNA sequences in the 3′-untranslated region (UTR). The unusually long 3′-UTR in rodent and human SCD1 cDNA contains several structural motifs characteristic of mRNA destabilization sequences. Whether these sequences are possible targets of PUFA merits further investigation.

While WAT SCD1 expression is positively regulated by glucose in murine adipocytes (Jones et al. 1998), the effect of high-carbohydrate diets on WAT SCD1 expression in vivo is inconsistent (Table 5.1). For example, while a 3-day high-carbohydrate diet increases the desaturation index in the venous blood draining SAT, WAT SCD1 expression remains unchanged (Chong et al. 2008). Others have reported positive associations between carbohydrate content in the diet and WAT SCD1 expression in humans (Mangravite et al. 2007). Low-fat high-carbohydrate diets also increase SCD2 expression in adipose tissue of mice (Kaestner et al. 1989).

Functions of WAT SCD

Regulation of Adipogenesis and Lipogenesis

In 3T3-L1 adipocytes, SCD1 desaturates palmitic (16:0) and stearic (18:0) acids to form palmitoleic (16:1 *n*-7) and oleic (18:1 *n*-9) acids, respectively. SCD2 appears to be more specific for the conversion of stearate to oleic acid (Kim et al. 2000).

SCD2 has an important function in preadipocyte differentiation. Knockdown of SCD2, but not SCD1, by small interfering RNA reduces adipogenesis in 3T3-L1 preadipocytes (Christianson et al. 2008). This could be possibly mediated via PPARγ, because SCD2 knockdown leads to reduced PPARγ transcription as well as translation (Christianson et al. 2008).

SCD and Obesity

SCD mRNA expression in WAT is greatly increased in genetic rodent models of obesity, such as the Zucker rat (Jones et al. 1996). In humans, BMI positively correlates with the adipose tissue desaturation index, suggesting an increased desaturase activity in obesity (Yee et al. 2012; Gong et al. 2011a; Okada et al. 2005). WAT SCD1 activity also closely correlates with insulin resistance (Sjogren et al. 2008). Further, large insulin-resistant adipocytes express higher levels of SCD1 compared to small insulin-sensitive ones (Matsubara et al. 2009). In conditions of chronic energy excess, there is a higher availability of SFAs to the WAT (either via dietary intake or due to de novo lipogenesis). Since MUFA are the preferred form of fatty acids to be stored in triglycerides via the action of SCD, it is plausible that the increase in SCD expression in obesity is a physiological adaptation to increased SFAs availability. This assertion is supported by the fact that intake of SFAs is associated with higher WAT SCD1 expression in rodents (Nikonova et al. 2008). Moreover, SCD1 knockout mice have low adiposity and reduced MUFA in adipocyte triglycerides (Liu et al. 2010).

However, other evidence suggests that SCD might have a regulatory role in energy balance and adiposity. For example, genetic polymorphisms in the SCD1 gene are associated with fat distribution and insulin sensitivity (Gong et al. 2011b). Moreover, products of reactions catalyzed by SCD such as palmitoleate (Cao et al. 2008) and oleate (Liu et al. 2010) are mediators of inflammation. Palmitoleate is considered to be an insulin-sensitizing lipokine in animal models (Cao et al. 2008). Interestingly, SAT from the lower body (gluteo-femoral) expresses higher SCD1 and produces more palmitoleate compared to SAT from anterior abdominal wall (Pinnick et al. 2012). Since lower-body adipose tissue expansion is associated with a favorable metabolic profile, it is possible that SCD and palmitoleate are possible mediators of these beneficial effects.

Effects of SCD on Adipose Tissue Inflammation

Obesity and increased adiposity lead to a chronic low-grade inflammation in the adipose tissue (described above). This adipose tissue inflammation is causally linked to the pathogenesis of obesity-associated insulin resistance and metabolic syndrome. Since adipose SCD expression is increased in obesity, it is plausible that SCD is a link between obesity and adipose inflammation. Indeed, loss of SCD in diet-induced and genetic models of obesity protects against adipose inflammation (Liu et al. 2010). Moreover, adipocyte-conditioned medium from SCD1 knockout mice does not induce pro-inflammatory cytokine gene expression in macrophages to the same extent as the ones from wild-type mice (Liu et al. 2010). However, SCD1 knockout mice (Ntambi et al. 2002) as well as mice with knockdown of SCD1 by SCD1-specific antisense oligonucleotide inhibitors (Jiang et al. 2005) have lower body weights, adiposity, and WAT lipogenic markers compared to wild-type counterparts. Moreover, these mouse models exhibit improved whole body insulin sensitivity. Therefore, reduced adipose inflammation in SCD1 knockout mice could be secondary to improvements in body weight and adiposity. Further, the less inflammatory nature of adipocyte-conditioned medium from SCD1 knockout mice could be due to their smaller adipocyte size.

To test a direct role of WAT SCD in modulating adipose inflammation, adipose tissue-specific SCD knockout has been employed. These mice were generated using the cre-loxp technology. Unlike the global SCD1 knockout mice, these adipose-specific SCD1 knockout mice (ASCD-KO) are not protected against high-fat diet-induced adiposity. Moreover, the ASCD-KO mice have comparable glucose tolerance to their wild-type counterparts. Taken together, this suggests that SCD expression in WAT per se does not regulate lipogenesis or adipose tissue inflammation and has a less important role in the regulation of energy balance. Interestingly, adipocytes from these ASCD-KO mice show increased GLUT1 expression with reduced adiponectin expression. Moreover, SCD inhibition in 3T3-L1 adipocytes leads to increased basal, but not insulin-stimulated glucose uptake. In contrast, global SCD1 KO mice exhibit increased GLUT4 and adiponectin expression (Hyun et al. 2010). This further supports the assertion that improvements of adipose tissue inflammation in global SCD1 knockout mice are secondary to reductions in adiposity.

Conclusion

SCD plays a pivotal role in metabolic function. In the WAT, SCD expression is regulated by dietary and hormonal factors. Hypercaloric diets, high-carbohydrate diets, and SFAs upregulate SCD expression, while PUFA and caloric restriction downregulates it. Of the hormonal factors, leptin is a potent inhibitor of SCD expression. WAT SCD upregulation in obesity appears to be an adaption to increased availability of carbohydrates and SFAs. In addition to facilitating lipogenesis via desaturation of fatty acids, WAT SCD also plays an important role in modulating metabolic functions. In this regard, palmitoleate and oleate, two products of the reaction catalyzed by this enzyme, have important metabolic functions. WAT SCD also appears to be important in modulating adipose tissue inflammation in obesity.

References

Boden G, Duan X, Homko C, Molina EJ, Song W, Perez O, Cheung P, Merali S (2008) Increase in endoplasmic reticulum stress-related proteins and genes in adipose tissue of obese, insulin-resistant individuals. Diabetes 57:2438–2444

Bryzgalova G, Lundholm L, Portwood N, Gustafsson JA, Khan A, Efendic S, Dahlman-Wright K (2008) Mechanisms of antidiabetogenic and body weight-lowering effects of estrogen in high-fat diet-fed mice. Am J Physiol Endocrinol Metab 295:E904–E912

Cao H, Gerhold K, Mayers JR, Wiest MM, Watkins SM, Hotamisligil GS (2008) Identification of a lipokine, a lipid hormone linking adipose tissue to systemic metabolism. Cell 134:933–944

Choi Y, Kim YC, Han YB, Park Y, Pariza MW, Ntambi JM (2000) The trans-10, cis-12 isomer of conjugated linoleic acid downregulates stearoyl-CoA desaturase 1 gene expression in 3T3-L1 adipocytes. J Nutr 130:1920–1924

Chong MF, Hodson L, Bickerton AS, Roberts R, Neville M, Karpe F, Frayn KN, Fielding BA (2008) Parallel activation of de novo lipogenesis and stearoyl-CoA desaturase activity after 3 d of high-carbohydrate feeding. Am J Clin Nutr 87:817–823

Christianson JL, Nicoloro S, Straubhaar J, Czech MP (2008) Stearoyl-CoA desaturase 2 is required for peroxisome proliferator-activated receptor gamma expression and adipogenesis in cultured 3T3-L1 cells. J Biol Chem 283:2906–2916

Cohen P, Miyazaki M, Socci ND, Hagge-Greenberg A, Liedtke W, Soukas AA, Sharma R, Hudgins LC, Ntambi JM, Friedman JM (2002) Role for stearoyl-CoA desaturase-1 in leptin-mediated weight loss. Science 297:240–243

Feuerer M, Herrero L, Cipolletta D, Naaz A, Wong J, Nayer A, Lee J, Goldfine AB, Benoist C, Shoelson S, Mathis D (2009) Lean, but not obese, fat is enriched for a unique population of regulatory T cells that affect metabolic parameters. Nat Med 15:930–939

Franck N, Gummesson A, Jernas M, Glad C, Svensson PA, Guillot G, Rudemo M, Nystrom FH, Carlsson LM, Olsson B (2011) Identification of adipocyte genes regulated by caloric intake. J Clin Endocrinol Metab 96:E413–E418

Frick F, Linden D, Ameen C, Eden S, Mode A, Oscarsson J (2002) Interaction between growth hormone and insulin in the regulation of lipoprotein metabolism in the rat. Am J Physiol Endocrinol Metab 283:E1023–E1031

Gallardo N, Bonzon-Kulichenko E, Fernandez-Agullo T, Molto E, Gomez-Alonso S, Blanco P, Carrascosa JM, Ros M, Andres A (2007) Tissue-specific effects of central leptin on the expression of genes involved in lipid metabolism in liver and white adipose tissue. Endocrinology 148:5604–5610

Gong J, Campos H, McGarvey S, Wu Z, Goldberg R, Baylin A (2011a) Adipose tissue palmitoleic acid and obesity in humans: does it behave as a lipokine? Am J Clin Nutr 93:186–191

Gong J, Campos H, McGarvey S, Wu Z, Goldberg R, Baylin A (2011b) Genetic variation in stearoyl-CoA desaturase 1 is associated with metabolic syndrome prevalence in Costa Rican adults. J Nutr 141:2211–2218

Hegele RA (2001) Molecular basis of partial lipodystrophy and prospects for therapy. Trends Mol Med 7:121–126

Hyun CK, Kim ED, Flowers MT, Liu X, Kim E, Strable M, Ntambi JM (2010) Adipose-specific deletion of stearoyl-CoA desaturase 1 up-regulates the glucose transporter GLUT1 in adipose tissue. Biochem Biophys Res Commun 399:480–486

Jaudszus A, Moeckel P, Hamelmann E, Jahreis G (2010) Trans-10, cis-12-CLA-caused lipodystrophy is associated with profound changes of fatty acid profiles of liver, white adipose tissue and erythrocytes in mice: possible link to tissue-specific alterations of fatty acid desaturation. Ann Nutr Metab 57:103–111

Jiang G, Li Z, Liu F, Ellsworth K, Dallas-Yang Q, Wu M, Ronan J, Esau C, Murphy C, Szalkowski D, Bergeron R, Doebber T, Zhang BB (2005) Prevention of obesity in mice by antisense oligonucleotide inhibitors of stearoyl-CoA desaturase-1. J Clin Invest 115:1030–1038

Johansson LE, Danielsson AP, Parikh H, Klintenberg M, Norstrom F, Groop L, Ridderstrale M (2012) Differential gene expression in adipose tissue from obese human subjects during weight loss and weight maintenance. Am J Clin Nutr 96:196–207

Jones BH, Maher MA, Banz WJ, Zemel MB, Whelan J, Smith PJ, Moustaid N (1996) Adipose tissue stearoyl-CoA desaturase mRNA is increased by obesity and decreased by polyunsaturated fatty acids. Am J Physiol 271:E44–E49

Jones BH, Standridge MK, Claycombe KJ, Smith PJ, Moustaid-Moussa N (1998) Glucose induces expression of stearoyl-CoA desaturase in 3T3-L1 adipocytes. Biochem J 335(Pt 2):405–408

Kaestner KH, Ntambi JM, Kelly TJ Jr, LANE MD (1989) Differentiation-induced gene expression in 3T3-L1 preadipocytes. A second differentially expressed gene encoding stearoyl-CoA desaturase. J Biol Chem 264:14755–14761

Kalupahana NS, Moustaid-Moussa N (2012a) The adipose tissue renin-angiotensin system and metabolic disorders: a review of molecular mechanisms. Crit Rev Biochem Mol Biol 47:379–390

Kalupahana NS, Moustaid-Moussa N (2012b) The renin-angiotensin system: a link between obesity, inflammation and insulin resistance. Obes Rev 13:136–149

Kalupahana NS, Claycombe KJ, Moustaid-Moussa N (2011) (n-3) Fatty acids alleviate adipose tissue inflammation and insulin resistance: mechanistic insights. Adv Nutr 2:304–316

Kalupahana NS, Massiera F, Quignard-Boulange A, Ailhaud G, Voy BH, Wasserman DH, Moustaid-Moussa N (2012a) Overproduction of angiotensinogen from adipose tissue induces adipose inflammation, glucose intolerance, and insulin resistance. Obesity (Silver Spring) 20:48–56

Kalupahana NS, Moustaid-Moussa N, Claycombe KJ (2012b) Immunity as a link between obesity and insulin resistance. Mol Aspects Med 33:26–34

Kim YC, Gomez FE, Fox BG, Ntambi JM (2000) Differential regulation of the stearoyl-CoA desaturase genes by thiazolidinediones in 3T3-L1 adipocytes. J Lipid Res 41:1310–1316

Kurebayashi S, Hirose T, Miyashita Y, Kasayama S, Kishimoto T (1997) Thiazolidinediones downregulate stearoyl-CoA desaturase 1 gene expression in 3T3-L1 adipocytes. Diabetes 46:2115–2118

Lin J, Choi YH, Hartzell DL, Li C, Della-Fera MA, Baile CA (2003) CNS melanocortin and leptin effects on stearoyl-CoA desaturase-1 and resistin expression. Biochem Biophys Res Commun 311:324–328

Liu X, Miyazaki M, Flowers MT, Sampath H, Zhao M, Chu K, Paton CM, Joo DS, Ntambi JM (2010) Loss of stearoyl-CoA desaturase-1 attenuates adipocyte inflammation: effects of adipocyte-derived oleate. Arterioscler Thromb Vasc Biol 30:31–38

Lundholm L, Zang H, Hirschberg AL, Gustafsson JA, Arner P, Dahlman-Wright K (2008) Key lipogenic gene expression can be decreased by estrogen in human adipose tissue. Fertil Steril 90:44–48

Macek Jilkova Z, Pavelka S, Flachs P, Hensler M, Kus V, Kopecky J (2010) Modulation of type I iodothyronine 5'-deiodinase activity in white adipose tissue by nutrition: possible involvement of leptin. Physiol Res 59:561–569

Mainieri D, Montani JP, Seydoux J, Giacobino JP, Boss O, Dulloo AG (2007) beta-Adrenergic control of stearoyl-CoA desaturase 1 repression in relation to sympathoadrenal regulation of thermogenesis. Int J Obes (Lond) 31:378–381

Mangravite LM, Dawson K, Davis RR, Gregg JP, Krauss RM (2007) Fatty acid desaturase regulation in adipose tissue by dietary composition is independent of weight loss and is correlated with the plasma triacylglycerol response. Am J Clin Nutr 86:759–767

Matsubara Y, Kano K, Kondo D, Mugishima H, Matsumoto T (2009) Differences in adipocytokines and fatty acid composition between two adipocyte fractions of small and large cells in high-fat diet-induced obese mice. Ann Nutr Metab 54:258–267

Mauvoisin D, Mounier C (2011) Hormonal and nutritional regulation of SCD1 gene expression. Biochimie 93:78–86

Nagare T, Sakaue H, Matsumoto M, Cao Y, Inagaki K, Sakai M, Takashima Y, Nakamura K, Mori T, Okada Y, Matsuki Y, Watanabe E, Ikeda K, Taguchi R, Kamimura N, Ohta S, Hiramatsu R, Kasuga M (2011) Overexpression of KLF15 transcription factor in adipocytes of mice results in down-regulation of SCD1 protein expression in adipocytes and consequent enhancement of glucose-induced insulin secretion. J Biol Chem 286:37458–37469

Nikonova L, Koza RA, Mendoza T, Chao PM, Curley JP, Kozak LP (2008) Mesoderm-specific transcript is associated with fat mass expansion in response to a positive energy balance. FASEB J 22:3925–3937

Nishimura S, Manabe I, Nagasaki M, Eto K, Yamashita H, Ohsugi M, Otsu M, Hara K, Ueki K, Sugiura S, Yoshimura K, Kadowaki T, Nagai R (2009) CD8+ effector T cells contribute to macrophage recruitment and adipose tissue inflammation in obesity. Nat Med 15:914–920

Ntambi JM, Buhrow SA, Kaestner KH, Christy RJ, Sibley E, Kelly TJ Jr, Lane MD (1988) Differentiation-induced gene expression in 3T3-L1 preadipocytes. Characterization of a differentially expressed gene encoding stearoyl-CoA desaturase. J Biol Chem 263:17291–17300

Ntambi JM, Miyazaki M, Stoehr JP, Lan H, Kendziorski CM, Yandell BS, Song Y, Cohen P, Friedman JM, Attie AD (2002) Loss of stearoyl-CoA desaturase-1 function protects mice against adiposity. Proc Natl Acad Sci U S A 99:11482–11486
Okada T, Furuhashi N, Kuromori Y, Miyashita M, Iwata F, Harada K (2005) Plasma palmitoleic acid content and obesity in children. Am J Clin Nutr 82:747–750
Pasarica M, Rood J, Ravussin E, Schwarz JM, Smith SR, Redman LM (2010) Reduced oxygenation in human obese adipose tissue is associated with impaired insulin suppression of lipolysis. J Clin Endocrinol Metab 95:4052–4055
Pinnick KE, Neville MJ, Fielding BA, Frayn KN, Karpe F, Hodson L (2012) Gluteofemoral adipose tissue plays a major role in production of the lipokine palmitoleate in humans. Diabetes 61:1399–1403
Sampath H, Miyazaki M, Dobrzyn A, Ntambi JM (2007) Stearoyl-CoA desaturase-1 mediates the pro-lipogenic effects of dietary saturated fat. J Biol Chem 282:2483–2493
Sessler AM, Kaur N, Palta JP, Ntambi JM (1996) Regulation of stearoyl-CoA desaturase 1 mRNA stability by polyunsaturated fatty acids in 3T3-L1 adipocytes. J Biol Chem 271:29854–29858
Shea J, French CR, Bishop J, Martin G, Roebothan B, Pace D, Fitzpatrick D, Sun G (2009) Changes in the transcriptome of abdominal subcutaneous adipose tissue in response to short-term overfeeding in lean and obese men. Am J Clin Nutr 89:407–415
Sjogren P, Sierra-Johnson J, Gertow K, Rosell M, Vessby B, de Faire U, Hamsten A, Hellenius ML, Fisher RM (2008) Fatty acid desaturases in human adipose tissue: relationships between gene expression, desaturation indexes and insulin resistance. Diabetologia 51:328–335
Theander-Carrillo C, Wiedmer P, Cettour-Rose P, Nogueiras R, Perez-Tilve D, Pfluger P, Castaneda TR, Muzzin P, Schurmann A, Szanto I, Tschop MH, Rohner-Jeanrenaud F (2006) Ghrelin action in the brain controls adipocyte metabolism. J Clin Invest 116:1983–1993
Turyn J, Stojek M, Swierczynski J (2010) Up-regulation of stearoyl-CoA desaturase 1 and elongase 6 genes expression in rat lipogenic tissues by chronic food restriction and chronic food restriction/refeeding. Mol Cell Biochem 345:181–188
Wang Z, Nakayama T (2010) Inflammation, a link between obesity and cardiovascular disease. Mediators Inflamm 2010:535918
Weiner FR, Smith PJ, Wertheimer S, Rubin CS (1991) Regulation of gene expression by insulin and tumor necrosis factor alpha in 3T3-L1 cells. Modulation of the transcription of genes encoding acyl-CoA synthetase and stearoyl-CoA desaturase-1. J Biol Chem 266:23525–23528
Yao-Borengasser A, Rassouli N, Varma V, Bodles AM, Rasouli N, Unal R, Phanavanh B, Ranganathan G, McGehee RE Jr, Kern PA (2008) Stearoyl-coenzyme A desaturase 1 gene expression increases after pioglitazone treatment and is associated with peroxisomal proliferator-activated receptor-gamma responsiveness. J Clin Endocrinol Metab 93:4431–4439
Yee JK, Phillips SA, Allamehzadeh K, Herbst KL (2012) Subcutaneous adipose tissue fatty acid desaturation in adults with and without rare adipose disorders. Lipids Health Dis 11:19

Chapter 6
Function and Regulation of Macrophage Stearoyl-CoA Desaturase in Metabolic Disorders

Nishan Sudheera Kalupahana, Shu Wang, Shaikh M. Rahman, and Naima Moustaid-Moussa

Introduction

Macrophages are key members of the innate immune system, which reside in most tissues of the body. The primary function of macrophages was traditionally considered to be phagocytosis and killing of foreign pathogens. More recently, a preponderance of evidence from cell to human studies suggests that macrophages play an important role in metabolic homeostasis as well (Bhargava and Lee 2012). For example, macrophages in adipose tissue and skeletal muscle regulate local inflammatory responses and insulin sensitivity, while macrophages in arterial walls are involved in atherosclerotic plaque formation. Further, macrophages in the liver, also known as Kupffer cells, are responsible for hepatic inflammation, steatosis, and insulin resistance. Therefore, macrophages play a key role in the pathogenesis of metabolic disorders such as type-2 diabetes and atherosclerosis.

Stearoyl-CoA desaturase (SCD) is a microsomal enzyme which facilitates the formation of monounsaturated fatty acids (MUFA) from saturated fatty acid (SFA) precursors (discussed in previous chapters). More specifically, SCD converts palmitoyl-CoA and stearoyl-CoA to palmitoleoyl-CoA and oleoyl-CoA, respectively (Mauvoisin and Mounier 2011). Of the two human SCD isoforms (SCD1 and SCD5), macrophages mainly express SCD1. While mice have four SCD isoforms, only SCD1 has been widely studied in macrophages, with a few studies reporting the expression of SCD2 as well (Saez et al. 2007; Sun et al. 2003).

Interestingly, macrophages isolated from subjects with type 2 diabetes have a higher linoleic acid content and fatty acid profiles consistent with increased SCD

N.S. Kalupahana (✉)
Department of Physiology, Faculty of Medicine, University of Peradeniya, Peradeniya, Sri Lanka
e-mail: skalupahana@pdn.ac.lk

S. Wang • S.M. Rahman • N. Moustaid-Moussa
Nutritional Sciences, Texas Tech University, Lubbock, TX, USA

J.M. Ntambi (ed.), *Stearoyl-CoA Desaturase Genes in Lipid Metabolism*, DOI 10.1007/978-1-4614-7969-7_6, © Springer Science+Business Media New York 2013

activity compared to control subjects (Senanayake et al. 2007). Moreover, incubation of human macrophages in serum isolated from dyslipidemic subjects leads to increased expression of SCD1 (Wong et al. 2011). Macrophages from obese animal models also express higher levels of SCD1 (Leroux et al. 2012). Therefore, it is likely that SCD plays an important role in the pathogenesis of macrophage-mediated metabolic disorders. This chapter discusses the functions of macrophages in relation to metabolic derangements, regulation of macrophage SCD expression, and effects of macrophage SCD expression in metabolic disorders.

Role of Macrophages in Metabolic Disorders

Macrophages and Obesity-Associated Insulin Resistance

Adipose tissue is considered as an endocrine organ (discussed in a separate chapter in this book), which consists of adipocytes and a stromal vascular fraction. The latter contains blood cells, endothelial cells, adipose tissue precursors, and macrophages (Kalupahana et al. 2012). Adipose tissue from a lean, normal-weight person contains macrophages which are of the M2 type. These are adipose tissue-resident macrophages (Rull et al. 2010), which have anti-inflammatory functions (Rull et al. 2010; Mosser and Edwards 2008). Adipose tissue of an obese individual is characterized by increased adipocyte (fat cell) numbers (hyperplasia) and cell size (hypertrophy). Obesity and adipocyte hypertrophy can cause lack of nutrients for cells, hypoxia, and endoplasmic reticulum stress, which can lead to adipocyte death via necrosis and apoptosis. Dead adipocytes are surrounded by macrophages, which can be identified microscopically as a "crown-like structure" (Yudkin 2007). The majority of macrophages in expanded adipose tissue are of the M1 type, which are mainly induced and recruited by proinflammatory cytokines. Expanded adipose tissue has an increased local inflammatory response. Monocyte chemoattractant protein 1 (MCP-1) is a chemokine secreted by adipocytes, macrophages, and activated endothelial cells in expanded adipose tissue. MCP-1 promotes the recruitment of monocytes into adipose tissue. Subsequent exposure of the monocytes to macrophage colony-stimulating factor causes differentiation to macrophages, which can lead to secretion of more MCP-1 (Kanda et al. 2006). M1 type macrophages in expanded adipose tissue secrete not only MCP-1 but also other proinflammatory cytokines including tumor necrosis factor α (TNFα), interleukin (IL)-6, IL-1, and IL-8, which can further amplify the local inflammatory response and decrease adipocyte insulin sensitivity (Coppack 2001; Zeyda and Stulnig 2009).

Obesity and increased fat mass are also accompanied with hepatic steatosis and inflammation. Since hypertrophic, insulin-resistant adipocytes have increased lipolytic rates, expanded adipose tissue continuously releases free fatty acids into the blood stream. These fatty acids can be taken up by the liver, skeletal muscle, and other tissues. Increased uptake of free fatty acids from adipose tissue and triglyceride-rich lipoproteins including chylomicrons and very low-density lipoproteins

(VLDLs) can increase fat accumulation in the liver, resulting in hepatic steatosis (Maher et al. 2008). Excess fat accumulation can recruit more inflammatory cells into the liver. These immune cells and Kupffer cells can secrete more inflammatory factors and amplify this inflammatory response, which can lead to hepatic insulin resistance (Huang et al. 2010; Neyrinck et al. 2009). Kupffer cells in obese rodents also express high levels of lipogenic genes such as SCD1, fatty acid synthase (FAS), and diacylglycerol acyltransferase 2 (DGAT2) (Leroux et al. 2012). As a consequence, these cells have elevated lipid content and recruit more immune cells, producing high levels of inflammatory factors.

Macrophages play an essential role in skeletal muscle repair during tissue damage and exercise (Chazaud et al. 2009). Increased blood free fatty acid concentrations can cause elevated fat content in skeletal muscle, which may recruit more immune cells including macrophages. This fat accumulation and inflammation can decrease insulin sensitivity in the muscle (Trayhurn et al. 2011).

Macrophages and Atherosclerosis

Macrophages play an important role in atherosclerotic lesion progression by facilitating cholesterol accumulation and amplifying the local inflammatory responses in blood vessel walls (Kunjathoor et al. 2002; Ludewig and Laman 2004). As early as in 1958, experimental data in rabbits suggested that the increase in white blood cells adhering to the vessel wall signaled an early stage of atherosclerotic lesion progression (Lusvarghi et al. 1958). It was likewise demonstrated that fatty streaks, the earliest detectable atherosclerotic lesions, contain macrophage-derived foam cells (MDFC) which were differentiated from blood monocytes. Monocytes adhere to the blood vessel wall by inflammatory signals such as vascular cell adhesion molecule-1 (VCAM-1), P-selectin, and E-selectin (Hope and Meredith 2003). Under normal conditions the endothelial monolayer of the vasculature resists adhesion of blood-borne monocytes. The presence of VCAM-1, P-selectin, and E-selectin increases the adhesion of monocytes to the blood vessel wall. Once adherent to the endothelial monolayer, the monocytes migrate between endothelial cells at their junction to penetrate the intimal layer (Parker et al. 2001). Modified lipoprotein particles, especially high levels of low-density lipoprotein (LDL) and low levels of high-density lipoprotein (HDL), have been shown to increase the expression of VCAM-1 (Lopez-Garcia et al. 2005; Baker et al. 1999). MCP-1, a chemoattractant cytokine (chemokine), also plays an important role in the recruitment of monocytes into the arterial intima (Sheikine and Hansson 2004).

After monocyte migration into the arterial intima, these monocytes differentiate into macrophages in response to macrophage colony-stimulating factor (Tojo et al. 1999). Macrophages express many scavenger receptors and membrane proteins, which can regulate cellular cholesterol and fat content. Macrophage scavenger receptor 1 (MSR1) and CD36 are membrane proteins involved in the uptake of cholesterol-rich modified lipoproteins (Kunjathoor et al. 2002; de Winther et al. 2000). Two important macrophage membrane proteins involved in cholesterol efflux are ATP-binding cassette transporter A1 (ABCA1) and scavenger receptor B class 1 (SR-B1). ABCA1 promotes cholesterol and phospholipid efflux from cells to HDL, while SR-B1 mediates the selective efflux of cellular cholesterol to HDL (Vedhachalam et al. 2004; Baranova et al. 2002). Cholesteryl ester (CE) in HDL is then selectively taken up by hepatocytes and used to synthesize bile acids or excreted into the bile completing a process termed reverse cholesterol transport. When macrophage cholesterol influx is greater than cholesterol efflux, cholesterol homeostasis in the macrophages is disturbed and CE accumulates in cytoplasmic droplets. These lipid-laden macrophages are known as MDFC. Both macrophage- and smooth muscle cell-derived foam cells characterize the early atherosclerotic lesion (Guyton and Klemp 1996). This cholesterol accumulation in the vessel intima leads to the formation of a cholesterol core, characteristic of advanced atherosclerotic lesions. MDFC in the atheroma secrete several inflammatory factors and proteases involved in lesion progression. Secretion of MCP-1 and matrix metalloproteinases promotes the progression of the atherosclerotic lesion formation.

Regulation of Macrophage SCD Expression

Only a few studies have investigated the regulation of SCD1 expression in macrophages. These studies indicate that macrophage SCD1 is regulated mainly by transcription factors and polyunsaturated fatty acids (PUFA).

Several transcription factors regulate SCD1 expression in macrophages. For example, activation of liver X receptor (LXR)/retinoid X receptor (RXR) via 22(R)-OH cholesterol and 9-*cis*-retinoic acid leads to increased expression of SCD1 and SCD2 in peritoneal macrophages (Martin-Fuentes et al. 2009). CCAAT/enhancer-binding protein (CEBP) β is a transcription factor which represses SCD1 expression (Zhang et al. 2012). The peroxisome proliferator-activated receptor (PPAR) γ agonist troglitazone suppresses macrophage SCD1 expression (Wang et al. 2004). Atorvastatin, an inhibitor of the enzyme 3-hydroxy-3-methylglutaryl-coenzyme A (HMG-CoA) reductase, also suppresses SCD1 expression in THP-1-derived macrophages. β-Amyloid peptide is an important mediator of Alzheimer's disease (AD), which induces inflammation in brain macrophages (microglia). One study has shown that this peptide can increase SCD1 expression in microglial cells, suggesting that it is a positive regulator of SCD1 in macrophages (Uryu et al. 2003).

Similar to its regulation in other tissues, macrophage SCD1 expression is repressed by PUFA. For example, alpha linolenic acid (ALA), an n-3 PUFA, reduces SCD1 expression in THP-1 MDFC (Zhang et al. 2011). The mechanism of ALA-induced SCD1 repression is as follows. ALA activates the nuclear receptor farnesoid-X-receptor (FXR), which increases the expression of small heterodimer partner (SHP), which in turn decreases LXR-dependent sterol regulatory element-binding protein 1c (SREBP1c) transcription, leading to repression of SCD1 expression (Zhang et al. 2012). Conjugated linoleic acid (CLA) also suppresses SCD1 in macrophages (Wang et al. 2004).

SCD and Macrophage Biology in Metabolic Disorders

Role of Macrophage SCD in the Pathogenesis of Obesity-Related Inflammation

Macrophages play a key role in the pathogenesis of adipose tissue inflammation and obesity-associated metabolic disorders such as type-2 diabetes and atherosclerosis. While obesity promotes a proinflammatory macrophage phenotype, the key molecules and the molecular signaling pathway (s) involved in this process are not clearly known. Interestingly, macrophages from subjects with type-2 diabetes exhibit higher accumulation of linoleic acid and fatty acid profiles consistent with increased SCD activity compared to control subjects (Senanayake et al. 2007). Recent studies have found that peritoneal macrophages from global C/EBPβ deletion exhibit significantly higher SCD1 gene expression and attenuated the expression of palmitate-induced inflammatory cytokines including TNFα (Rahman et al. 2012). Thus, it is reasonable to propose that forced expression of macrophage SCD1 might be protective against high-fat diet-induced macrophage lipotoxicity.

The assertion that macrophage SCD1 expression is unlikely to promote an inflammatory phenotype is also supported by findings from SCD1 knockout mice. While SCD1 deficiency in obese mouse models such as Agouti and high-fat diet-fed mice prevents inflammation in white adipose tissue and adipocytes, it does not prevent the inflammatory activation of peritoneal macrophages (Liu et al. 2010). Moreover, adipocytes isolated from SCD1 knockout adipose tissue show a blunted response to lipopolysaccharide (LPS) stimulation. However, peritoneal macrophages isolated from SCD1 knockout and wild-type (WT) mice show similar expression of inflammatory genes when challenged with LPS. These results suggest that the modulation of inflammation by SCD1 is cell type-specific. In order to assess a paracrine connection between adipocytes and macrophages, Liu et al. collected conditioned medium (CM) from WT and SCD1 knockout adipocytes and tested its effects on inflammation in RAW264.7 macrophage cells. The induction of proinflammatory genes TNFα and IL-1β was significantly lower in RAW264.7 macrophages treated with SCD1-deficient conditioned medium (CM), compared with the treatment with WT CM. These data indicate that adipocyte-derived soluble factors are the likely mediators of inflammation in macrophages and that the levels of these factors are lower in SCD1-deficient CM.

Macrophage SCD as a Mediator of Atherosclerosis

Macrophage-mediated inflammation also plays a critical role in the pathogenesis of atherosclerosis, a leading cause of cardiovascular morbidity and mortality in the world. Macrophages play a central role in atherogenesis through regulation of accumulation of cholesterol and inflammation (discussed above). As discussed earlier in

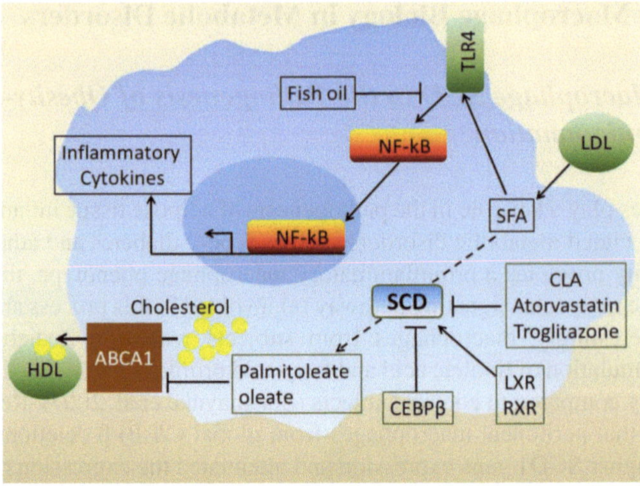

Fig. 6.1 Regulation and functions of macrophage SCD. Macrophage SCD expression is induced by activation of LXR/RXR transcription factors, while repressed by CEBPβ. SCD activity is also inhibited by PUFA, Atorvastatin, and PPARγ agonist troglitazone. SCD catalyzes MUFA formation from SFA, which in turn destabilize ABCA1 leading to reductions in cholesterol efflux. Therefore, SCD inhibition leads to increased cholesterol efflux. However, SCD inhibition leads to accumulation of SFA, which can activate TLR4 and thereby the NF-kB signaling pathway, leading to production of proinflammatory cytokines, which can aggravate atherogenesis. Fish oil can prevent this effect, probably via inhibition of the NF-kB pathway. *SCD* stearoyl-CoA desaturase, *LXR* liver X receptor, *RXR* retinoid X receptor, *CEBP* CCAAT/enhancer-binding protein, *PUFA* polyunsaturated fatty acids, *PPAR* peroxisome proliferator-activated receptor, *MUFA* monounsaturated fatty acids, *SFA* saturated fatty acids, *ABCA1* ATP-binding cassette transporter A1, *TLR4* toll-like receptor-4, *NF-kB* nuclear factor-kB

this chapter, ABCA1 is an important cholesterol efflux receptor, which promotes cholesterol and phospholipid efflux from macrophages to HDL. When macrophage cholesterol influx is greater than cholesterol efflux, cholesterol homeostasis in the macrophage is disturbed and cholesteryl ester accumulates in cytoplasmic droplets. These lipid-laden macrophages are known as MDFC, which characterize the early atherosclerotic lesion (Vedhachalam et al. 2004). Cholesterol efflux from these foam cells provides protection against the progression of the plaque.

Because of the promising role of SCD in the regulation of metabolic functions, much effort has been directed in elucidating the role of SCD in atherosclerosis. A summary is given in Fig. 6.1. Sun et al. have shown that SCD1 and SCD2 are induced in peritoneal macrophages after treatment with LXR/RXR activators, 22(R)-OH cholesterol and 9-*cis*-retinoic acid. This induction is associated with inhibition of ABCA1 (Sun et al. 2003). Other studies also report the LXR-mediated induction of SCD gene expression and subsequent reduction of ABCA1 in macrophages (Wang et al. 2004). Conversely, SCD inhibitors, CLA and troglitazone attenuate the ABCA1 destabilization induced by palmitate and stearate but not by linoleate. These results suggest that LXR/RXR ligands activate SCD, which in turn promote the production of MUFA from SFA precursors, which destabilize ABCA1. However, there is a body of literature showing that LXR agonists increase both ABCA1 and ATP-binding cassette subfamily G member 1 (ABCG1) gene transcription and cholesterol efflux in macrophages (Larrede et al. 2009; Zhou et al. 2010). These studies, however, do not clarify whether LXR activation-mediated increase of ABCA1 and ABCG1 gene expression and therefore cholesterol efflux in macrophages is dependent on SCD expression (Beyea et al. 2007; Teupser et al. 2008). Thus, the role of SCD in LXR activation-mediated induction of ABCA1 and ABCG1 remains unclear. Further work in this area to determine the exact relationship between LXR agonism, SCD1 expression, and cholesterol efflux is certainly warranted.

Consistent with the effect of SCD inhibitors on attenuating SFA-induced ABCA1 destabilization, SCD inhibition also promotes cholesterol efflux from foam cells. For example, ALA treatment and siRNA-mediated knockdown of SCD1 significantly reduce SCD1 expression in macrophage foam cells with concomitant increases in cholesterol efflux and reductions in intracellular cholesterol storage in these cells (Zhang et al. 2012). SCD1 knockdown also reduces fatty acid content, increases ABCA1 protein levels, and increases cholesterol efflux in cultured human macrophages (Wong et al. 2011). Taken together, this suggests a negative effect of macrophage SCD expression on cholesterol efflux, thereby potentially promoting atherogenesis. Interestingly, the anti-atherogenic pharmacological agent atorvastatin is also known to suppress SCD gene expression in macrophages. In one such study, Martín-Fuentes et al. assessed the effects of atorvastatin on oxidized low-density lipoprotein (oxLDL)-loaded human macrophage cells (THP-1). They found a significant reduction in the percentage of palmitoleic and oleic acids in THP-1 cells incubated with atorvastatin, which was associated with marked reduction of SCD and sterol regulatory element-binding protein 1 (SREBP1) gene expression (Martin-Fuentes et al. 2009).

Since increased macrophage SCD expression is associated with reduced cholesterol efflux, SCD inhibition could possibly have a favorable effect on atherogenesis. Paradoxically, rodent studies have indicated otherwise. Brown et al. used antisense oligonucleotides (ASO) to inhibit SCD1 in a mouse model of hyperlipidemia and atherosclerosis (LDLr−/−Apob100/100) and showed that inhibition of SCD1 attenuated diet-induced obesity, insulin resistance, and hepatic steatosis but promoted aortic atherosclerosis, which was not reversed by dietary oleate (Brown et al. 2008). Furthermore, these mice had increased levels of SFA in plasma and tissues, reduced plasma triglycerides, and normal LDL cholesterol levels. Next, they examined whether increased SFA availability in this model affected toll-like receptor-4 (TLR4) activation. As expected, mice with SCD1 inhibition exhibited hypersensitivity to TLR4 agonists. It is possible that inhibition of SCD1 in the liver leads to secretion of VLDL particles that are enriched in SFA-rich cholesteryl esters (CE), giving rise to SFA-CE-rich LDL particles. Since the macrophages also have reduced SCD1 expression, they are unable to convert the SFA delivered to them via LDLs to MUFA, and therefore are exposed to high levels of SFA. This in turn will activate TLR4 which can further activate the nuclear factor-kB (NF-kB) pathway, leading to transcription of proinflammatory cytokines (Fig. 6.1). This proinflammatory environment promotes atherogenesis, especially in hyperlipidemic conditions.

This model is further supported by the finding that the atherosclerosis in LDLr–/–Apob100/100 mice with SCD1 knockdown can be prevented by the co-treatment with fish oil (Brown et al. 2010). We and others have previously shown that eicosapentaenoic acid (EPA), an n-3 PUFA found in fish oil, can inhibit the NF-kB pathway, possibly via activation of the G protein-coupled receptor 120 (Siriwardhana et al. 2012; Oh et al. 2010). It is plausible that fish oil attenuates the SFA-induced activation of TLR4 and subsequent activation of the NF-kB pathway and transcription of proinflammatory cytokines in the LDLr–/–Apob100/100 mice treated with SCD1 ASO, providing protection against the development of atherosclerosis.

Other studies have also shown that SCD1 deficiency promotes atherosclerosis under hypercholesterolemic conditions. For example, low-density lipoprotein receptor (LDLr)-deficient mice with SCD1 deficiency have increased atherosclerosis when fed a Western-type diet (Macdonald et al. 2009). In addition, SCD1 deletion in bone marrow cells of LDLr–/– mice also does not prevent atherosclerotic lesion formation. However, SCD1 knockdown in the chronic intermittent hypoxia mouse model, a model of obstructive sleep apnea which exhibits atherosclerosis, is protected from atherosclerosis. This discrepancy of the effects of SCD1 knockdown on atherosclerosis could indicate that under hypercholesterolemic conditions, SCD might protect against SFA-mediated inflammation (Liu et al. 2011). Taken together, this indicates that while macrophage SCD deficiency seems to promote cholesterol efflux, the proinflammatory environment induced by SFA-mediated TLR4 activation promotes atherogenesis.

Conclusion

SCD plays a pivotal role in metabolic function and exerts diverse cell-specific patterns of function and regulation. SCD is nutritionally and hormonally regulated and is altered in various metabolic and inflammatory disorders including obesity and diabetes. The role of SCD in macrophages in obesity and atherosclerosis is not fully understood and very limited studies have addressed its regulation in immune cells. Most of the studies were conducted either in RAW macrophages or in peritoneal macrophages isolated from global SCD1–/– mice. Several isoforms of SCD exist (Liu et al. 2011) and their comparative role in macrophage inflammation and lipotoxicity in response to obesity and its associated disorders have not been addressed. Future studies with isoform-specific roles of SCDs in macrophages might shed new insights into SCD's role in obesity-induced macrophage lipotoxicity and metabolic disorders.

References

Baker PW, Rye KA, Gamble JR, Vadas MA, Barter PJ (1999) Ability of reconstituted high density lipoproteins to inhibit cytokine-induced expression of vascular cell adhesion molecule-1 in human umbilical vein endothelial cells. J Lipid Res 40:345–353

Baranova I, Vishnyakova T, Bocharov A, Chen Z, Remaley AT, Stonik J, Eggerman TL, Patterson AP (2002) Lipopolysaccharide down regulates both scavenger receptor B1 and ATP binding cassette transporter A1 in RAW cells. Infect Immun 70:2995–3003

Beyea MM, Heslop CL, Sawyez CG, Edwards JY, Markle JG, Hegele RA, Huff MW (2007) Selective up-regulation of LXR-regulated genes ABCA1, ABCG1, and APOE in macrophages through increased endogenous synthesis of 24(S),25-epoxycholesterol. J Biol Chem 282:5207–5216

Bhargava P, Lee CH (2012) Role and function of macrophages in the metabolic syndrome. Biochem J 442:253–262

Brown JM, Chung S, Sawyer JK, Degirolamo C, Alger HM, Nguyen T, Zhu X, Duong MN, Wibley AL, Shah R, Davis MA, Kelley K, Wilson MD, Kent C, Parks JS, Rudel LL (2008) Inhibition of stearoyl-coenzyme A desaturase 1 dissociates insulin resistance and obesity from atherosclerosis. Circulation 118:1467–1475

Brown JM, Chung S, Sawyer JK, Degirolamo C, Alger HM, Nguyen TM, Zhu X, Duong MN, Brown AL, Lord C, Shah R, Davis MA, Kelley K, Wilson MD, Madenspacher J, Fessler MB, Parks JS, Rudel LL (2010) Combined therapy of dietary fish oil and stearoyl-CoA desaturase 1 inhibition prevents the metabolic syndrome and atherosclerosis. Arterioscler Thromb Vasc Biol 30:24–30

Chazaud B, Brigitte M, Yacoub-Youssef H, Arnold L, Gherardi R, Sonnet C, Lafuste P, Chretien F (2009) Dual and beneficial roles of macrophages during skeletal muscle regeneration. Exerc Sport Sci Rev 37:18–22

Coppack SW (2001) Pro-inflammatory cytokines and adipose tissue. Proc Nutr Soc 60:349–356

de Winther MP, van Dijk KW, Havekes LM, Hofker MH (2000) Macrophage scavenger receptor class A: a multifunctional receptor in atherosclerosis. Arterioscler Thromb Vasc Biol 20:290–297

Guyton JR, Klemp KF (1996) Development of the lipid-rich core in human atherosclerosis. Arterioscler Thromb Vasc Biol 16:4–11

Hope SA, Meredith IT (2003) Cellular adhesion molecules and cardiovascular disease. Part II. Their association with conventional and emerging risk factors, acute coronary events and cardiovascular risk prediction. Intern Med J 33:450–462

Huang W, Metlakunta A, Dedousis N, Zhang P, Sipula I, Dube JJ, Scott DK, O'Doherty RM (2010) Depletion of liver Kupffer cells prevents the development of diet-induced hepatic steatosis and insulin resistance. Diabetes 59:347–357

Kalupahana NS, Moustaid-Moussa N, Claycombe KJ (2012) Immunity as a link between obesity and insulin resistance. Mol Aspects Med 33:26–34

Kanda H, Tateya S, Tamori Y, Kotani K, Hiasa K, Kitazawa R, Kitazawa S, Miyachi H, Maeda S, Egashira K, Kasuga M (2006) MCP-1 contributes to macrophage infiltration into adipose tissue, insulin resistance, and hepatic steatosis in obesity. J Clin Invest 116:1494–1505

Kunjathoor VV, Febbraio M, Podrez EA, Moore KJ, Andersson L, Koehn S, Rhee JS, Silverstein R, Hoff HF, Freeman MW (2002) Scavenger receptors class A-I/II and CD36 are the principal receptors responsible for the uptake of modified low density lipoprotein leading to lipid loading in macrophages. J Biol Chem 277:49982–49988

Larrede S, Quinn CM, Jessup W, Frisdal E, Olivier M, Hsieh V, Kim MJ, van Eck M, Couvert P, Carrie A, Giral P, Chapman MJ, Guerin M, le Goff W (2009) Stimulation of cholesterol efflux by LXR agonists in cholesterol-loaded human macrophages is ABCA1-dependent but ABCG1-independent. Arterioscler Thromb Vasc Biol 29:1930–1936

Leroux A, Ferrere G, Godie V, Cailleux F, Renoud ML, Gaudin F, Naveau S, Prevot S, Makhzami S, Perlemuter G, Cassard-Doulcier AM (2012) Toxic lipids stored by Kupffer cells correlates with their pro-inflammatory phenotype at an early stage of steatohepatitis. J Hepatol 57:141–149

Liu X, Miyazaki M, Flowers MT, Sampath H, Zhao M, Chu K, Paton CM, Joo DS, Ntambi JM (2010) Loss of stearoyl-CoA desaturase-1 attenuates adipocyte inflammation: effects of adipocyte-derived oleate. Arterioscler Thromb Vasc Biol 30:31–38

Liu X, Strable MS, Ntambi JM (2011) Stearoyl CoA desaturase 1: role in cellular inflammation and stress. Adv Nutr 2:15–22

Lopez-Garcia E, Schulze MB, Meigs JB, Manson JE, Rifai N, Stampfer MJ, Willett WC, Hu FB (2005) Consumption of trans fatty acids is related to plasma biomarkers of inflammation and endothelial dysfunction. J Nutr 135:562–566

Ludewig B, Laman JD (2004) The in and out of monocytes in atherosclerotic plaques: balancing inflammation through migration. Proc Natl Acad Sci U S A 101:11529–11530

Lusvarghi E, Bellesia L, Mucci P (1958) Osmotic resistance of white blood cells in different circulatory districts (renal, intestinal, hepatic and peripheral) in the rabbit. Blut 4:254–260

Macdonald ML, van Eck M, Hildebrand RB, Wong BW, Bissada N, Ruddle P, Kontush A, Hussein H, Pouladi MA, Chapman MJ, Fievet C, van Berkel TJ, Staels B, McManus BM, Hayden MR (2009) Despite antiatherogenic metabolic characteristics, SCD1-deficient mice have increased inflammation and atherosclerosis. Arterioscler Thromb Vasc Biol 29:341–347

Maher JJ, Leon P, Ryan JC (2008) Beyond insulin resistance: innate immunity in nonalcoholic steatohepatitis. Hepatology 48:670–678

Martin-Fuentes P, Garcia-Otin AL, Calvo L, Gomez-Coronado D, Civeira F, Cenarro A (2009) Atorvastatin decreases stearoyl-CoA desaturase gene expression in THP-1 macrophages incubated with oxidized LDL. Lipids 44:115–123

Mauvoisin D, Mounier C (2011) Hormonal and nutritional regulation of SCD1 gene expression. Biochimie 93:78–86

Mosser DM, Edwards JP (2008) Exploring the full spectrum of macrophage activation. Nat Rev Immunol 8:958–969

Neyrinck AM, Cani PD, Dewulf EM, de Backer F, Bindels LB, Delzenne NM (2009) Critical role of Kupffer cells in the management of diet-induced diabetes and obesity. Biochem Biophys Res Commun 385:351–356

Oh DY, Talukdar S, Bae EJ, Imamura T, Morinaga H, Fan W, Li P, Lu WJ, Watkins SM, Olefsky JM (2010) GPR120 is an omega-3 fatty acid receptor mediating potent anti-inflammatory and insulin-sensitizing effects. Cell 142:687–698

Parker C III, Vita JA, Freedman JE (2001) Soluble adhesion molecules and unstable coronary artery disease. Atherosclerosis 156:417–424

Rahman SM, Janssen RC, Choudhury M, Baquero KC, Aikens RM, de la Houssaye BA, Friedman JE (2012) CCAAT/enhancer-binding protein beta (C/EBPbeta) expression regulates dietary-induced inflammation in macrophages and adipose tissue in mice. J Biol Chem 287:34349–34360

Rull A, Camps J, Alonso-Villaverde C, Joven J (2010) Insulin resistance, inflammation, and obesity: role of monocyte chemoattractant protein-1 (or CCL2) in the regulation of metabolism. Mediators Inflamm 2010

Saez F, Chabory E, Cadet R, Vernet P, Baron S, Lobaccaro JM, Drevet JR (2007) Liver X receptors and epididymal epithelium physiology. Asian J Androl 9:574–582

Senanayake S, Brownrigg LM, Panicker V, Croft KD, Joyce DA, Steer JH, Puddey IB, Yeap BB (2007) Monocyte-derived macrophages from men and women with Type 2 diabetes mellitus differ in fatty acid composition compared with non-diabetic controls. Diabetes Res Clin Pract 75:292–300

Sheikine Y, Hansson GK (2004) Chemokines and atherosclerosis. Ann Med 36:98–118

Siriwardhana N, Kalupahana NS, Fletcher S, Xin W, Claycombe KJ, Quignard-Boulange A, Zhao L, Saxton AM, Moustaid-Moussa N (2012) n-3 and n-6 polyunsaturated fatty acids differentially regulate adipose angiotensinogen and other inflammatory adipokines in part via NF-kappaB-dependent mechanisms. J Nutr Biochem 23:1661–1667

Sun Y, Hao M, Luo Y, Liang CP, Silver DL, Cheng C, Maxfield FR, Tall AR (2003) Stearoyl-CoA desaturase inhibits ATP-binding cassette transporter A1-mediated cholesterol efflux and modulates membrane domain structure. J Biol Chem 278:5813–5820

Teupser D, Kretzschmar D, Tennert C, Burkhardt R, Wilfert W, Fengler D, Naumann R, Sippel AE, Thiery J (2008) Effect of macrophage overexpression of murine liver X receptor-alpha (LXR-alpha) on atherosclerosis in LDL-receptor deficient mice. Arterioscler Thromb Vasc Biol 28:2009–2015

Tojo N, Asakura E, Koyama M, Tanabe T, Nakamura N (1999) Effects of macrophage colony-stimulating factor (M-CSF) on protease production from monocyte, macrophage and foam cell in vitro: a possible mechanism for anti-atherosclerotic effect of M-CSF. Biochim Biophys Acta 1452:275–284

Trayhurn P, Drevon CA, Eckel J (2011) Secreted proteins from adipose tissue and skeletal muscle—adipokines, myokines and adipose/muscle cross-talk. Arch Physiol Biochem 117:47–56

Uryu S, Tokuhiro S, Oda T (2003) beta-Amyloid-specific upregulation of stearoyl coenzyme A desaturase-1 in macrophages. Biochem Biophys Res Commun 303:302–305

Vedhachalam C, Liu L, Nickel M, Dhanasekaran P, Anantharamaiah GM, Lund-Katz S, Rothblat GH, Phillips MC (2004) Influcncc of ApoA-I structure on the ABCA1-mediated efflux of cellular lipids. J Biol Chem 279:49931–49939

Wang Y, Kurdi-Haidar B, Oram JF (2004) LXR-mediated activation of macrophage stearoyl-CoA desaturase generates unsaturated fatty acids that destabilize ABCA1. J Lipid Res 45:972–980

Wong BX, Kyle RA, Myhill PC, Croft KD, Quinn CM, Jessup W, Yeap BB (2011) Dyslipidemic diabetic serum increases lipid accumulation and expression of stearoyl-CoA desaturase in human macrophages. Lipids 46:931–941

Yudkin JS (2007) Inflammation, obesity, and the metabolic syndrome. Horm Metab Res 39:707–709

Zeyda M, Stulnig TM (2009) Obesity, inflammation, and insulin resistance—a mini-review. Gerontology 55:379–386

Zhang J, Grieger JA, Kris-Etherton PM, Thompson JT, Gillies PJ, Fleming JA, Vanden Heuvel JP (2011) Walnut oil increases cholesterol efflux through inhibition of stearoyl CoA desaturase 1 in THP-1 macrophage-derived foam cells. Nutr Metab (Lond) 8:61

Zhang J, Kris-Etherton PM, Thompson JT, Hannon DB, Gillies PJ, Heuvel JP (2012) Alpha-linolenic acid increases cholesterol efflux in macrophage-derived foam cells by decreasing stearoyl CoA desaturase 1 expression: evidence for a farnesoid-X-receptor mechanism of action. J Nutr Biochem 23:400–409

Zhou X, Yin Z, Guo X, Hajjar DP, Han J (2010) Inhibition of ERK1/2 and activation of liver X receptor synergistically induce macrophage ABCA1 expression and cholesterol efflux. J Biol Chem 285:6316–6326

Chapter 7
Stearoyl-CoA Desaturase-1 in the Regulation of Toll-Like Receptor Signaling and Endoplasmic Reticulum Stress Signaling

Tabitha C. Ting and Makoto Miyazaki

SCD1 and Lipid Biosynthesis

Stearoyl-CoA desaturase 1 (SCD1) is an enzyme involved in the de novo lipogenesis of fatty acids (Fig. 7.1). In the first step of this biosynthetic pathway, acetyl-CoA carboxylase converts acetyl-CoA to malonyl-CoA. Next, malonyl-CoA is converted into the saturated fatty acid (SFA) palmitate (C16:0), through a multiple-step process catalyzed by fatty acid synthase (FAS). Elongases convert palmitate into stearate (C18:0), the SFA substrate for SCD. Following elongation, SCD catalyzes the conversion of stearate into the monounsaturated fatty acid (MUFA) oleate (C18:1). MUFAs such as oleate serve as major substrates for the synthesis of complex lipids such as phospholipids, cholesterol esters, and wax esters. Since SCD1 controls the delicate balance between SFAs and MUFAs, SCD1 plays an important role in determining the composition of fatty acids and complex lipids in the cell. An SFA to MUFA ratio that is either excessive or insufficient can lead to severe metabolic consequences, including atherosclerosis, obesity, and type II diabetes (Sampath et al. 2007). To study global and tissue-specific effects of SCD1 on these diseases, researchers have developed several models that have a deletion or decreased expression of SCD1. These include conditional knockout mouse models ($SCD1^{-/-}$) or transient knockdown mouse models using antisense oligonucleotide technology (SCD1 ASO). Inhibition of SCD1 in these models leads to an increase in SFAs such as stearate and a decrease in MUFAs such as oleate (Sampath et al. 2007). These are therefore very useful models to study the physiological role of SFAs and MUFAs.

T.C. Ting • M. Miyazaki (✉)
Division of Renal Diseases and Hypertension, University of Colorado Denver, Aurora, CO 80045, USA
e-mail: Makoto.Miyazaki@ucdenver.edu

J.M. Ntambi (ed.), *Stearoyl-CoA Desaturase Genes in Lipid Metabolism*, DOI 10.1007/978-1-4614-7969-7_7, © Springer Science+Business Media New York 2013

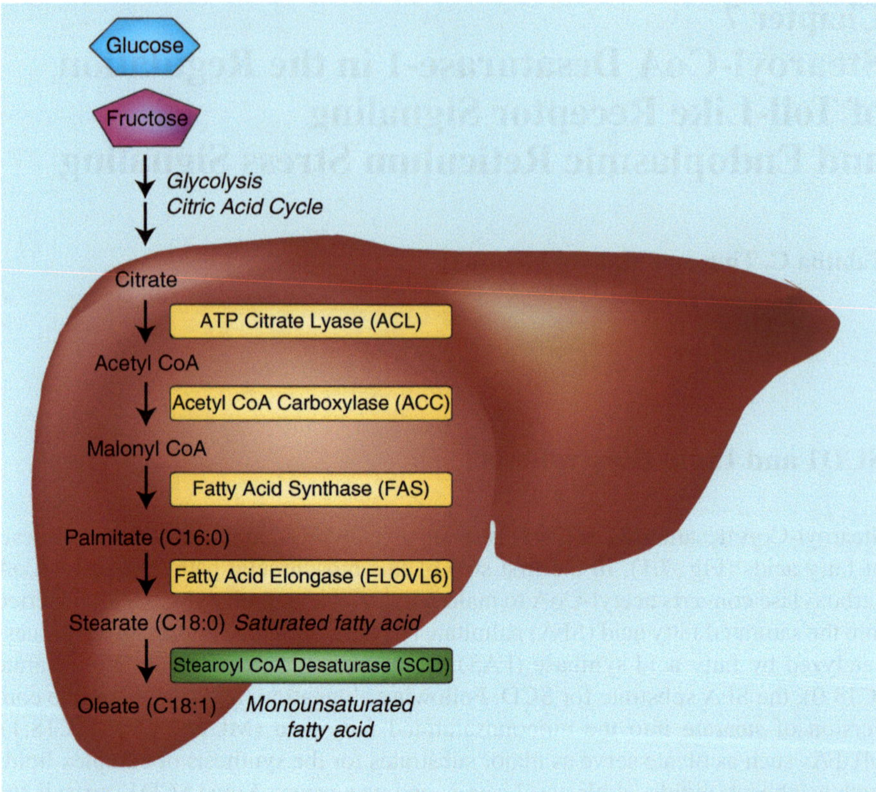

Fig. 7.1 SCD1 in de novo lipogenesis. Dietary carbohydrates such as glucose and fructose are major substrates of hepatic de novo lipogenesis. The carbohydrates are converted to citrate through glycolysis and citric acid cycle. Citrate is converted back to acetyl-CoA in the reaction of ATP citrate lyase (ACL). Acetyl-CoA carboxylase (ACC) catalyzes the irreversible conversion of the 2-carbon acetyl-CoA to the 3-carbon intermediate, malonyl-CoA. Malonyl-CoA serves as the precursor for the endogenous synthesis of fatty acids via the fatty acid synthase (FAS) multienzyme complex. FAS catalyzes seven cycles of sequential condensation, reduction, and dehydration reactions to form the 16-carbon saturated fatty acid (SFA), palmitate (16:0). Further elongation generally occurs through the actions of microsomal elongase (Elovl6) to form stearate (18:0). Cellular levels of stearate are regulated by a lipogenic enzyme, stearoyl-CoA desaturase, which catalyzes the conversion of stearate to oleate. More than 60 % of stearate derived from foods or de novo lipogenesis is converted into oleate. The reaction of stearoyl-CoA desaturase involves the introduction of the first *cis*-double bond in the Δ9 position in a spectrum of saturated fatty acyl-CoAs

SCD1 and Inflammation

Inflammation is a key event in the pathogenesis of many vascular diseases, including atherosclerosis. SCD1 may have an intriguing role in macrophage-mediated inflammation. Studies on β amyloid peptide and toll-like receptor 4 (TLR4)/NF-κB signaling demonstrate that SCD1 inhibition accelerates inflammation, whereas other studies report that SCD1 inhibition exerts no effect on inflammation (Macdonald et al. 2009a, b; Brown et al. 2008; Uryu et al. 2003; Liu et al. 2010,

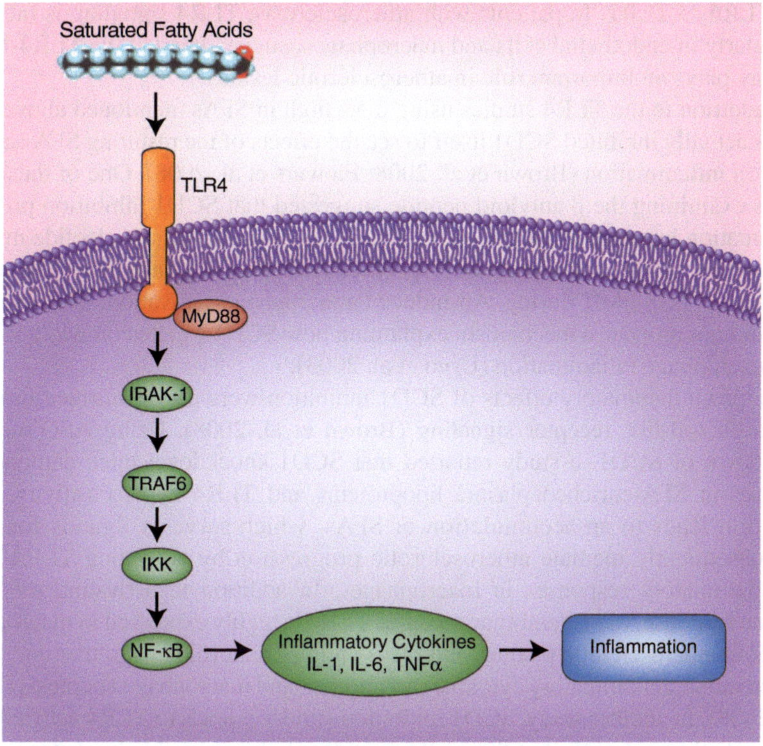

Fig. 7.2 SFAs in TLR4 signaling. SFAs directly bind and activate TLR4, resulting in inflammation. *TLR4* toll-like receptor-4, *MyD88* myeloid differentiation primary response gene-88, *IRAK-1* interleukin-1 receptor-associated kinase 1, *TRAF6* TNF receptor-associated factor-6, *IKK* IκB kinase, *NF-κB* nuclear factor kappa-light-chain-enhancer of activated B cells

2011). In studies supporting the pro-inflammatory role of fatty acids, accumulation of SFAs was shown to directly induce inflammation through TLR4 and NF-κB signaling (Lee et al. 2001, 2003a, b, 2004). In this pathway, SFAs activate TLR4, a pattern-recognition receptor that plays a role in activating innate immunity and inflammation (Fig. 7.2). TLR4 then induces NF-κB, a protein complex involved in cellular responses to harmful stimuli such as stress, free radicals, and toxic lipids. NF-κB activation increases the transcription of most enzymes in the de novo synthesis of ceramides, which are sphingosines covalently linked to a fatty acid. De novo ceramide synthesis uses SFAs, but not unsaturated fatty acids. Therefore, an increase in SFAs due to SCD1 inhibition leads to ceramide accumulation. Ceramides potently inhibit Akt, a serine/threonine kinase that upregulates nutrient storage and inhibits apoptosis. Increased ceramide and the resulting Akt inhibition lead to the activation of pro-apoptotic pathways through enzymes such as caspase-9 (Holland et al. 2007; Summers 2006). This pathway seems to be differentially activated in various tissues, since it has been shown that stearate activates inflammatory genes in macrophages but not in adipose tissue. In addition to TLR4, it has also been shown that SFAs induce macrophage inflammation through TLR2 when the receptor is dimerized with TLR6 or TLR1. In patients with atherosclerosis, TLR4 signaling is increased particularly in endothelial cells and macrophages, suggesting that the TLR4-NF-κB pathway plays an important role in atherosclerotic lesions.

In addition to the TLR4 studies using diets high in SFAs mentioned above, other studies actually inhibited SCD1 itself to see the effects of the resulting SFA accumulation on inflammation (Brown et al. 2008; Flowers et al. 2008). One of these early studies examining the β amyloid peptide suggested that SCD1 inhibition promotes inflammation in macrophages (Uryu et al. 2003). Using an oligonucleotide microarray analysis, SCD1 was found to be specifically and significantly upregulated by β amyloid peptide (Aβ) during Aβ-induced macrophage activation. However, this study did not propose a mechanism explaining how SCD1 expression was correlated with macrophage inflammation (Uryu et al. 2003).

The pro-inflammatory effects of SCD1 inhibition were also confirmed in related studies on toll-like receptor signaling (Brown et al. 2008). Using ASO-mediated knockdown of SCD1, a study reported that SCD1-knockdown mice demonstrated increases in SFA-enriched plasma lipoproteins and TLR4 hypersensitivity. SCD1 inhibition leads to an accumulation of SFAs, which serve as ligands for TLR4 and consequently mediate atherosclerotic progression by activating TLR4-driven pro-inflammatory responses in macrophages. In addition to activating the innate immune response, transmembrane receptor TLR4 is highly expressed in macrophages present in atherosclerotic plaques and plays an important role in vascular endothelial cell activation, inflammatory cytokine recruitment, and macrophage apoptosis (Brown et al. 2008). In another study, SCD1 inhibition similarly leads to TLR4 activation and increased atherosclerosis, but these effects appear to be reversible by dietary supplementation of ω-3 polyunsaturated fatty acids (PUFAs) from fish oil. Using a hyperlipidemic mouse model, a combination of SCD1 ASO and fish oil treatments decreased both metabolic syndrome and atherosclerosis (Brown et al. 2010).

Although the mechanisms by which SFAs induce TLR4 have yet to be elucidated, one putative explanation is that SFAs change lipid and protein composition of microdomains on raft membranes. Changes in membrane composition will affect TLR signaling since activated TLR4 is thought to translocate to these domains. SFAs will allow for easier TLR4 trafficking to domains, while PUFAs will disturb lipid composition and raft order thus interfering with TLR4 recruitment to rafts. As seen in the study using SCD1 ASO and fish oil-derived ω-3 PUFAs, dual therapy of SCD1 inhibitors and other anti-inflammatory agents may be effective in reducing both metabolic syndrome and atherosclerosis (Brown et al. 2010). Although TLR4 antagonists appear to be a putative treatment for atherosclerotic progression in mice models, the role of TLR4 in human atherosclerosis has been debated. Therefore, further studies will be required to determine whether TLR4 activation is necessary for SFA-induced atherosclerosis and whether these results can be applied clinically.

In contrast to the Aβ and TLR4 studies, others reported that SCD1 inhibition exerts no effect on inflammation. A recent study compared peritoneal macrophages from SCD1-deficient mice and wild-type mice but reported no differences between groups when the macrophages were treated with LPS (Liu et al. 2010). In a similar study using peritoneal macrophages from SCD1-deficient mice, macrophage inflammation also did not differ from the control, although inflammatory changes in the skin and plasma were observed (MacDonald et al. 2009b). One plausible explanation for this discrepancy may be due to the presence of more than one SCD isoform. SCD2 is more highly expressed in macrophages than SCD1 and therefore might be compensating for the low expression of SCD1 in macrophages. The expression of SCD1 and SCD2 in different cell types may be responsible for differences in inflammatory responses (Liu et al. 2010, 2011; Liu and Ntambi 2009). Although it is well-known that SCD1 deficiency exerts harmful pro-inflammatory effects in several cell types including skin and β-cells (Zheng et al. 1999; Flowers et al. 2007), the role of SCD1 in macrophage inflammation is still ambiguous. Further investigation will be necessary to determine the complex regulation of SCD1 in macrophage inflammation.

SCD and ER Stress

A second method by which SCD1 inhibition and SFAs promote lipid-related diseases is by inducing stress in the endoplasmic reticulum (ER), a central organelle for protein processing and lipid synthesis (Fig. 7.3). ER stress is an accumulation of misfolded or unfolded proteins in the endoplasmic reticulum due to adverse conditions such as ischemia, hypoxia, heat shock, oxidative stress, or depletion of stored ER calcium. ER stress triggers a restorative/corrective pathway known as the unfolded protein response (UPR), which attempts to restore proper ER function. The UPR alleviates ER stress through three main compensatory mechanisms: (1) decreasing the load of proteins that enter the ER by suppressing translation, (2) raising protein-folding capacity by increasing the number of available chaperones, or (3) increasing degradation of misfolded proteins through a ubiquitin-proteasome pathway. If all three compensatory mechanisms fail and homeostasis cannot be reestablished, prolonged ER stress and UPR activation can trigger apoptosis. The UPR comprises three branches mediated by ER-resident transmembrane proteins: PKR-like ER kinase (PERK), inositol requiring enzyme 1 (IRE1), and activating transcription factor-6 (ATF6). These three ER-resident signaling proteins are activated either by binding to a protein chaperone called BiP or by directly sensing the presence of misfolded proteins in the ER lumen. Following activation, these ER-resident proteins induce downstream effectors such as XBP-1 (X-box binding protein 1) and CHOP (C/EBP-homologous protein) (Flowers et al. 2007; Tabas and Ron 2011; Walter and Ron 2011).

Recent studies have shown that abnormal lipid metabolism activates ER stress and prolonged UPR in atherosclerotic plaques. In normal, healthy cells, macrophages transport lipoprotein-cholesterol to the ER, where the cholesterol is esterified (Scull and Tabas 2011). However, in atherosclerotic lesions, this process is interrupted and massive amounts of free cholesterol accumulate. SCD can regulate intracellular free cholesterol levels, since its product oleate is a preferred substrate for cholesteryl ester synthesis. SCD1$^{-/-}$ mice have higher levels of free cholesterol in the skin, liver, and aorta compared to wild-type mice. In addition to this increase in free

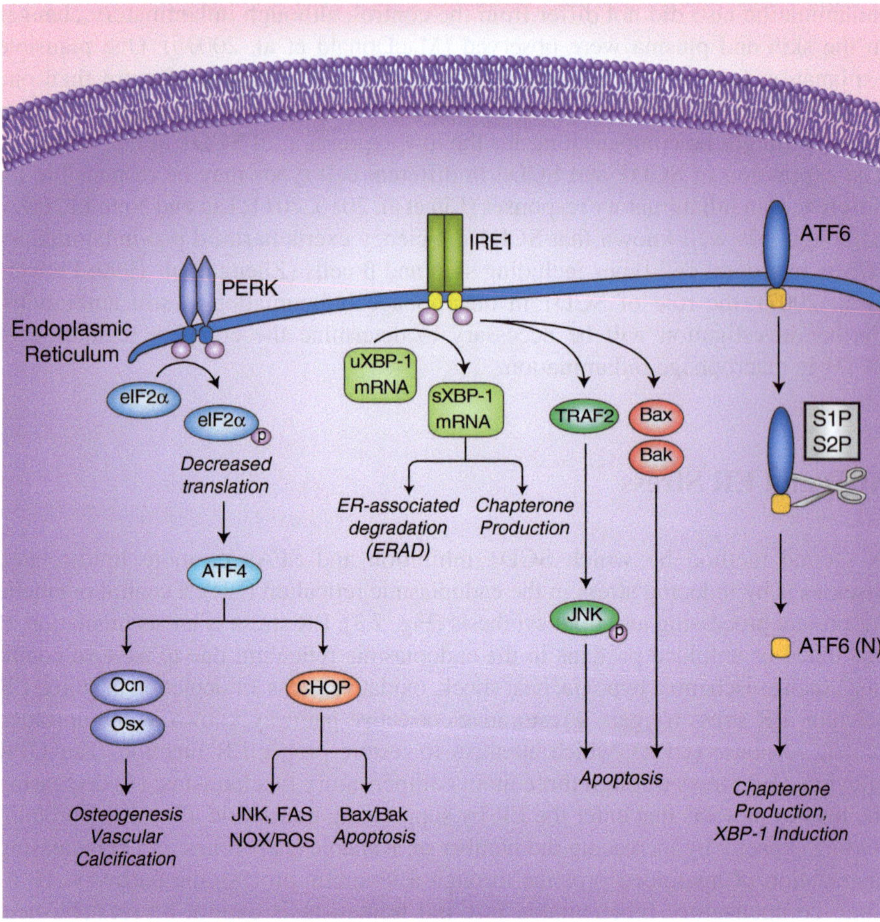

Fig. 7.3 ER stress signaling. Upon accumulation of unfolded proteins and lipids such as SFAs in the ER, three ER stress sensors are activated and initiate signal transduction events that control cell survival or death. *PERK* PKR-like ER kinase, *IRE1* inositol requiring enzyme 1, *ATF6* activating transcription factor-6, *eIF2α* eukaryotic translation initiation factor 2α, *ATF4* activating transcription factor-4, *CHOP* C/EBP-homologous protein, *Ocn* Osteocalcin, *Osx* Osterix, *uXBP-1* unspliced X-box binding protein 1, *sXBP-1* spliced X-box binding protein 1, *TRAF2* TNF receptor-associated factor 2, *JNK* Jun N-terminal kinase, *Bax* Bcl-2-associated X protein, *Bak* Bcl-2 homologous antagonist/killer, *S1P* site-1 protease, *S2P* site-2 protease

cholesterol, lesion exposure to SFAs such as stearic acid induces ER stress, leading to macrophage apoptosis in atherosclerotic plaques (Anderson et al. 2012). Several ER stress transducers and their downstream effectors have been implicated in the development of atherosclerosis. However, better understanding of the molecular mechanisms behind this induction is required. Expanding knowledge of how lipids induce ER stress and atherosclerosis may lead to better treatment of atherosclerosis and associated vascular calcification.

The PERK branch of the UPR is activated by SCD1 inhibition and the resulting accumulation of stearate (Masuda et al. 2012). In this UPR branch, PERK oligomerizes and phosphorylates itself before phosphorylating the α subunit of eukaryotic initiation factor 2 (eIF2). Phosphorylation of this α subunit renders eIF2α inactive, resulting in the inhibition of 80S ribosomal assembly and translation as well as a reduction in the protein load entering the ER. Although most protein translation is reduced when eIF2α is phosphorylated, translation of activating transcription factor-4 (ATF4) increases because ATF4 possesses upstream open reading frames that are bypassed only when eIF2α is phosphorylated. ATF4 activates its targets including the pro-apoptotic CHOP and proteins involved in osteoblast differentiation such as Osteocalcin (Ocn) and Osterix (Osx) (Masuda et al. 2012). During prolonged ER stress, CHOP induces apoptosis through BH3-only proteins, Bcl-2-associated X protein (Bax), and Bcl-2 homologous antagonist/killer (Bak). Using in vitro models, our group has previously shown that SCD1 inhibition increases stearate, leading to increased ATF4 and CHOP expression through the PERK-eIF2α pathway. We found that stearate most potently induces vascular calcification in preference to other fatty acids. ATF4 knockdown inhibits stearate-induced vascular calcification. We therefore concluded that UPR activation of the PERK-eIF2α-ATF4 branch contributes to stearate-mediated vascular calcification (Masuda et al. 2012).

A second branch of the UPR, the IRE1 branch, has also been shown to be activated by SFA accumulation (Wei et al. 2006). In this branch, IRE1 cleaves XBP-1, which then activates the transcription of UPR genes for chaperones, lipid synthesis, and ER-associated degradation of misfolded proteins. IRE1 can also recruit TRAF2, leading to the activation of caspase-12 and Jun N-terminal kinase (JNK). Apoptosis can then be activated either by the TRAF2-JNK pathway or by pro-apoptotic proteins Bax and Bak, which are targets of IRE1 as well as CHOP (Choi et al. 2011). Our study showed that SCD1 inhibition induces the expression of other ER stress targets including spliced XBP-1 in vascular smooth muscle cells (Masuda et al. 2012). In studies on β cells, SFAs such as palmitate were shown to induce the IRE1 pathway, leading to an increase in the IRE1-dependent JNK response (Ron and Walter 2007). Other studies have also confirmed that both stearate and palmitate can induce the IRE1 branch of the UPR, leading to an increase in pro-apoptotic proteins such as caspase-3 (Wei et al. 2006). These changes were accompanied by increases in other UPR-related proteins including ATF4, CHOP, chaperone GRP78, and growth arrest and DNA damage-inducible protein (GADD34). Although β-cell ceramide synthesis has been correlated with fatty acid-induced apoptosis, this study showed that UPR activation by stearate and palmitate could occur independently of de novo ceramide synthesis. Taken together, these studies demonstrate that another UPR branch, the IRE1 pathway, can also mediate the effects of SFAs on apoptosis.

PERK and IRE1 branches of the UPR have been found to link lipotoxic signals to atherosclerosis and other diseases (Scull and Tabas 2011; Erbay et al. 2009). ER stress signaling is significantly activated in animal models of atherosclerosis and other lipid-mediated disorders and human atherosclerotic lesions (Myoishi et al. 2007; Duan et al. 2009). CHOP deficiency decreases atherosclerotic plaque lesions, cell death lesions, and vascular remodeling in hyperlipidemic ApoE$^{-/-}$ and

$LDLR^{-/-}$ mice (Thorp et al. 2009; Gao et al. 2011). Chemical chaperones such as 4-phenyl butyrate and taurine-conjugated chenodeoxycholic acid were shown to reduce atherosclerotic lesion area, accompanied with reduced ER stress (Erbay et al. 2009). In a study examining macrophage ER stress, toxic lipids such as oxidized LDL and palmitate (C16:0) induced ER stress through a macrophage lipid chaperone called fatty acid-binding protein-4 (aP2). Their results suggested that toxic lipids accumulate and upregulate aP2, which then inhibits de novo lipogenesis and leads to activation of UPR-mediated apoptosis (Erbay et al. 2009). Other studies have observed the induction of ER chaperone GRP78, PERK, and CHOP, following lipid accumulation in mouse and rat models on high fat diets (Brookheart et al. 2009). In rat models of nonalcoholic fatty liver disease, high SFA diets and prolonged ER stress have also been shown to activate caspase-3, a crucial death protease that mediates apoptosis. Collectively, these studies suggest that dyslipidemia plays a pivotal role in inducing ER stress and apoptosis. Although links between dyslipidemia and ER stress have been found, mechanisms by which SFAs induce ER stress have not been fully elucidated. Several studies have proposed potential mechanisms. In a study using $SCD1^{-/-}$ mice on very low fat diets, SCD1 loss led to increases in spliced XBP-1, CHOP, and ATF3 (another ER stress-induced transcription factor) in the liver. In this study, Flowers et al. postulated that SCD1 inhibition activates ER stress by altering hepatic fatty acid composition of cellular membranes, leading to impaired function of membrane transport proteins (Flowers et al. 2008). Another explanation attributes the effects of SFAs on ER stress to unsaturated fatty acid depletion, which may contribute to oxidative stress (Flowers et al. 2008). Our studies have shown that the SFA stearate must be converted to its CoA conjugated form (stearoyl-CoA) in order to activate ER stress and vascular calcification (Masuda et al. 2012). We hypothesize that stearoyl-CoA is incorporated into an ER membrane lipid. This altered membrane composition is then sensed by ER stress transducers, which activate the UPR. Although current evidence links SCD1 deficiency to ER stress, the necessity for defining mechanisms behind ER stress induction warrants further investigation.

In addition to SFAs themselves, downstream products of fatty acids such as phospholipids have been shown to regulate ER stress. For example, previous studies have demonstrated that phospholipids play a role in activating ER stress (Testerink et al. 2009; Van Der Sanden et al. 2003; Fu et al. 2011). The two most abundant phospholipids in the ER membrane are phosphatidylcholine (PC) and phosphatidylethanolamine (PE). When lipogenesis or dietary lipid intake is increased, PC is the most common phospholipid component for packaging and storing lipid droplets and lipoproteins. In addition, SCD inhibition changes intracellular PC and PE levels (Dobrzyn et al. 2005). PC synthesis is catalyzed by choline-phosphate cytidylyltransferase A (PCTY1a), while PC to PE conversion is regulated by phosphatidylethanolamine N-methyl transferase (PEMT) (Fu et al. 2011). In ER samples isolated from the obese liver tissue of leptin-deficient mice, the PC/PE ratio was higher than in the control. Since the PC/PE ratio in lean controls was the same as that found in their diet, the increased PC/PE ratio in obese hepatic ER seemed to result from de novo lipogenesis rather than from dietary sources (Fu et al. 2011). Similar to free cholesterol, higher PC levels in ER membranes were shown to inhibit the activity of sarco/endoplasmic reticulum calcium ATPase (SERCA), a transport protein that maintains calcium homeostasis. Altered calcium levels caused by SERCA dysfunction impaired protein-folding chaperones such as BiP and calnexin, and thus activate ER stress. This upregulation of ER stress was evidenced by IRE1α and eIF2α phosphorylation, accompanied by increased expression of ER chaperones GRP78 and GRP94. Conversely, a decreased PC/PE ratio caused by PEMT suppression also led to IRE1α and eIF2α phosphorylation and the induction of CHOP, homocysteine-inducible ER stress-inducible protein (HERP), and Der1-like domain family member 2 (DERL2). Other studies have also implicated the role of PC in the activation of ER stress. The active spliced form of an IRE1 target, XBP-1, was found to be associated with ER expansion and increased PC synthesis in fibroblasts in vitro. Collectively, these studies seem to indicate that an overabundance of PC is integral to the development of ER stress (Testerink et al. 2009; Van der Sanden et al. 2003). These studies suggest that an alteration of PC/PE ratio in the ER membrane contributes to ER stress mediated by SCD inhibition and SFA overload.

Conclusion

In this chapter, we have discussed specific pathways by which SCD1 and its substrates, SFAs, contribute to inflammation and apoptosis (Fig. 7.4). The SCD1 activity is tightly regulated mostly at the transcription level. However, once this tight regulation is interrupted, the induction of SCD1 expression increases levels of oleate, which is a preferred substrate for triglyceride and cholesteryl ester synthesis. Conversely, substantial reduction of SCD1 increases levels of intracellular SFAs such as stearate and palmitate, leading to TLR4 and ER stress activation. ER stress is induced by abnormal ratios of lipid species such as saturated/unsaturated fatty acids, free/esterified cholesterol, and PC/PE, leading to apoptosis frequently found in atherosclerotic lesions. These ratios can be regulated through the modulation of SCD1 activity. Thus, composition and ratio of lipid species contributing to SFA-mediated ER stress remain to be determined through further investigation. Turning to the broader perspective of disease pathology, we see that the development of atherosclerosis is characterized by all of these factors: abnormal lipid accumulation, TLR4-mediated inflammation, and ER stress-mediated apoptosis. These factors clearly suggest that SCD1, its SFA substrates, and its MUFA products play a multi-faceted role in the development of vascular diseases such as atherosclerosis and vascular calcification.

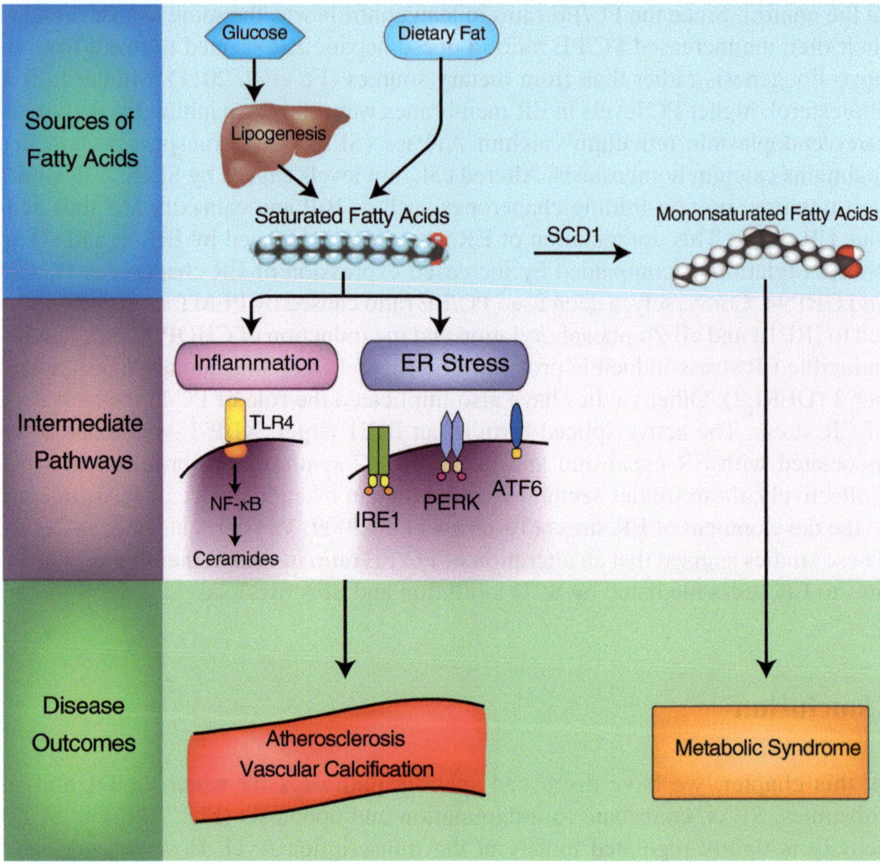

Fig. 7.4 SCD1 in inflammation and ER stress. Loss of SCD1 and the resulting accumulation of SFAs activate TLR4 and ER stress signaling, resulting in inflammation and cell death leading to vascular diseases such as atherosclerosis and vascular calcification. *TLR4* toll-like receptor-4, *NF-κB* nuclear factor kappa-light-chain-enhancer of activated B cells, *PERK* PKR-like ER kinase, *IRE1* inositol requiring enzyme 1, *ATF6* activating transcription factor-6

References

Anderson EK, Hill AA, Hasty AH (2012) Stearic acid accumulation in macrophages induces toll-like receptor 4/2-independent inflammation leading to endoplasmic reticulum stress-mediated apoptosis. Arterioscler Thromb Vasc Biol 32:1687–1695

Brookheart RT, Michel CI, Schaffer JE (2009) As a matter of fat. Cell Metab 10:9–12

Brown JM, Chung S, Sawyer JK, Degirolamo C, Alger HM, Nguyen T, Zhu X, Duong MN, Wibley AL, Shah R, Davis MA, Kelley K, Wilson MD, Kent C, Parks JS, Rudel LL (2008) Inhibition of stearoyl-coenzyme A desaturase 1 dissociates insulin resistance and obesity from atherosclerosis. Circulation 118:1467–1475

Brown JM, Chung S, Sawyer JK, Degirolamo C, Alger HM, Nguyen TM, Zhu X, Duong MN, Brown AL, Lord C, Shah R, Davis MA, Kelley K, Wilson MD, Madenspacher J, Fessler MB, Parks JS, Rudel LL (2010) Combined therapy of dietary fish oil and stearoyl-CoA desaturase 1 inhibition prevents the metabolic syndrome and atherosclerosis. Arterioscler Thromb Vasc Biol 30:24–30

Choi SE, Jung IR, Lee YJ, Lee SJ, Lee JH, Kim Y, Jun HS, Lee KW, Park CB, Kang Y (2011) Stimulation of lipogenesis as well as fatty acid oxidation protects against palmitate-induced INS-1 beta-cell death. Endocrinology 152:816–827

Dobrzyn A, Dobrzyn P, Miyazaki M, Sampath H, Chu K, Ntambi JM (2005) Stearoyl-CoA desaturase 1 deficiency increases CTP:choline cytidylyltransferase translocation into the membrane and enhances phosphatidylcholine synthesis in liver. J Biol Chem 280:23356–23362, Epub 2005 Apr 13

Duan X, Zhou Y, Teng X, Tang C, Qi Y (2009) Endoplasmic reticulum stress-mediated apoptosis is activated in vascular calcification. Biochem Biophys Res Commun 387:694–699

Erbay E, Babaev VR, Mayers JR, Makowski L, Charles KN, Snitow ME, Fazio S, Wiest MM, Watkins SM, Linton MF, Hotamisligil GS (2009) Reducing endoplasmic reticulum stress through a macrophage lipid chaperone alleviates atherosclerosis. Nat Med 15:1383–1391

Flowers JB, Rabaglia ME, Schueler KL, Flowers MT, Lan H, Keller MP, Ntambi JM, Attie AD (2007) Loss of stearoyl-CoA desaturase-1 improves insulin sensitivity in lean mice but worsens diabetes in leptin-deficient obese mice. Diabetes 56:1228–1239

Flowers MT, Keller MP, Choi Y, Lan H, Kendziorski C, Ntambi JM, Attie AD (2008) Liver gene expression analysis reveals endoplasmic reticulum stress and metabolic dysfunction in SCD1-deficient mice fed a very low-fat diet. Physiol Genomics 33:361–372

Fu S, Yang L, Li P, Hofmann O, Dicker L, Hide W, Lin X, Watkins SM, Ivanov AR, Hotamisligil GS (2011) Aberrant lipid metabolism disrupts calcium homeostasis causing liver endoplasmic reticulum stress in obesity. Nature 473:528–531

Gao J, Ishigaki Y, Yamada T, Kondo K, Yamaguchi S, Imai J, Uno K, Hasegawa Y, Sawada S, Ishihara H, Oyadomari S, Mori M, Oka Y, Katagiri H (2011) Involvement of endoplasmic stress protein C/EBP homologous protein in arteriosclerosis acceleration with augmented biological stress responses. Circulation 124:830–839

Holland WL, Brozinick JT, Wang LP, Hawkins ED, Sargent KM, Liu Y, Narra K, Hoehn KL, Knotts TA, Siesky A, Nelson DH, Karathanasis SK, Fontenot GK, Birnbaum MJ, Summers SA (2007) Inhibition of ceramide synthesis ameliorates glucocorticoid-, saturated-fat-, and obesity-induced insulin resistance. Cell Metab 5:167–179

Lee JY, Sohn KH, Rhee SH, Hwang D (2001) Saturated fatty acids, but not unsaturated fatty acids, induce the expression of cyclooxygenase-2 mediated through Toll-like receptor 4. J Biol Chem 276:16683–16689

Lee JY, Plakidas A, Lee WH, Heikkinen A, Chanmugam P, Bray G, Hwang DH (2003a) Differential modulation of Toll-like receptors by fatty acids: preferential inhibition by n-3 polyunsaturated fatty acids. J Lipid Res 44:479–486

Lee JY, Ye J, Gao Z, Youn HS, Lee WH, Zhao L, Sizemore N, Hwang DH (2003b) Reciprocal modulation of Toll-like receptor-4 signaling pathways involving MyD88 and phosphatidylinositol 3-kinase/AKT by saturated and polyunsaturated fatty acids. J Biol Chem 278:37041–37051

Lee JY, Zhao L, Youn HS, Weatherill AR, Tapping R, Feng L, Lee WH, Fitzgerald KA, Hwang DH (2004) Saturated fatty acid activates but polyunsaturated fatty acid inhibits Toll-like receptor 2 dimerized with Toll-like receptor 6 or 1. J Biol Chem 279:16971–16979

Liu X, Ntambi JM (2009) Atherosclerosis: keep your macrophages in shape. Nat Med 15: 1357–1358

Liu X, Miyazaki M, Flowers MT, Sampath H, Zhao M, Chu K, Paton CM, Joo DS, Ntambi JM (2010) Loss of Stearoyl-CoA desaturase-1 attenuates adipocyte inflammation: effects of adipocyte-derived oleate. Arterioscler Thromb Vasc Biol 30:31–38

Liu X, Strable MS, Ntambi JM (2011) Stearoyl CoA desaturase 1: role in cellular inflammation and stress. Adv Nutr 2:15–22

Macdonald ML, Bissada N, Vallance BA, Hayden MR (2009a) Absence of stearoyl-CoA desaturase-1 does not promote DSS-induced acute colitis. Biochim Biophys Acta 1791:1166–1172

MacDonald ML, van Eck M, Hildebrand RB, Wong BW, Bissada N, Ruddle P, Kontush A, Hussein H, Pouladi MA, Chapman MJ, Fievet C, van Berkel TJ, Staels B, McManus BM, Hayden MR (2009b) Despite antiatherogenic metabolic characteristics, SCD1-deficient mice have increased inflammation and atherosclerosis. Arterioscler Thromb Vasc Biol 29:341–347

Masuda M, Ting TC, Levi M, Saunders SJ, Miyazaki-Anzai S, Miyazaki M (2012) Activating transcription factor 4 regulates stearate-induced vascular calcification. J Lipid Res 53:1543–1552

Myoishi M, Hao H, Minamino T, Watanabe K, Nishihira K, Hatakeyama K, Asada Y, Okada K, Ishibashi-Ueda H, Gabbiani G, Bochaton-Piallat ML, Mochizuki N, Kitakaze M (2007) Increased endoplasmic reticulum stress in atherosclerotic plaques associated with acute coronary syndrome. Circulation 116:1226–1233

Ron D, Walter P (2007) Signal integration in the endoplasmic reticulum unfolded protein response. Nat Rev Mol Cell Biol 8:519–529

Sampath H, Miyazaki M, Dobrzyn A, Ntambi JM (2007) Stearoyl CoA desaturase-1 mediates the pro-lipogenic effects of dietary saturated fat. J Biol Chem 282(4):2483–93

Scull CM, Tabas I (2011) Mechanisms of ER stress-induced apoptosis in atherosclerosis. Arterioscler Thromb Vasc Biol 31:2792–2797

Summers SA (2006) Ceramides in insulin resistance and lipotoxicity. Prog Lipid Res 45:42–72

Tabas I, Ron D (2011) Integrating the mechanisms of apoptosis induced by endoplasmic reticulum stress. Nat Cell Biol 13:184–190

Testerink N, van der Sanden MH, Houweling M, Helms JB, Vaandrager AB (2009) Depletion of phosphatidylcholine affects endoplasmic reticulum morphology and protein traffic at the Golgi complex. J Lipid Res 50:2182–2192

Thorp E, Li G, Seimon TA, Kuriakose G, Ron D, Tabas I (2009) Reduced apoptosis and plaque necrosis in advanced atherosclerotic lesions of Apoe–/– and Ldlr–/– mice lacking CHOP. Cell Metab 9:474–481

Uryu S, Tokuhiro S, Oda T (2003) beta-Amyloid-specific upregulation of stearoyl coenzyme A desaturase-1 in macrophages. Biochem Biophys Res Commun 303:302–305

Van der Sanden MH, Houweling M, van Golde LM, Vaandrager AB (2003) Inhibition of phosphatidylcholine synthesis induces expression of the endoplasmic reticulum stress and apoptosis-related protein CCAAT/enhancer-binding protein-homologous protein (CHOP/GADD153). Biochem J 369:643–650

Walter P, Ron D (2011) The unfolded protein response: from stress pathway to homeostatic regulation. Science 334:1081–1086

Wei Y, Wang D, Topczewski F, Pagliassotti MJ (2006) Saturated fatty acids induce endoplasmic reticulum stress and apoptosis independently of ceramide in liver cells. Am J Physiol Endocrinol Metab 291:E275–281

Zheng Y, Eilertsen KJ, Ge L, Zhang L, Sundberg JP, Prouty SM, Stenn KS, Parimoo S (1999) Scd1 is expressed in sebaceous glands and is disrupted in the asebia mouse. Nat Genet 23:268–270

Chapter 8
Stearoyl-CoA Desaturase in the Control of Heart Metabolism

Pawel Dobrzyn and Agnieszka Dobrzyn

Introduction

Stearoyl-CoA desaturase (SCD) is a rate-limiting enzyme that catalyzes the synthesis of monounsaturated fatty acids, mainly oleate and palmitoleate, from saturated fatty acyl-CoAs. The preferred substrates for SCD are palmitoyl- and stearoyl-CoAs (Dobrzyn and Ntambi 2004). Four isoforms of SCD have been identified in the mouse (SCD1-4) (Kaestner et al. 1989; Zheng et al. 2001; Miyazaki et al. 2003) and two (SCD-1 and SCD-5) in the human genome (Zhang et al. 1999; Beiraghi et al. 2003). Human SCD1 shows 85 % homology to murine SCD1 (Zhang et al. 1999). In an adult mouse, under normal dietary conditions, SCD1 mRNA is highly expressed in white adipose tissue, brown adipose tissue, meibomian glands, Harderian and preputial glands (Miyazaki et al. 2001). The mRNA is dramatically induced in the liver in response to a high-carbohydrate diet (Ntambi and Miyazaki 2004). SCD2 is expressed predominantly in the brain (Kaestner et al. 1989) but also, to a lesser extent, in the kidney, spleen, skeletal muscles, heart, and lungs, where its expression is induced in response to a high-carbohydrate diet (Kaestner et al. 1989; Zheng et al. 2001). SCD3 is expressed mainly in the differentiated sebocytes (Zheng et al. 2001), and SCD4 is expressed exclusively in the heart (Miyazaki et al. 2003). The physiological role of each SCD isoform and the reason for having multiple SCD gene isoforms that share considerable sequence homology and catalyze the same biochemical reaction are still under investigation.

P. Dobrzyn (✉)
Laboratory of Molecular and Medical Biochemistry, Nencki Institute of Experimental Biology, 3 Pasteur Street, 02-093 Warsaw, Poland
e-mail: p.dobrzyn@nencki.gov.pl

A. Dobrzyn
Laboratory of Cell Signaling and Metabolic Disorders, Nencki Institute of Experimental Biology, Warsaw, Poland

J.M. Ntambi (ed.), *Stearoyl-CoA Desaturase Genes in Lipid Metabolism*,
DOI 10.1007/978-1-4614-7969-7_8, © Springer Science+Business Media New York 2013

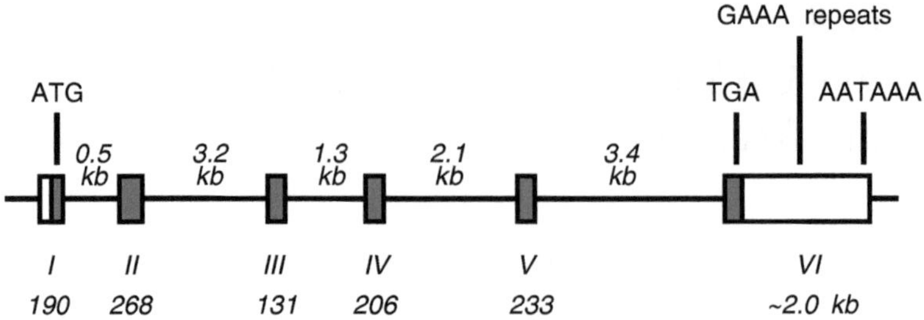

Fig. 8.1 Structure of mouse SCD4 gene. SCD4 is expressed exclusively in the heart. Adopted from Miyazaki et al. (2003)

In the heart, three SCD isoforms (SCD1, SCD2, and SCD4) are expressed. The expression of SCD1 is several times higher in the murine heart compared with that of SCD4 (Miyazaki et al. 2003; Dobrzyn et al. 2010a). However, SCD4 expression is significantly increased after a high-fat diet (Miyazaki et al. 2003), in obese leptin-deficient mice (Miyazaki et al. 2003; Dobrzyn et al. 2010a), and in the state of SCD1 deficiency (Miyazaki et al. 2003; Dobrzyn et al. 2008b). The cardiac SCD4 cDNA encodes a 353-amino acid residue protein with four transmembrane regions that is more than 80 % identical to the other three mouse SCD genes (Miyazaki et al. 2003). SCD4 also contains three conserved histidine-rich motifs that are present in the other SCD isoforms and essential for Δ9-desaturase function (Ntambi et al. 1988; Kaestner et al. 1989; Zheng et al. 2001). This enzyme desaturates both stearate and palmitate to the corresponding monounsaturated fatty acids (Miyazaki et al. 2003). SCD4 maps to mouse chromosome 19 D2, where three other SCD genes are located within a 200-kb region. SCD4 lies between the SCD1 and SCD2 genes, indicating that this region is composed of a cluster of Δ9-desaturases. The 3.1-kb transcript is the predominant species; thus, the SCD4 mRNA is smaller than the 4.9 kb of the other SCD isoforms because of a shorter 3′-untranslated region (Miyazaki et al. 2003) (Fig. 8.1). While in the liver, SCD1 is a major target gene of leptin, in the heart, leptin represses the expression of SCD4 but not that of SCD1 or SCD2 (Miyazaki et al. 2003). It is important to note that the expression of cardiac SCD4 is induced by a high-carbohydrate diet and LXRα agonists; however, unlike SCD1, it is not repressed by polyunsaturated fatty acids (PUFA) (Miyazaki et al. 2003).

The heart has a limited capacity for lipogenesis and de novo lipid synthesis. Why, then, is cardiac SCD such a highly expressed and regulated enzyme? Many studies underline the important role of lipogenic enzymes in proper heart function and suggest that the role of lipogenic genes in the heart may be distinct from other tissues. Cardiac sterol regulatory element-binding protein 1 was shown to activate the G-protein coupled inwardly reflecting K^+ channel, leading to enhanced acetylcholine-sensitive K^+ currents and reduced arrhythmias post-myocardial infraction (Park et al. 2008). It has also been shown that (1) the heart-specific knockdown of peroxisome proliferator-activated receptors γ (PPARγ) (Duan et al. 2005) and acyl-CoA synthase 1 (Ellis et al. 2011) induces cardiac hypertrophy; (2) transgenic overexpression of fatty acid transport protein 1 in the heart causes lipotoxic cardiomyopathy

(Chiu et al. 2005) and (3) fatty acid synthase (FAS), the enzyme that catalyzes de novo fatty acids (FA), is involved in the regulation of the ability of the heart to respond to stress through the activation of Ca^{2+}/calmodulin-dependent protein kinase II (Razani et al. 2011). Studies on mouse strains that have a mutation in the SCD1 gene provided evidence that SCD also plays an important role in the regulation of cardiac metabolism and function (Dobrzyn et al. 2008b, 2010a, b; Paton and Ntambi 2009). The lack of SCD1 expression decreases FA uptake and oxidation and increases glucose transport and oxidation in the heart (Dobrzyn et al. 2008b). Disruption of the SCD1 gene improves cardiac function in obese leptin-deficient ob/ob mice by correcting the systolic and diastolic dysfunction (Dobrzyn et al. 2010a). The improvement is associated with the reduced expression of the genes involved in FA transport and lipid synthesis within the heart along with decreased cardiac free fatty acid (FFA), diacylglycerol, trigliceride (TG), and ceramide levels, and reduced cardiomyocyte apoptosis (Dobrzyn et al. 2010a). Here, we review recent advances made in understanding the role of SCD in the control of heart metabolism and its involvement in the pathogenesis of lipotoxic cardiomyopathy.

Role of SCD in the Regulation of Cardiac Substrate Utilization

The heart has a major energy requirement based on its large work output. Although the cardiomyocytes are capable of using carbohydrates and ketone bodies as energy sources, FA are considered the preferred fuel. Under aerobic conditions, the heart derives 60–90 % of the energy necessary for contractile function from FA oxidation, while the rest is obtained mainly from carbohydrates (glucose and lactate) (Sambandam and Lopaschuk 2003; Stanley et al. 2005). There is evidence to suggest that impaired substrate metabolism contributes to contractile dysfunction and to the progressive left ventricular remodeling that is characteristic of heart failure. In disease states such as ischemia-reperfusion, diabetes, or obesity, cardiac substrate utilization is shifted into an excessive use of FA in place of glucose (Sambandam and Lopaschuk 2003; Chaitman et al. 2004; Stanley et al. 2005). It has been suggested that this shift in metabolism plays a role in the development of cardiomyopathy, which leads to both impaired contractile function and ischemic injury (Stanley et al. 2005; Lopaschuk et al. 2010). Thus, the regulation of substrate metabolism in the heart plays a fundamental role in the control of normal cardiac function.

One of the key control points of FA β-oxidation is the rate of FA transfer into the mitochondria through carnitine palmitoyltransferase 1 (CPT1) (Awan and Saggerson 1993). CPT1 activity and the rate of FA β-oxidation are increased in liver, skeletal muscle, and brown adiposse tissue of SCD1−/− compared with SCD1+/+ mice (Dobrzyn et al. 2004, 2005; Lee et al. 2004a). Interestingly, however, both CPT1 mRNA and protein levels as well as its activity are decreased in the heart of SCD1−/− mice (Fig. 8.2). As a result of lower FA availability and uptake and decreased rate of FA transport into mitochondria by CPT1, the rate of mitochondrial FA oxidation is decreased in the heart of SCD1−/− animals (Fig. 8.2). Remarkably, there was no

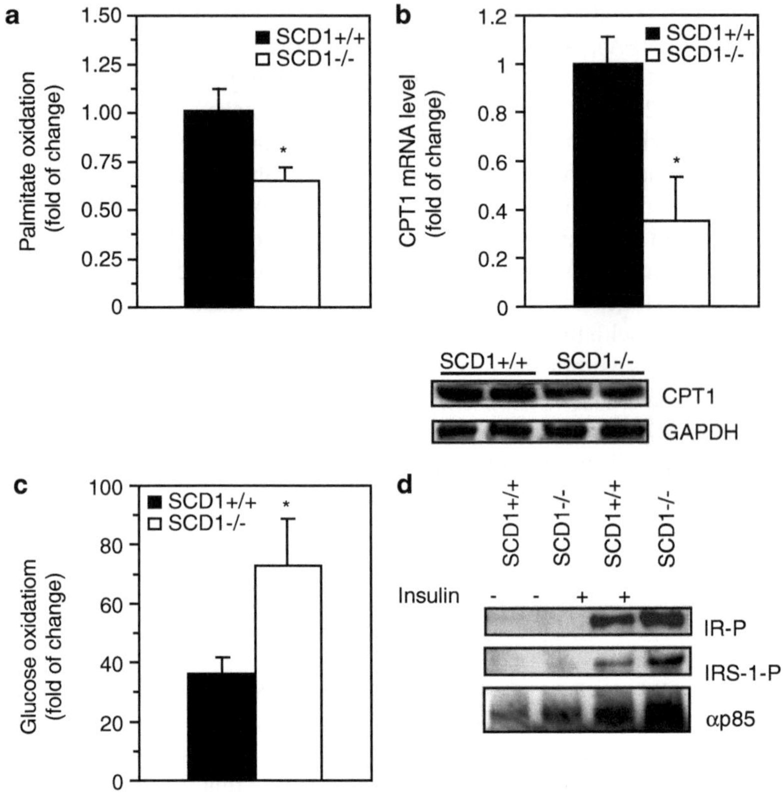

Fig. 8.2 The effect of SCD1 deficiency on the rate of fatty acid and glucose oxidation in the heart. (**a**) The rate of palmitoyl-CoA oxidation in SCD1+/+ and SCD1–/– mice. (**b**) CPT1 mRNA and protein levels. (**c**) The rate of glucose oxidation. (**d**) IR and IRS-1 phosphorylation, and association of IRS-1 with the αp85 subunit of PI3-kinase in basal condition and after insulin administration. *$P<0.05$ vs. SCD1+/+ mice. Adopted from Dobrzyn et al. (2008b)

significant difference in SCD enzyme activity in either group of mice (Dobrzyn et al. 2008b). Furthermore, there were no significant differences in levels of palmitoleic acid, a product of de novo desaturation by SCD, in cardiac lipids between SCD1+/+ and SCD1–/– mice. These results, examined together with increased SCD4 mRNA level in SCD1–/– mice, indicate that the SCD4 compensates for the lack of SCD1 in the heart (Miyazaki et al. 2003; Dobrzyn et al. 2008b). It is possible that this relative lack of a change in cardiac Δ9-desaturase activity may be accountable for the tissue-specific differences in fat oxidation observed in the heart, as opposed to other SCD1 deficient tissues, such as skeletal muscle, liver, or brown adiposse tissue. However, the potential role of SCD4 in the regulation of FA oxidation requires further studies.

The utilization of FA and glucose is tightly coupled in the myocardium (Sambandam and Lopaschuk 2003; Stanley et al. 2005). When FA are unavailable as a source of ATP, the heart extends its use of carbohydrates as an energy supply. This effect (i.e., increased glucose oxidation) was observed in the heart of SCD1−/− mice to compensate for ATP supply (Fig. 8.2). It was accompanied by increased tyrosine phosphorylation of the insulin receptor (IR) and IR substrate (IRS)-1 and greater IRS association with the αp85 subunit of PI 3-kinase in the heart (Fig. 8.2), despite lower levels of plasma insulin (Dobrzyn et al. 2008b). The increased insulin signaling is responsible for enhanced glucose uptake in the heart of SCD1−/− mice. A markedly increased insulin-stimulated glucose flux in the heart of SCD1−/− mice was also observed by Flowers et al. (2007) in studies using a hyperinsulinemic-euglycemic clamp. The decrease in fat oxidation coupled with increased insulin sensitivity leads to a shift in substrate utilization from FA to glucose in the SCD1-deficient heart (Dobrzyn et al. 2008b). The important role of SCD1 in regulation of cardiac myocyte substrate utilization was also shown in the studies of neonatal cardiomyocytes, where SCD1 overexpression attenuated palmitic acid oxidation and restored palmitate-induced suppression of glucose oxidation in the heart (Matsui et al. 2012).

AMP-activated protein kinase (AMPK) is an important factor that regulates metabolic pathways such as FA oxidation, glucose transporter translocation, glucose uptake, and glycolysis (Towler and Hardie 2007). Thus, activation of AMPK in the heart can potentially increase both fat and glucose metabolism (Coven et al. 2003; Xing et al. 2003; Stanley et al. 2005). SCD1 deficiency is known to activate AMPK in liver and skeletal muscle (Dobrzyn et al. 2004, 2005). However, in the heart, AMPK phosphorylation and protein levels were not affected by SCD1 deficiency, indicating that AMPK is unlikely to play a role in the shift in substrate oxidation in the myocardium of SCD1−/− mice (Dobrzyn et al. 2008b).

The transcription factor PPARα is highly expressed in tissues with a high capacity for FA oxidation, including hepatocytes, cardiomyocytes, the renal cortex, and skeletal muscles. Its activation promotes FA oxidation, ketone body synthesis, and glucose sparing (Ferre 2004). Cardiac-specific overexpression of PPARα has been shown to cause insulin resistance and increased FA oxidation in the heart (Park et al. 2005), whereas ablation of fatty acid translocase/CD36 (FAT/CD36) in the context of PPARα cardiac overexpression largely reverses these effects and promotes glucose uptake and oxidation in the heart (Yang et al. 2007). These studies underscore the role of PPARα in regulating cardiac substrate utilization. The expression of PPARα is decreased significantly in the myocardium of SCD1−/− mice. Therefore, it is possible that the decreased expression of genes encoding proteins involved in FA oxidation such as CPT1 and acyl-CoA oxidase is mediated by decreased PPARα activity in the heart of SCD1−/− mice (Dobrzyn et al. 2008b). PGC1α and nonesterified PUFA are two main regulators of PPARα activity (Sampath and Ntambi 2005; Finck and Kelly 2006). Although cardiac PGC1α expression is not affected by SCD1 deficiency, the intracellular PUFA contents are reduced by 30 % in the heart of SCD1−/− mice, suggesting a viable mechanism for the decreased PPARα activity observed in SCD1-deficient hearts (Dobrzyn et al. 2008b).

SCD1 and Cardiac Function

A shift in cardiac substrate utilization may lead to cardiac structural and/or functional abnormalities (Stanley et al. 2005). Transthoracic echocardiography with Doppler flow analysis was used to assess the structure and function of an SCD1−/− heart (Dobrzyn et al. 2008b). Left ventricle (LV) weights normalized to body weight measured by both wet weights obtained at necropsy and echocardiography were ~15 % higher in SCD1−/− compared with SCD1+/+ mice (Table 8.1). The LV diameter in the heart of SCD1−/− mice was significantly larger; however, the LV posterior and anterior wall thicknesses were similar in SCD1-deficient and wild-type hearts. The structural changes in the heart of SCD1−/− mice were not accompanied by functional abnormalities. The heart rate (beats/min) measured during echocardiography was similar in SCD1−/− and SCD1+/+ mice. LV fractional shortening, which is used as a measure of systolic function, and the isovolumetric relaxation time, which is used as a measure of diastolic function, were not significantly affected by SCD1 deficiency (Table 8.1). Also, the myocardial performance index, a Doppler-based measure of left ventricular function and the velocity of blood flow across the mitral valve in early diastole were unchanged in SCD1−/− compared with SCD1+/+ mice (Dobrzyn et al. 2008b).

Ectopic deposition of lipids in the myocardium may lead to functional impairments as observed in obese leptin-resistant Zucker diabetic rats (ZDF) (Zhou et al. 2000) and db/db mice as well as leptin-deficient ob/ob mice (Barouch et al. 2006). Ob/ob mice develop pathologic LV hypertrophy along with elevated TG content and

Table 8.1 Cardiovascular measurements of heart function and structure of SCD1+/+, SCD1−/−, ob/ob and ob/ob;SCD1−/− mice (Dobrzyn et al. 2010a)

	SCD1+/+	SCD1−/−	ob/ob	ob/ob;SCD1−/−
HR (bpm)	453.13 ± 23.1	473.15 ± 28.4	515.13 ± 36.0	481.33 ± 67.1
AWd (mm)	0.83 ± 0.2	0.85 ± 0.2	0.89 ± 0.1	0.85 ± 0.1
PWd (mm)	0.82 ± 0.1	0.86 ± 0.2	0.92 ± 0.1	0.87 ± 0.1
LVDd (mm)	3.27 ± 0.2	3.56 ± 0.5	3.79 ± 0.3[a]	3.34 ± 0.2[c]
LV mass (mg)	100.01 ± 4.4	110.64 ± 7.2	128.64 ± 3.9[a,b]	107.60 ± 3.4[c]
LV mass/BW (mg/g)	3.51 ± 0.3	3.98 ± 0.6	2.01 ± 0.2[a,b]	2.69 ± 0.4[a,b,c]
% FS	52.33 ± 4.6	55.02 ± 5.7	43.1 ± 3.7[a,b]	55.3 ± 7.8[c]
IVRT (s)	0.014 ± 0.006	0.016 ± 0.004	0.020 ± 0.002	0.020 ± 0.003
MPI	0.44 ± 0.2	0.49 ± 0.1	0.57 ± 0.2	0.50 ± 0.1
Ea/Aa	1.78 ± 0.5	1.64 ± 0.2	1.33 ± 0.2	1.07 ± 0.2[a,b,c]
E/Ea	29.03 ± 4.3	27.36 ± 3.6	17.6 ± 5.1[a,b]	22.3 ± 2.4[a]

HR heart rate in beats per minute, *AWd* anterior wall in diastole, *PWd* posterior wall in diastole, *LVDd* left ventricular diameter in diastole, *LV mass* left ventricular mass, *FS* fractional shortening ((LVDd−LVDs)/LVDd), *IVRT* isovolumic relaxation time in seconds, *MPI* myocardial performance index = the ratio of isovolumic contraction and relaxation to ejection time, $\mathrm{MPI} = (a-b)/b$, where a = the time of mitral value closure and b = aortic ejection time, *Ea* early diastolic maximal velocity from tissue Doppler, *Aa* late diastolic maximal velocity from tissue Doppler, *E* transmitral early filling velocity

[a] $P < 0.05$ vs. SCD1+/+ mice

[b] $P < 0.05$ vs. SCD1−/− mice

[c] $P < 0.05$ vs. ob/ob mice

increased myocyte apoptosis (Barouch et al. 2003; Miyazaki et al. 2003). A number of studies suggest an important role of SCD in the pathogenesis of lipid-induced heart disease. SCD1 expression and activity are significantly increased in the myocardium of ob/ob and db/db mice (Ge et al. 2012). The activity of SCD is also significantly induced in the hearts of obese diabetic rats fed a high-sucrose diet, and SCD1 overexpression markedly induces lipid droplet accumulation in isolated neonatal rat cardiac myocytes incubated with palmitate (Matsui et al. 2012). Furthermore, cardiac SCD1 and SCD4 are overexpressed in acyl-CoA synthase transgenic mice, which develop severe lipotoxic cardiomyopathy (Lee et al. 2004b). The expression of SCD1 appears to be more abundant in the hearts of diabetic patients than in those from lean subjects and nondiabetics (Matsui et al. 2012).

It has been proposed that SCD1 repression may be a viable approach for decreasing FA uptake and oxidation in the heart and thereby aid in the prevention and treatment of lipotoxic cardiomyopathy observed in the diabetic and obese states. To test this hypothesis, an ob/ob;SCD1−/− double knockout mouse model was generated (Dobrzyn et al. 2010a). Transthoracic echocardiography analysis showed that ob/ob;SCD1−/− mice are characterized by a smaller LV diameter and significantly reduced LV mass compared to the ob/ob mice (Table 8.1), though LV wall thickness was not affected by SCD1 deficiency (Dobrzyn et al. 2010a). Because the body mass of the ob/ob mice was much larger than that of either the wild type or ob/ob;SCD1−/− counterparts, the LV mass/body weight ratio was significantly increased in ob/ob;SCD1−/− mice. Ob/ob mice appeared to have a significant reduction in systolic function, as demonstrated by the impaired fractional shortening (Dobrzyn et al. 2010a). SCD1 deficiency increased the percent of fractional shortening by 28 % in the heart of ob/ob;SCD1−/− compared with ob/ob mice (Table 8.1). The percent of fractional shortening in ob/ob;SCD1−/− mice was not different from that of wild-type animals (Dobrzyn et al. 2010a). The myocardial performance index, a Doppler-based measure of LV function, was marginally different between ob/ob and ob/ob;SCD1−/− mice. The Doppler flow analysis demonstrated that the E/Ea ratio was reduced by 57 % in ob/ob mice compared with the wild-type mice, indicating a diastolic dysfunction in ob/ob mice (Dobrzyn et al. 2010a). Although the E/Ea ratio in the heart of ob/ob;SCD1−/− mice was still significantly lower than that in the wild-type mice, SCD1 deficiency increased the E/Ea ratio by 27 % in the heart of these mice compared with ob/ob mice (Table 8.1). These results showed that disruption of the SCD1 gene improves cardiac function in obese leptin-deficient ob/ob mice.

Reduction of SCD1 activity was also shown to exert a positive effect on heart function in the animal models of diet-induced obesity. Rats fed a high-carbohydrate, high-fat diet developed eccentric hypertrophy, a characteristic of increased preload, defined by increased left ventricular internal diameter in diastole without any changes to relative wall thickness. Consequently, these rats showed impaired systolic function seen as decreased fractional shortening, increased wall stress, increased diastolic stiffness, decreased developed pressure, and decreased LV contractility. Additionally, diastolic, systolic, stroke volumes, and cardiac output were all elevated in high-carbohydrate, high-fat fed rats without any changes in heart rate (Poudyal et al. 2012a). These negative changes in cardiac function were accompanied by increased SCD activity. A number of studies showed that increased amounts of α-linoleic acid (ALA), docosahexaenoic acid (DHA), and eicosapentaenoic acid (EPA) suppress SCD1 expression (Hofacer et al. 2012; Poudyal and Brown 2011). ALA, DHA, and EPA supplementation in high-carbohydrate, high-fat diet-induced metabolic syndrome rats caused lipid redistribution away from the abdominal area and favorably improved glucose tolerance, insulin sensitivity, dyslipidemia, hypertension and left ventricular dimensions, contractility, volumes, and stiffness. These effects were associated with complete suppression of SCD activity (Poudyal et al. 2013), further emphasizing the importance of SCD for heart function regulation.

SCD1 and Cardiomyocyte Apoptosis

Excessive deposition of intramyocardial TG enlarges the intracellular pool of fatty acyl-CoA, thereby providing a substrate for nonoxidative metabolic pathways, such as ceramide synthesis (Mengi and Dhalla 2004). Increased levels of ceramide cause apoptosis of cardiac myocytes, which can result in LV chamber expansion, contractile dysfunction, and impaired diastolic filling contributing to the cardiomyopathy observed in the settings of obesity and diabetes (Zhou et al. 2000). This phenomenon, broadly referred to as "cardiac lipotoxicity," has been observed in several genetic models in mice and rats. Obese ZDF rats develop cardiac dilatation and reduced contractility at 14 weeks of age, effects that are associated with elevated intramyocardial TG, ceramide levels, and increased apoptosis (Zhou et al. 2000). Troglitazone therapy of ZDF rats decreased cardiac TG and ceramide levels, which was associated with the complete prevention of apoptosis and restoration of cardiac function.

Accumulation of intramyocardial ceramide and TG, LV chamber enlargement, and impaired contractile function were also observed in obese leptin-deficient ob/ob mice (Dobrzyn et al. 2010a). Interestingly, the knockout of SCD1 gene resulted in decreased TG and ceramide contents in the heart of ob/ob;SCD1−/− mice in comparison to control ob/ob mice. Notably, the ceramide level in the hearts of ob/ob;SCD1−/− mice was comparable to values noted in the myocardium of the wild-type counterparts (Dobrzyn et al. 2010a). Ceramide may be formed by hydrolysis of sphingomyelin, de novo synthesis via condensation of palmitoyl-CoA and serine, glycosphingolipid breakdown, or conversion of other sphingolipids (Hanada 2003). The decrease in ceramide content in SCD1-deficient hearts appears to be due to decreased de novo synthesis, as evidenced by decreased SPT activity and gene expression, and reduced incorporation of [^{14}C]palmitate into ceramide (Dobrzyn et al. 2010a). The reduced intracellular palmitate level in the heart of ob/ob;SCD1−/− mice resulting from decreased lipogenesis and reduced FA uptake may also be one rate-limiting factor in de novo ceramide synthesis. Similar effects, including reduced SCD activity and decreased intramuscular palmitoyl-CoA content, concomitant with the downregulation of SPT activity and reduction in ceramide synthesis, were observed in the oxidative skeletal muscles of SCD1−/− and SCD1-deficient ob/ob mice (Dobrzyn et al. 2005).

Because the ceramide pathway is the most important of the lipoapoptotic routes in cardiomyocytes (Shimabukuro et al. 1998), decreased ceramide content due to SCD1 deficiency resulted in a reduced rate of apoptosis in the heart of ob/ob mice (Fig. 8.3). It has been shown that two key markers of ceramide-induced apoptosis, NO production (measured by iNOS activity) and caspase-3 activity, were significantly reduced in the hearts of ob/ob;SCD1−/− double mutant mice (Dobrzyn et al. 2010a). Decreased activities of iNOS and caspase-3 might result from inhibition of de novo ceramide synthesis due to SCD1 deficiency (Fig. 8.3). Bielawska and coworkers (1997) proposed that ceramide upregulates iNOS expression and increases NO production, which causes an increase in peroxynitrite and apoptosis. Additionally, downregulation of caspase-3 was often linked with ceramide action (Ravid et al. 2003; Castillo et al. 2007). Ravid et al. (2003) showed that ceramide-mediated apoptosis is blocked by a general caspase inhibitor, Boc-D-fluoromethylketone. Ruvolo et al. (2002) established that exogenous ceramide can downregulate antiapoptotic factor Bcl-2 expression and phosphorylation, thereby activating caspase-3 and apoptosis. It has also been suggested that the antiapoptotic effect of Bcl-2 may occur via the modulation of ceramide production and the prevention of ceramide-mediated caspase activation (Ruvolo et al. 2002; Sawada et al. 2000). Increased mRNA level of Bcl-2 was observed in the heart of ob/ob;SCD1−/− mice compared with ob/ob controls. Thus, the increased gene expression of Bcl-2 could be another factor contributing to the downregulation of caspase-3 activity and the reduction in the rate of apoptosis in the heart of ob/ob;SCD1−/− double knockout mice (Fig. 8.3).

Furthermore, it has been suggested that increased oxidation of palmitate through CPT1 is involved in the occurrence of apoptosis in cardiomyocytes. Previous studies established that palmitate-induced cell death was enhanced by carnitine, a cofactor needed for palmitate transport into mitochondria via CPT1 (Kong and Rabkin 2002). CPT1 mRNA levels were decreased in the heart of ob/ob;SCD1−/− mice compared with ob/ob mice (Dobrzyn et al. 2010a). Therefore, decreased mitochondrial FA oxidation may account for decreased apoptosis in the heart of SCD1-deficient ob/ob mice. As described above, knockout of the SCD1 gene significantly improves diastolic and systolic LV function in ob/ob mice (Table 8.1).

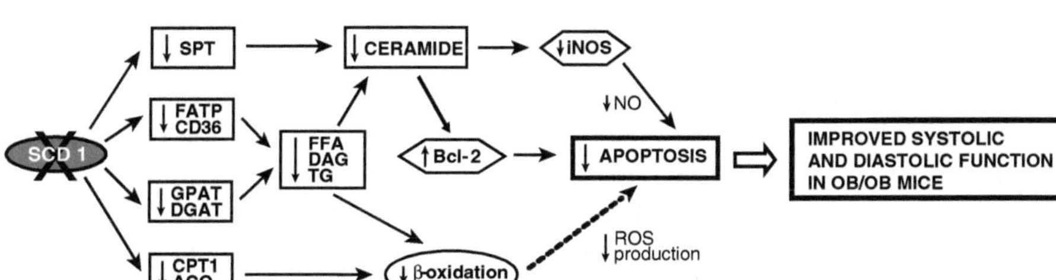

Fig. 8.3 Proposed model for the effect of SCD1 gene deletion on heart lipid metabolism and left ventricle function in leptin deficiency. Reduction in myocardial lipid accumulation and inhibition of lipid-induced apoptosis appears to be the main mechanisms responsible for improved cardiac function in leptin-deficient ob/ob mice caused by lack of SCD1 function. Adopted from Dobrzyn et al. (2010a)

Because reduced apoptosis in cardiomyocytes was shown to improve cardiac function (Barouch et al. 2006), inhibition of lipid-induced apoptosis due to SCD1 deficiency might be directly responsible for the improvement of heart function in ob/ob mice (Fig. 8.3).

Studies performed on isolated cardiomyocytes showed that SCD1 has protective effects against saturated FA-induced apoptosis (Matsui et al. 2012). Palmitic acid or stearic acid increased the activities of caspases 3 and 7 and the number of TUNEL-positive stained cells in neonatal rat cardiomyocytes. In contrast, SCD1 overexpression significantly attenuated saturated FA-induced apoptotic changes. These results suggest that SCD1 suppresses saturated FA-induced apoptosis in cardiac myocytes, as shown first by Listenberger et al. (2003) in CHO cells; however, this effect was observed in studies performed in vitro only.

The Role of Oleate in Regulation of Cardiac Metabolism

Dietary stearic acid intake ranks second among the SFA consumed in the United States, accounting for 25.8 % of SFA intake and 2.9 % of total kcal (Kris-Etherton et al. 2005). Stearate is a precursor to endogenously synthesized oleate by SCD (Sampath and Ntambi 2005). Oleic acid is the most relevant intermediate between saturated FA and polyunsaturated FA, and it is best suited for being stored or incorporated into glycerolipids. It also modulates the basic features of biomembranes (Kamp et al. 2003). Due to its unique features, oleate plays an important metabolic role in the cell (e.g., it directly affects signal transduction by activating the isoenzymes of protein kinase C), regulating the phosphorylation of proteins and modulating the expression of genes (Kennedy et al. 2009; Dziewulska et al. 2012). In the heart, oleate plays an important role in the regulation of substrate metabolism and seems to be involved in the pathogenesis of lipotoxic cardiomyopathy (Posner et al. 1991; Xu et al. 2006; Dobrzyn et al. 2012). It has been shown that the majority of FA undergoing β-oxidation in the heart are not saturated FA but, rather, monounsaturated FA, with oleic acid being the preferred substrate (Lopaschuk et al. 2010). An increased rate of FA β-oxidation was found in the heart of rats fed trioleate- (TO) or tristearate (TS)-supplemented diets, suggesting that both dietary and endogenously synthesized oleate are able to activate oxidative pathways (Dobrzyn et al. 2012). Very little is known about the potential mechanisms involved in the regulation of cardiac FA oxidation by oleate. One of the possible mechanisms that link oleate and FA oxidation may involve the activation of PPARα. Transgenic mice with an overexpression of cardiac PPARα have increased myocardial fatty acid uptake, cardiac TG and ceramide accumulation, cardiac hypertrophy, and left ventricular dilation (Finck et al. 2002, 2003) as well as clear adverse effects on cardiac function and structure (Chiu et al. 2005; Finck et al. 2002). PPARα activation increases fatty acid oxidation and protects the heart from substrate-induced contractile dysfunction when there is an oversupply of FA and TG (Morgan et al. 2006). Gilde et al. (2003) showed that the administration of long-chain saturated and monounsaturated FA results in moderate increases in the mRNA expression of several PPARα-regulated genes; however, this effect was only seen in isolated cardiomyocytes (Gilde et al. 2003). Recent studies demonstrate that dietary or de novo synthesized oleate increases the expression and activity of PPARα (Dobrzyn et al. 2012). Furthermore, feeding with TS- and TO-supplemented diets results in increased phosphorylation and activity of AMPK in the heart, another important factor for regulating cardiac FA oxidation and glucose transporter 4 (GLUT4) membrane translocation (Lopaschuk et al. 2010; Towler and Hardie 2007). Analyses performed on the perfused heart show that oleic acid activates AMPK in the myocardium (Clark et al. 2004). Thus, it is possible that the activation of both the AMPK and PPARα pathways by oleate is a viable mechanism for increasing mitochondrial FA oxidation in response to increased oleate availability.

As mentioned above, oleate can be easily incorporated into glycerolipids for storage or into phospholipids, leading to changes in plasma membrane properties. As shown using TS- and TO-fed rats, augmentation of cardiac oleate is accompanied by an increased protein level of FAT/CD36 (Dobrzyn et al. 2012), the major protein responsible for membrane FA transport (Carley and Severson 2005). This change could be especially valuable for mitochondrial FA oxidation because overexpression of FAT/CD36 in the skeletal muscle leads to an enhanced level of mitochondrial FA oxidation (Nickerson et al. 2009). An increase in oleate availability also results in increased TG accumulation in the myocardium (Dobrzyn et al. 2012). The increased TG level in the hearts of TO- and TS-fed rats was likely related to the fact that oleate is the preferred substrate for TG esterification. However, both TO and TS feedings resulted in an upregulation of the cardiac lipogenic genes, i.e., FAS, acyl-CoA synthetase, and glycerol-3-phosphate transferase (GPAT) in the heart. The parallel upregulation of SCD1/oleate availability, FAS, and GPAT in conditions of increased cardiac FA oxidation suggests that these proteins may cooperate in maintaining the intracellular oleate level within a narrow range and thus may play a role in the regulation of cardiac energy metabolism.

Increased levels of dietary or endogenously synthesized oleate may be responsible for not only increasing the rate of FA oxidation but also for reducing glucose uptake in the heart (Dobrzyn et al. 2012). Recent studies showed that long-chain FA affect GLUT4 expression and membrane translocation as well as Akt phosphorylation in cardiac tissue. Reporter studies in H9C2 cardiomyotubes have shown that hyperlipidemia in vitro, which is induced by high levels of stearic and oleic acids, repressed transcription of the GLUT4 promoter (Armoni et al. 2005). Hommelberg et al. (2009) observed a reduced rate of GLUT4 translocation and deoxyglucose uptake in L6 skeletal muscle cells incubated with palmitate and stearate. Oleate was shown to reduce Akt phosphorylation in perfused rat livers (Anderwald et al. 2007) and in primary hepatocytes (Liu et al. 2007). Furthermore, prolonged exposure of hepatocytes to oleate decreased the insulin-induced tyrosine phosphorylation of IRS1/2 while slightly increasing the serine phosphorylation of IRS (Liu et al. 2007). Therefore, it is possible that the dietary or de novo synthesized oleate from stearic acid decreases the rate of glucose uptake in the heart by the reduction of Akt phosphorylation and GLUT4 membrane translocation, as was found in TS- and TO-fed rats (Dobrzyn et al. 2012). These data are in agreement with results obtained in SCD1−/− mice, where increased cardiac glucose transport and oxidation were accompanied by decreased oleic acid content (Dobrzyn et al. 2008b).

Plasma Desaturation Index as Predictor of Cardiac Health

Recent studies establish the important role of SCD activity and plasma fatty acid desaturation on the human heart function and cardiometabolic risks. First, Warensjö et al. (2008a) showed that the plasma SCD desaturation index is elevated by a diet high in saturated fat, compared with a diet rich in unsaturated fat. Next, the same investigators suggested that endogenous fatty acid desaturation, which is partly independent of diet, is associated with and might even contribute to mortality; thus, desaturase indexes may be used to predict cardiovascular mortality risk (Warensjö et al. 2008b). These results were confirmed by Djoussé et al. (2012), who showed that plasma SCD activity (16:1n-7/16:0 ratio) was positively associated with the risk of heart failure, whereas oleic acid and *cis*-vaccenic acid concentrations were not related to heart failure risk. Moreover, the plasma SCD desaturation index was positively associated with heart rate (Ebbesson et al. 2012). Additionally, a study including 93 healthy volunteers showed that SCD desaturation index is closely associated with the features of cardiometabolic risks (i.e., increased body mass index and TG content as well as decreased HDL-cholesterol level) in Koreans (Do et al. 2011). All together, these results suggest that the level of fatty acid desaturation in blood plasma is related to heart rate and function and is possibly related to arrhythmia and sudden death. This would partially explain the observed association between cardiovascular mortality risk and SCD activity.

Conclusion

Research over the past decade has identified SCD as an important regulator of physiological processes and body adiposity. Using knockout mouse models, we have learned that SCD1 is a critical control point of lipid partitioning. While high SCD activity favors fat storage, suppression of the enzyme activates metabolic pathways that promote the burning of fat and decrease lipid synthesis in white adipose tissue and the liver. SCD1 deficiency upregulates insulin-signaling components and affects glycogen metabolism in insulin-sensitive tissues. In the heart, the lack of SCD1 decreases FA uptake and oxidation while increasing glucose transport and oxidation. These metabolic changes cause a decrease in the accumulation of neutral lipids and ceramide thereby inhibiting apoptotic pathways in the LV of leptin-deficient ob/ob mice and improve systolic and diastolic function. The findings on SCD1 point to a potentially novel strategy for the treatment of lipotoxic heart disease. However, the potential use of an SCD inhibitor as a therapeutic agent awaits a more complete understanding of the mechanism underlying the effects of SCD deficiency and an indication that inhibition of this enzyme is both safe and efficacious.

Acknowledgements P. Dobrzyn is supported by grants from by the National Science Centre (UMO-2011/01/D/NZ3/04777) and the National Centre for Research and Development (LIDER/19/2/L-2/10/NCBiR/2011); A. Dobrzyn is supported by grant from the Polish Science Foundation (TEAM/2010-5/2).

References

Anderwald C, Brunmair B, Stadlbauer K, Krebs M, Furnsinn C, Roden M (2007) Effects of free fatty acids on carbohydrate metabolism and insulin signalling in perfused rat liver. Eur J Clin Invest 37:774–782

Armoni M, Harel C, Bar-Yoseph F, Milo S, Karnieli E (2005) Free fatty acids repress the GLUT4 gene expression in cardiac muscle via novel response elements. J Biol Chem 280:34786–34795

Awan MM, Saggerson ED (1993) Malonyl-CoA metabolism in cardiac myocytes and its relevance to the control of fatty acid oxidation. Biochem J 295:61–66

Barouch LA, Berkowitz DE, Harrison RW, O'Donnell CP, Hare JM (2003) Disruption of leptin signaling contributes to cardiac hypertrophy independently of body weight in mice. Circulation 108:754–759

Barouch LA, Gao D, Chen L, Miller KL, Xu W, Phan AC, Kittleson MM, Minhas KM, Berkowitz DE, Wei C, Hare JM (2006) Cardiac myocyte apoptosis is associated with increased DNA damage and decreased survival in murine models of obesity. Circ Res 98:119–124

Beiraghi S, Zhou M, Talmadge CB, Went-Sumegi N, Davis JR, Huang D, Saal H, Seemayer TA, Sumegi J (2003) Identification and characterization of a novel gene disrupted by a pericentric inversion inv(4)(p13.1q21.1) in a family with cleft lip. Gene 309:11–21

Bielawska AE, Shapiro JP, Jiang L, Melkonyan HS, Piot C, Wolfe CL, Tomei LD, Hannun YA, Umansky SR (1997) Ceramide is involved in triggering of cardiomyocyte apoptosis induced by ischemia and reperfusion. Am J Pathol 151:1257–1263

Carley AN, Severson DL (2005) Fatty acid metabolism is enhanced in type 2 diabetic hearts. Biochim Biophys Acta 1734:112–126

Castillo SS, Levy M, Wang C, Thaikoottathil JV, Khan E, Goldkorn T (2007) Nitric oxide-enhanced caspase-3 and acidic sphingomyelinase interaction: a novel mechanism by which airway epithelial cells escape ceramide-induced apoptosis. Exp Cell Res 313:816–823

Chaitman BR, Pepine CJ, Parker JO, Skopal J, Chumakova G, Kuch J, Wang W, Skettino SL, Wolff AA (2004) Effects of ranolazine with atenolol, amlodipine, or diltiazem on exercise tolerance and angina frequency in patients with severe chronic angina: a randomized controlled trial. JAMA 291:309–316

Chiu HC, Kovacs A, Blanton RM, Han X, Courtois M, Weinheimer CJ, Yamada KA, Brunet S, Xu H, Nerbonne JM, Welch MJ, Fettig NM, Sharp TL, Sambandam N, Olson KM, Ory DS, Schaffer JE (2005) Transgenic expression of fatty acid transport protein 1 in the heart causes lipotoxic cardiomyopathy. Circ Res 96:225–233

Clark H, Carling D, Saggerson D (2004) Covalent activation of heart AMP-activated protein kinase in response to physiological concentrations of long-chain fatty acids. Eur J Biochem 271:2215–22124

Coven DL, Hu X, Cong L, Bergeron R, Shulman GI, Hardie DG, Young LH (2003) Physiological role of AMP-activated protein kinase in the heart: graded activation during exercise. Am J Physiol Endocrinol Metab 285:E629–E636

Djoussé L, Weir NL, Hanson NQ, Tsai MY, Gaziano JM (2012) Plasma phospholipid concentration of cis palmitoleic acid and risk of heart failure. Circ Heart Fail 5:703–709

Do HJ, Chung HK, Moon J, Shin MJ (2011) Relationship between the estimates of desaturase activities and cardiometabolic phenotypes in Koreans. J Clin Biochem Nutr 49:131–135

Dobrzyn A, Ntambi JM (2004) The role of stearoyl-CoA desaturase in body weight regulation. Trends Cardiovasc Med 14:77–81

Dobrzyn P, Dobrzyn A, Miyazaki M, Cohen P, Asilmaz E, Hardie DG, Friedman JM, Ntambi JM (2004) Stearoyl-CoA desaturase 1 deficiency increases fatty acid oxidation by activating AMP-activated protein kinase in liver. Proc Natl Acad Sci U S A 101:6409–6414

Dobrzyn A, Dobrzyn P, Lee SH, Miyazaki M, Cohen P, Asilmaz E, Hardie DG, Friedman JM, Ntambi JM (2005) Stearoyl-CoA desaturase-1 deficiency reduces ceramide synthesis by down-regulating serine palmitoyltransferase and increasing beta-oxidation in skeletal muscle. Am J Physiol Endocrinol Metab 288:E599–E607

Dobrzyn P, Sampath H, Dobrzyn A, Miyazaki M, Ntambi JM (2008b) Loss of stearoyl-CoA desaturase 1 inhibits fatty acid oxidation and increases glucose utilization in the heart. Am J Physiol Endocrinol Metab 294:E357–E364

Dobrzyn P, Dobrzyn A, Miyazaki M, Ntambi JM (2010a) Loss of stearoyl-CoA desaturase 1 rescues cardiac function in obese leptin-deficient mice. J Lipid Res 51:2202–2210

Dobrzyn P, Jazurek M, Dobrzyn A (2010b) Stearoyl-CoA desaturase and insulin signaling -what is the molecular switch? Biochim Biophys Acta 1797:1189–1194

Dobrzyn P, Pyrkowska A, Jazurek M, Dobrzyn A (2012) Increased availability of endogenous and dietary oleic acid contributes to the upregulation of cardiac fatty acid oxidation. Mitochondrion 12:132–137

Duan SZ, Ivashchenko CY, Russell MW, Milstone DS, Mortensen RM (2005) Cardiomyocyte-specific knockout and agonist of peroxisome proliferator-activated receptor-gamma both induce cardiac hypertrophy in mice. Circ Res 97:372–379

Dziewulska A, Dobrzyn P, Jazurek M, Pyrkowska A, Ntambi JM, Dobrzyn A (2012) Monounsaturated fatty acids are required for membrane translocation of protein kinase C-theta induced by lipid overload in skeletal muscle. Mol Membr Biol 29:309–320

Ebbesson SO, Lopez-Alvarenga JC, Okin PM, Devereux RB, Tejero ME, Harris WS, Ebbesson LO, MacCluer JW, Wenger C, Laston S, Fabsitz RR, Kennish J, Howard WJ, Howard BV, Umans J, Comuzzie AG (2012) Heart rate is associated with markers of fatty acid desaturation: the GOCADAN study. Int J Circumpolar Health 71:17343

Ellis JM, Mentock SM, Depetrillo MA, Koves TR, Sen S, Watkins SM, Muoio DM, Cline GW, Taegtmeyer H, Shulman GI, Willis MS, Coleman RA (2011) Mouse cardiac acyl coenzyme a synthetase 1 deficiency impairs fatty acid oxidation and induces cardiac hypertrophy. Mol Cell Biol 31:1252–1262

Ferre P (2004) The biology of peroxisome proliferator-activated receptors: relationship with lipid metabolism and insulin sensitivity. Diabetes 53(Suppl 1):S43–S50

Finck BN, Kelly DP (2006) PGC-1 coactivators: inducible regulators of energy metabolism in health and disease. J Clin Invest 116:615–622

Finck BN, Lehman JJ, Leone TC, Welch MJ, Bennett MJ, Kovacs A, Han X, Gross RW, Kozak R, Lopaschuk GD, Kelly DP (2002) The cardiac phenotype induced by PPARα overexpression mimics that caused by diabetes mellitus. J Clin Invest 109:121–130

Finck BN, Han X, Courtois M, Aimond F, Nerbonne JM, Kovacs A, Gross RW, Kelly DP (2003) A critical role for PPAR-mediated lipotoxicity in the pathogenesis of diabetic cardiomyopathy: modulation by dietary fat content. Proc Natl Acad Sci U S A 100:1226–1231

Flowers JB, Rabaglia ME, Schueler KL, Flowers MT, Lan H, Keller MP, Ntambi JM, Attie AD (2007) Loss of stearoyl-CoA desaturase-1 improves insulin sensitivity in lean mice but worsens diabetes in leptin-deficient obese mice. Diabetes 56:1228–1239

Ge F, Hu C, Hyodo E, Arai K, Zhou S, Lobdell Iv H, Walewski JL, Homma S, Berk PD (2012) Cardiomyocyte triglyceride accumulation and reduced ventricular function in mice with obesity reflect increased long chain fatty acid uptake and de novo fatty acid synthesis. J Obes 2012:205648

Gilde AJ, Van Der Lee KA, Willemsen PH, Chinetti G, Van Der Leij FR, van der Vusse GJ, Staels B, van Bilsen M (2003) Peroxisome proliferator-activated receptor (PPAR)-α and PPARβ/γ, but not PPARγ, modulate the expression of genes involved in cardiac lipid metabolism. Circ Res 92:518–524

Hanada K (2003) Serine palmitoyltransferase, a key enzyme of sphingolipid metabolism. Biochim Biophys Acta 1632:16–30

Hofacer R, Magrisso IJ, Jandacek R, Rider T, Tso P, Benoit SC, MacNamara MK (2012) Omega-3 fatty acid deficiency increases stearoyl-CoA desaturase expression and activity indices in rat liver: positive association with non-fasting plasma triglyceride levels. Prostaglandins Leukot Essent Fatty Acids 86:71–77

Hommelberg PP, Plat J, Langen RC, Schols AM, Mensink RP (2009) Fatty acid-induced NF-kappaB activation and insulin resistance in skeletal muscle are chain length dependent. Am J Physiol Endocrinol Metab 296:E114–E120

Kaestner KH, Ntambi JM, Kelly TJ, Lane MD (1989) Differentiation-induced gene expression in 3T3-L1 preadipocytes. A second differentially expressed gene encoding stearoyl-CoA desaturase. J Biol Chem 264:14755–14761

Kamp F, Guo W, Souto R, Pilch PF, Corkey CE, Hamilton JA (2003) Rapid flip-flop of oleic acid across the plasma membrane of adipocytes. J Biol Chem 278:7988–7995

Kennedy A, Martinez K, Chuang CC, LaPoint K, McIntosh M (2009) Saturated fatty acid-mediated inflammation and insulin resistance in adipose tissue: mechanisms of action and implications. J Nutr 139:1–4

Kong JY, Rabkin SW (2002) Palmitate-induced cardiac apoptosis is mediated through CPT-1 but not influenced by glucose and insulin. Am J Physiol Heart Circ Physiol 282:H717–H725

Kris-Etherton PM, Griel AE, Psota TL, Gebauer SK, Zhang J, Etherton TD (2005) Dietary stearic acid and risk of cardiovascular disease: intake, sources, digestion, and absorption. Lipids 40:1193–1200

Lee SH, Dobrzyn A, Dobrzyn P, Rahman SM, Miyazaki M, Ntambi JM (2004a) Lack of stearoyl-CoA desaturase 1 upregulates basal thermogenesis but causes hypothermia in a cold environment. J Lipid Res 45:1674–1682

Lee Y, Naseem RH, Duplomb L, Park BH, Garry DJ, Richardson JA, Schaffer JE, Unger RH (2004b) Hyperleptinemia prevents lipotoxic cardiomyopathy in acyl CoA synthase transgenic mice. Proc Natl Acad Sci U S A 101:13624–13629

Listenberger LL, Han X, Lewis SE, Cases S, Farese RV Jr, Ory DS, Schaffer JE (2003) Triglyceride accumulation protects against fatty acid-induced lipotoxicity. Proc Natl Acad Sci U S A 100:3077–3082

Liu HY, Collins QF, Xiong Y, Moukdar F, Lupo EG, Liu Z, Cao W (2007) Prolonged treatment of primary hepatocytes with oleate induces insulin resistance through p38 mitogen-activated protein kinase. J Biol Chem 282:14205–14212

Lopaschuk GD, Ussher JR, Folmes CD, Jaswal JS, Stanley WC (2010) Myocardial fatty acid metabolism in health and disease. Physiol Rev 90:207–258

Matsui H, Yokoyama T, Sekiguchi K, Iijima D, Sunaga H, Maniwa M, Ueno M, Iso T, Arai M, Kurabayashi M (2012) Stearoyl-CoA desaturase-1 (SCD1) augments saturated fatty acid-induced lipid accumulation and inhibits apoptosis in cardiac myocytes. PLoS One 7:e33283

Mengi SA, Dhalla NS (2004) Carnitine palmitoyltransferase-I, a new target for the treatment of heart failure: perspectives on a shift in myocardial metabolism as a therapeutic intervention. Am J Cardiovasc Drugs 4:201–209

Miyazaki M, Kim HJ, Man WC, Ntambi JM (2001) Oleoyl-CoA is the major de novo product of stearoyl-CoA desaturase 1 gene isoform and substrate for the biosynthesis of the Harderian gland 1-alkyl-2,3-diacylglycerol. J Biol Chem 276:39455–39461

Miyazaki M, Jacobson MJ, Man WC, Cohen P, Asilmaz E, Friedman JM, Ntambi JM (2003) Identification and characterization of murine SCD4, a novel heart-specific stearoyl-CoA desaturase isoform regulated by leptin and dietary factors. J Biol Chem 278:33904–33911

Morgan EE, Dennison JH, Young ME, McElfresh TA, Kung TA, Tserng KY, Hoit BD, Stanley WC, Handler MP (2006) Effects of chronic activation of peroxisome proliferator-activated receptor-alpha or high-fat feeding in a rat infarct model of heart failure. Am J Physiol Heart Circ Physiol 290:H1899–H1904

Nickerson JG, Alkhateeb H, Benton CR, Lally J, Nickerson J, Han XX, Wilson MH, Jain SS, Snook LA, Glatz JF, Chabowski A, Luiken JJ, Bonen A (2009) Greater transport efficiencies of the membrane fatty acid transporters FAT/CD36 and FATP4 compared with FABPpm and FATP1 and differential effects on fatty acid esterification and oxidation in rat skeletal muscle. J Biol Chem 284:16522–16530

Ntambi JM, Miyazaki M (2004) Regulation of stearoyl-CoA desaturases and role in metabolism. Prog Lipid Res 43:91–104

Ntambi JM, Buhrow SA, Kaestner KH, Christy RJ, Sibley E, Kelly TJ, Lane MD (1988) Differentiation-induced gene expression in 3T3-L1 preadipocytes. Characterization of a differentially expressed gene encoding stearoyl-CoA desaturase. J Biol Chem 263:17291–17300

Park SY, Cho YR, Finck BN, Kim HJ, Higashimori T, Hong EG, Lee MK, Danton C, Deshmukh S, Cline GW, Wu JJ, Bennett AM, Rothermel B, Kalinowski A, Russell KS, Kim YB, Kelly DP, Kim JK (2005) Cardiac-specific overexpression of peroxisome proliferator-activated receptor-alpha causes insulin resistance in heart and liver. Diabetes 54:2514–2524

Park HJ, Georgescu SP, Du C, Madias C, Aronovitz MJ, Welzig CM, Wang B, Begley U, Zhang Y, Blaustein RO, Patten RD, Karas RH, Van Tol HH, Osborne TF, Shimano H, Liao R, Link MS, Galper JB (2008) Parasympathetic response in chick myocytes and mouse heart is controlled by SREBP. J Clin Invest 118:259–271

Paton CM, Ntambi JM (2009) Biochemical and physiological function of stearoyl-CoA desaturase. Am J Physiol Endocrinol Metab 297:E28–E37

Posner BM, Cobb JL, Belanger AJ, Cupples LA, D'Agostino RB, Stokes J (1991) Dietary lipid predictors of coronary heart disease in men. The Framingham Study. Arch Intern Med 151:1181–1187

Poudyal P, Brown L (2011) Stearoyl-CoA desaturase: a vital checkpoint in the development and progression of obesity. Endocr Metab Immune Disord Drug Targets 11:217–231

Poudyal H, Panchal SK, Waanders J, Ward L, Brown L (2012a) Lipid redistribution by α-linolenic acid-rich chia seed inhibits stearoyl-CoA desaturase-1 and induces cardiac and hepatic protection in diet-induced obese rats. J Nutr Biochem 23:153–162

Poudyal H, Panchal SK, Ward LC, Brown L (2013) Effects of ALA, EPA and DHA in high-carbohydrate, high-fat diet-induced metabolic syndrome in rats. J Nutr Biochem 24:1041–1052

Ravid T, Tsaba A, Gee P, Rasooly R, Medina EA, Goldkorn T (2003) Ceramide accumulation precedes caspase-3 activation during apoptosis of A549 human lung adenocarcinoma cells. Am J Physiol Lung Cell Mol Physiol 284:L1082–L1092

Razani B, Zhang H, Schulze PC, Schilling JD, Verbsky J, Lodhi IJ, Topkara VK, Feng C, Coleman T, Kovacs A, Kelly DP, Saffitz JE, Dorn GW 2nd, Nichols CG, Semenkovich CF (2011) Fatty acid synthase modulates homeostatic responses to myocardial stress. J Biol Chem 286:30949–30961

Ruvolo PP, Clark W, Mumby M, Gao F, May WS (2002) A functional role for the B56 alpha-subunit of protein phosphatase 2A in ceramide-mediated regulation of Bcl2 phosphorylation status and function. J Biol Chem 277:22847–22852

Sambandam N, Lopaschuk GD (2003) AMP-activated protein kinase (AMPK) control of fatty acid and glucose metabolism in the ischemic heart. Prog Lipid Res 42:238–256

Sampath H, Ntambi JM (2005) Polyunsaturated fatty acid regulation of genes of lipid metabolism. Annu Rev Nutr 25:317–340

Sawada M, Nakashima S, Banno Y, Yamakawa H, Takenaka K, Shinoda J, Nishimura Y, Sakai N, Nozawa Y (2000) Influence of Bax or Bcl-2 overexpression on the ceramide-dependent apoptotic pathway in glioma cells. Oncogene 19:3508–3520

Shimabukuro M, Higa M, Zhou YT, Wang MY, Newgard CB, Unger RH (1998) Lipoapoptosis in beta-cells of obese prediabetic fa/fa rats. Role of serine palmitoyltransferase overexpression. J Biol Chem 273:32487–32490

Stanley WC, Recchia FA, Lopaschuk GD (2005) Myocardial substrate metabolism in the normal and failing heart. Physiol Rev 85:1093–1129

Towler MC, Hardie DG (2007) AMP-activated protein kinase in metabolic control and insulin signaling. Circ Res 100:328–341

Warensjö E, Risérus U, Gustafsson IB, Mohsen R, Cederholm T, Vessby B (2008a) Effects of saturated and unsaturated fatty acids on estimated desaturase activities during a controlled dietary intervention. Nutr Metab Cardiovasc Dis 18:683–690

Warensjö E, Sundström J, Vessby B, Cederholm T, Risérus U (2008b) Markers of dietary fat quality and fatty acid desaturation as predictors of total and cardiovascular mortality: a population-based prospective study. Am J Clin Nutr 88:203–209

Xing Y, Musi N, Fujii N, Zou L, Luptak I, Hirshman MF, Goodyear LJ, Tian R (2003) Glucose metabolism and energy homeostasis in mouse hearts overexpressing dominant negative alpha2 subunit of AMP-activated protein kinase. J Biol Chem 278:28372–28377

Xu J, Eilat-Adar S, Loria C, Goldbourt U, Howard BV, Fabsitz RR, Zephier EM, Mattil C, Lee ET (2006) Dietary fat intake and risk of coronary heart disease: the Strong Heart Study. Am J Clin Nutr 84:894–902

Yang J, Sambandam N, Han X, Gross RW, Courtois M, Kovacs A, Febbraio M, Finck BN, Kelly DP (2007) CD36 deficiency rescues lipotoxic cardiomyopathy. Circ Res 100:1208–1217

Zhang L, Ge L, Parimoo S, Stenn K, Prouty SM (1999) Human stearoyl-CoA desaturase: alternative transcripts generated from a single gene by usage of tandem polyadenylation sites. Biochem J 340:255–264

Zheng Y, Prouty SM, Harmon A, Sundberg JP, Stenn KS, Parimoo S (2001) Scd3: a novel gene of the stearoyl-CoA desaturase family with restricted expression in skin. Genomics 71:182–191

Zhou YT, Grayburn P, Karim A, Shimabukuro M, Higa M, Baetens D, Orci L, Unger RH (2000) Lipotoxic heart disease in obese rats: implications for human obesity. Proc Natl Acad Sci U S A 97:1784–1789

[illegible]

Chapter 9
Stearoyl-CoA Desaturase-1 Activity in Skeletal Muscle: Is It Good or Bad?

Joseph Stevens and Matthew W. Hulver

Introduction

Skeletal muscle is known to be an important site for metabolic processes such as glucose disposal and fatty acid oxidation, and, as such, dysregulation of these processes in muscle is associated with and may play a causative role in many disease states and comorbidities including obesity, hypertension, insulin resistance, and hypertriglyceridemia (Lee et al. 2003; McGarry 2002; Petersen and Shulman 2002; Scheuermann-Freestone et al. 2003). Specifically, these disease/co-morbid states are associated with dysregulated glucose and fatty acid metabolism and excess lipid accumulation in skeletal muscle (Flowers and Ntambi 2009; Kahn and Flier 2000). Studies in humans show that SCD1 is associated with insulin resistance, increased intramuscular triacylglycerol (IMTG) content, and reduced fatty acid oxidation (Dubé et al. 2011; Hulver et al. 2005). Studies in rodents clearly show that genetic deletion of SCD1 prevents high-fat diet-induced weight gain and insulin resistance (Ntambi et al. 2002). Conversely, others have suggested that heightened SCD1 activity in skeletal muscle may be protective (Dobrzyn et al. 2010; Schenk and Horowitz 2007). To date, the role of SCD1 in dysregulated metabolism, specifically in skeletal muscle, is not definitively known, as there are conflicting reports. The purpose of this chapter is to highlight the available evidence for both the deleterious and protective roles of SCD1 in the context of metabolic deranged states such as insulin resistance and obesity. The role of SCD1 in the context of inflammatory signaling and adaptation to exercise will also be touched upon.

J. Stevens • M.W. Hulver, Ph.D. (✉)
Department of Human Nutrition, Foods, and Exercise, Virginia Tech, Integrated Life Sciences Building, 1981 Kraft Drive, Blacksburg, VA 24073, USA
e-mail: hulvermw@vt.edu

J.M. Ntambi (ed.), *Stearoyl-CoA Desaturase Genes in Lipid Metabolism*,
DOI 10.1007/978-1-4614-7969-7_9,

A Causative Role for SCD1 in Dysregulated Metabolism

Data from experiments using mouse models with whole-body deletion, a naturally occurring mutation in the gene, or through limited cell culture models with a gain or loss of SCD1 function have all provided insight into this topic; however, there is still much to be learned. Tissue specific deletion of SCD1 has contributed a great deal to the current understanding of the role of SCD1 in the modulation of whole-body metabolism (Liu et al. 2011). Currently, no data from skeletal muscle specific knockout of SCD1 has been published. Cell culture models, where studies in which SCD1 function has been interrupted in myotubes have shown a direct effect on fatty acid metabolism (Hulver et al. 2005; Kim et al. 2011; Peter et al. 2009), suggest a causative role for SCD1 in dysregulated substrate metabolism.

The Role of SCD1 in the Modulation of Substrate Metabolism and Insulin Sensitivity

Obesity is associated with ectopic lipid accumulation and insulin resistance, both of which are risk factors for the development of the metabolic syndrome and type 2 diabetes (Virtue and Vidal-Puig 2010). A majority of research has shown a causative role for SCD1 in the development of obesity, inflammation, and insulin resistance (Kim et al. 2011; Miyazaki et al. 2002; Ntambi et al. 2002; Pinnamaneni et al. 2006; Rahman et al. 2003). However, this causality as it relates to skeletal muscle has been difficult to discern. SCD1 expression is relatively low in skeletal muscle compared to other tissues such as the liver and sebaceous glands of the skin (Paton and Ntambi 2009). However, under high-fat fed conditions, rodent models of SCD1 deletion are protected from insulin resistance and possess reduced levels of intramyocellular lipid intermediates, which are known to disrupt insulin-stimulated glucose uptake (Rahman et al. 2003). Skeletal muscle is an important tissue for postprandial glucose homeostasis; 80 % of insulin-stimulated glucose uptake is accounted for by muscle tissue (Baron et al. 1988). The importance of skeletal muscle in glucose clearance taken along with current research showing a role for SCD1 in contributing to defects in insulin-signaling and glucose uptake suggest a causative role for SCD1 in mediating the alterations in metabolic function. Interestingly, mouse models with liver and/or adipose tissue specific deletion of SCD1 are not protected from the diet-induced obesity observed with the whole-body deletion of SCD1 (Flowers et al. 2012). Because the liver and adipose tissue are paramount to normal whole-body metabolic regulation and substrate homeostasis and the protective effects of high-fat feeding of SCD1 deletion at these sites are not observed, this points to SCD1 function in some other metabolically-important tissue (e.g., skeletal muscle) as the culprit for whole-body metabolic dysregulation.

SCD1 mRNA levels in skeletal muscle are elevated in the obese state and associated with dysregulated lipid metabolism (Hulver et al. 2005). Specifically, a hallmark feature of obesity is blunted fatty acid oxidation and elevated triacylglycerol accumulation in skeletal muscle, both of which are associated with increased levels of monounsaturated fatty acids (Aguilera et al. 2008). Hulver et al. showed higher SCD1 expression in skeletal muscle of obese humans, relative to non-obese controls, which was associated with decreased AMP-activated kinase (AMPK) phosphorylation and increases in acetyl-CoA carboxylase (ACC) beta levels, both of which occurred in the context of decreased FA oxidation. Consequences of whole-body SCD1 deficiency include increased rates of fatty acid oxidation as well as a reduction in triglyceride synthesis and storage in various tissues including liver, BAT, and skeletal muscle (Flowers et al. 2008; Ntambi et al. 2002; Sampath and Ntambi 2011). SCD1 deficiency appears to increase the rates of FA oxidation in a muscle-fiber type-dependent manner as rates of FA oxidation were increased with SCD1 deficiency in soleus and gastrocnemius, oxidative and oxidative/glycolytic fiber types, respectively (Dobrzyn et al. 2005). Conversely, rates of fatty acid oxidation were not observed in white gastrocnemius or extensor digitorum longus, which are primarily glycolytic fiber types (Dobrzyn and Dobrzyn 2006).

The mechanism by which SCD1 deletion increases FA oxidation is thought to be dependent on AMPK signaling (Dobrzyn et al. 2004a; Kim et al. 2011). It was shown by Dobrzyn et al. that increased AMPK signaling caused a decrease in ACC activity with subsequent reductions in cytosolic malonyl-CoA levels. Malonyl CoA inhibits carnitine palmitoyl transferase 1 (CPT1), which has a primary function of shuttling FA into the mitochondria for oxidation (Hardie and Pan 2002). These findings in SCD1 KO mice are similar to other models in which AMPK signaling is up-regulated (Tomas et al. 2002; Yoon et al. 2006). A similar study in which C2C12 myotubes were treated with the SCD1 inhibitor, 4-(2-cholorophenoxy)-*N*-(3-(methylcarbamoyl)-phenyl) piperidine-1-carboxamide, showed a decrease in ACC and fatty acid synthase (FAS) expression and increase in expression of (CPT1), alternative oxidase (AOX), and peroxisome proliferator-activated receptor gamma coactivator 1-alpha (PGC1-α) (Kim et al. 2011).

Altered levels of SCD1 expression in skeletal muscle can also have affects on the synthesis sphingolipids molecule, ceramide. Ceramide synthesis occurs through either a de-novo process involving esterification of palmitoyl-CoA and serine by serine palmitoyl transferase (SPT) or through the hydrolysis of sphingomyelin (Menaldino et al. 2003; Perry 2002). Ceramide is known to play a role in apoptosis and insulin resistance through the modulation of intracellular signaling processes (Blázquez et al. 2001; Shimabukuro et al. 1998). In particular, increases in substrates for ceramide synthesis such as FFAs have been shown to contribute to β-cell apoptosis (Shimabukuro et al. 1998). SCD1 KO mice show a 42 % and (Hayashi et al. 1998) 48 % decrease in ceramide content in soleus and gastrocnemius. In addition, sphingomyelin content was significantly reduced (~30 %) in the soleus and gastrocnemius muscles of SCD1 KO compared to their WT counterparts, with no changes in the mRNA levels of the shingomyelinases. This suggests a limited role for SCD1 specifically in sphingomyelin hydrolysis (Dobrzyn and Dobrzyn 2006). Expression of SPT and ceramide are regulated by many factors including intracellular FA content. As such, SCD1 activity may be very important in the modulation of ceramide-mediated effects on apoptotic processes and the inhibition of signaling events that contribute to insulin resistance.

One factor shown to down-regulate ceramide synthesis is AMPK signaling. This was shown through the use of the ribonucleoside AICAR that stimulates AMPK activity (Hayashi et al. 1998). As stated previously, AMPK signaling has been shown to be responsible for the up-regulation of FA oxidation shown with SCD1 KO mice. Treatment with AICAR resulted in significant decreases in palmitate-induced ceramide synthesis (Blázquez et al. 2001). Another FA and product of SCD1, oleate, has been shown to down-regulate ceramide levels, and this appears to occur through prevention of the generation and/or the scavenging of ceramide molecules (Listenberger et al. 2003). These data suggest that deletion of SCD1 has beneficial effects on cellular signaling processes through increases in AMPK activity and decreases in transcriptional events regulating lipogenic processes.

A consequence of abnormalities in lipid metabolism such as intramyocellular fat accumulation and increased FFA flux in skeletal muscle is insulin resistance (Kahn and Flier 2000; Kelley and Goodpaster 2001; Liu et al. 2007; Roden 2004). Dysregulated lipid metabolism disrupts insulin-signaling and the normal process of glucose transporter 4 (Glut4) translocation to the cellular membrane, and thus, reduces insulin-stimulated glucose uptake (Savage et al. 2007). Lipid metabolites such as ceramide and diacylglycerol can lead to endoplasmic reticulum stress that can cause serine/threonine phosphorylation of the insulin receptor as well as activation of inflammatory pathways (Könner and Brüning 2011; Peter et al. 2009). Decreases in expression of protein tyrosine phosphatase 1-beta (PTP-1b) expression can cause dephosphorylation of the insulin receptor. It can also lead to an inability of the insulin receptor to bind substrates, and through changes in membrane fluidity, it can affect aggregation of the insulin receptor. These effects have been shown to contribute to insulin insensitivity (Dobrzyn and Dobrzyn 2006). Increased SCD1 expression has been shown to induce the changes in FA composition that can affect insulin sensitivity particularly through an increase in the lipid products of SCD1 activity (Attie et al. 2002; Houdali et al. 2003; Warensjo et al. 2009). These lipid products are able to exert deleterious effects on the insulin-signaling machinery that can cause decreases in insulin sensitivity.

As previously stated, models with a disruption in whole-body or skin SCD1 gene expression show improvements in whole-body glucose and insulin tolerance when compared to WT counterparts on the same high-fat diet (Brown and Goldstein 1998; Flowers et al. 2008; Miyazaki et al. 2009; Ntambi et al. 2002). Specifically, KO mice in these studies show improvements in their fasting insulin levels and when put on a high-fat diet, KO mice have an increased ability to clear glucose compared to their WT littermates (Ntambi and Miyazaki 2004). These mice are also resistant high-fat diet induced obesity, which is related to increases in FA oxidation. Increases in FA oxidation lead to decreases in lipid metabolites such as ceramide and fatty acyl CoAs, metabolites that are known to activate protein kinase C (PKC), which has negative effects on the insulin-signaling machinery such as decreased AKT and IRS1 phosphorylation (Geraldes and King 2010; Turban and Hajduch 2011). Along the same line of thought, insulin-resistant, ob/ob mice given the SCD1 inhibitor sterculic acid, showed improvements in glucose tolerance, which was associated with decreases in both 16:0/16:1 and 18:0/18:1 desaturase indices (Ortinau et al. 2013).

The mechanism by which SCD1 is able to exert these effects on glucose tolerance appears to be through changes in the cell's normal response to glucose. This appears to include, in addition to the above-mentioned mechanisms, a negative influence on the intracellular Glut4 machinery responsible for glucose uptake, particularly under insulin-stimulated conditions (Rahman et al. 2003; Voss et al. 2005). When subjected to over-expression of SCD1, rat myoblasts show a 1.75-fold decrease in total Glut4 levels. These decreases track with suppressed basal and insulin-stimulated glucose uptake measured by 2-deoxyglucose uptake experiments (Voss et al. 2005). Additionally, mice with a deletion of SCD1 show increases in Glut4 levels that occur along with increases in glucose uptake compared to WT mice (Rahman et al. 2003). Whereas SCD1 can negatively influence the insulin receptor and its ability to bind substrate, it can also affect important downstream molecules such as Glut4. Most likely these negative influences on the downstream molecules such as Glut4 are occurring through a mechanism by which upstream molecules such as AKT are inhibited (Dobrzyn et al. 2005).

Another molecule thought to be modulated by SCD1 activity in skeletal muscle is the PTP-1b. This molecule is responsible for the rapid dephosphorylation and subsequent down-regulation of insulin receptor function as well as that of the insulin receptor substrates 1 and 2 (Rahman et al. 2003, 2005). This has been shown through the use of PTP-1b KO mice that have an increase in IR tyrosine phosphorylation in muscle (Elchebly et al. 1999). PTP-1b is down-regulated with SCD1 deletion, and this leads to the autophosphorylation and sustained activity of the insulin receptor. This was associated with increases in insulin-stimulated glucose uptake, and these changes were shown to be independent of circulating insulin levels (Rahman et al. 2003). As of yet, it isn't understood whether the changes in PTP-1b are a direct consequence of decreases in SCD1 activity or a consequence of the alterations in lipid metabolism caused by a lack of SCD1.

There are data showing up-regulation of SCD1 in the presence of certain dietary factors other than fatty acids. Diseases such as type 2 diabetes have been shown to be caused by diets high in carbohydrates (Flowers and Ntambi 2009). High levels of dietary carbohydrates including glucose, sucrose, and fructose significantly up-regulate the expression and activity of SCD1 (Hasty et al. 2000; Matsuzaka et al. 2004; Miyazaki et al. 2004). In the case of a 6-month diet consisting of high sucrose levels, there is a significant increase in levels of oleic acid, the product for SCD1 (Brenner 2003). Excess carbohydrates are metabolized and acetyl-CoA levels rise beyond that needed for ATP production. This results in increased de novo lipid synthesis, which is transcriptionally controlled by lipogenic transcription factors such as SREBP1c and positive modulation SCD1 expression (Hasty et al. 2000; Paton and Ntambi 2009). Rats fed a diet high in glucose show an up-regulation of SCD1 levels (Houdali et al. 2003), and muscle cells treated with high levels of glucose show a very similar effect (Voss et al. 2005).

Another mechanism by which altered SCD1 activity can lead to changes in insulin sensitivity is through its regulation of membrane fluidity; this occurs through the alteration of the FA composition of the cell membrane (Agatha et al. 2001; Ntambi 1999; Storlien et al. 1996). The ratio of saturated to monounsaturated fatty acids is the major determining factor of membrane fluidity and has been shown to influence insulin signaling. This is related to the amount of polyunsaturated FAs present (Ntambi and Miyazaki 2004). In SCD1 KO models, a decrease in monounsaturated FA content of membranes is met with an increase in polyunsaturated FAs in order to maintain membrane integrity (Ntambi and Miyazaki 2004). This causes increased membrane fluidity, which can lead to aggregation of the IR and subsequent hyperphosphorylation of the IR in response to insulin (Frangioudakis et al. 2010; Lamping et al. 2013) Therefore, data has shown a direct correlation between increases in PUFA content of membranes and the degree of insulin resistance in these same tissues (Dobrzyn and Dobrzyn 2006).

SCD1 and Inflammation

SCD1 has also been shown to play a role in the pro and anti-inflammatory processes that are associated with metabolic disease. Free fatty acids are known to activate pro-inflammatory pathways (Boden et al. 2005; Shi et al. 2006). The substrates as well as the products of SCD1 have been shown to both stimulate and inhibit an inflammatory response (Fessler et al. 2009; Kim and Sears 2010). Specifically, the fatty acid product of SCD1, palmitoleic acid, has been shown to bind to and activate inflammatory signaling pathways (Staiger et al. 2006; Weigert et al. 2004). Palmitoleic acid initiates an inflammatory response through binding of the toll-like receptor 4 and the subsequent activation of the nuclear factor kappa beta (NFκB) transcription factor and stimulation of the machinery involved in initiating inflammatory pathways (Schäffler et al. 2006). NFκB causes the production and activation of inflammatory cytokines, chemokines, and other various inflammatory molecules that can function to inhibit insulin-signaling processes (Hotamisligil 2006; Shoelson et al. 2006; Wellen and Hotamisligil 2005). Free fatty acids such as palmitoleate can also bind to various G-protein coupled receptors that can cause an inflammatory response and downregulation of the insulin-signaling machinery (Talukdar et al. 2011). In humans, increases in SCD1 activity measured by increases in plasma stearic to oleic acid ratios have been positively correlated with elevated whole-body C-reactive protein, a marker for inflammation (Black et al. 2004; Stryjecki et al. 2012). Research pertaining to an anti-inflammatory role for these substrates and products of SCD1 centers around studies showing that treatment with these molecules can have protective roles against an inflammatory response (Grimble and Tappia 1998; Vassiliou et al. 2009). Similarly, the substrates for SCD1 such as palmitate show a pro-inflammatory role that also appears to function through a TLR4-dependent mechanism (Kim et al. 2007). Mice given diets high in saturated fats such as palmitate show increases in various inflammatory markers including NFκB (Kim et al. 2007; Suganami et al. 2007).

The increases in plasma fatty acids shown with obesity have been shown to cause defects in insulin signaling (Roden et al. 1996). Excess FFAs are able to affect insulin signaling through their ability to elicit an inflammatory response (Eizirik et al. 2008). Both stearate and palmitate are strong inducers of an inflammatory response and are tightly controlled by SCD1 as they are it's primary substrates (Miyazaki and Ntambi 2003). There appears to be an interesting phenomena relating to the role of SCD1 in modulating inflammation. Evidence exists that SCD1 activity can control the activation of inflammatory pathways that can be causative of problems, especially insulin resistance. The question remains whether this controlling force is by activation or deactivation of these pathways. Mice treated with antisense oligonucleotides (ASO) to inhibit SCD1 show marked increases in saturated fatty acid accumulation in plasma and tissues when fed a westernized high-fat diet. These increases are also demonstrated in macrophages, and these show concomitant increases in TLR4 inflammatory gene expression (Brown et al. 2008). Another model in which mice are deficient in both SCD1 and the low-density lipoprotein receptor shows improvements over control littermates in their metabolic characteristics when given a westernized diet; however, these mice also show a similar increase in inflammation to the Brown et al. study (MacDonald et al. 2009). Saturated FAs have been shown to induce macrophage inflammation through a TLR4-dependent mechanism and ASO treated macrophages show a marked hypersensitivity to TLR4 agonist (Brown et al. 2008; Shi et al. 2006; Suganami et al. 2007). It has been demonstrated previously that macrophage infiltration into adipose and skeletal muscle, and subsequent activation by FFAs can mediate the inflammatory response that can be causative of skeletal muscle insulin resistance (Hevener et al. 2007; Odegaard et al. 2007). This would support a protective role for increases in SCD1 activity. Evidence also exists showing a role for DAG and ceramide in mediating an inflammatory response (Bilan et al. 2009; Holland et al. 2011). Along with this, research using exercise and models with over-expression of SCD1 in skeletal muscle has suggested a role for SCD1 in protecting the body against many of the negative effects of lipid intermediates and alterations in lipid metabolism as they relate to inflammatory pathway activation (Dobrzyn et al. 2010; Schenk and Horowitz 2007). This includes research showing improvements in metabolic outcomes that track with increases in SCD1 levels and activity (Dubé et al. 2008; Peter et al. 2009).

Interestingly, these findings have not been universal as certain models of SCD1 deficiency have shown improvements in inflammatory and stress responses (Liu et al. 2011). Those models that show deleterious effects SCD1 include studies of mice fed a high saturated diet where SCD1 expression and activity correlate with decreased insulin responsiveness (Lee et al. 2006). However, this was shown without studying activation of inflammatory markers. When SCD1 expression was knocked out in two mouse models of obesity, there were marked reductions in adipose tissue-derived inflammatory markers when compared to control mice. This was associated with improvements in basal insulin signaling (Liu et al. 2011). This suggests a possible mechanism by which SCD1 is influencing circulating FA composition and thus contributing to the chronic low-grade inflammation that is characteristic in the obese insulin-resistant state.

With all of this taken into consideration, it is possible that the involvement of SCD1 in these inflammatory processes is cell and model dependent. In models where the substrates for SCD1 are causing pro-inflammatory responses, SCD1 could have a protective effect by removing these substrates. In the cases where 16:1 or 18:1 fatty acids are causing a similar response, it is possible that SCD1 is contributing to the defect.

A Protective Role for SCD1

While there is considerable evidence for a causative role of SCD1 in the metabolic syndrome and its related comorbidities, there appears to be a collection of research showing SCD1 as a much needed and protective molecule. Both over-expression and exercise models have demonstrated a protective capacity for SCD1 (Dubé et al. 2011; Listenberger et al. 2003; Pinnamaneni et al. 2006). Research in this area has shown, both in rodent and human models, that there are improvements in FA partitioning and oxidation that are dependent on increases in SCD1, and that these lead to positive changes in glucose clearance (Amati et al. 2011; Dobrzyn et al. 2010; Dubé et al. 2008). This leads to the belief that SCD1 may be a double-edged sword capable of both protective and harmful effects. As with the causative mechanisms shown for SCD1 in skeletal muscle, the protective mechanisms seem to differ depending on the specific comorbidity.

Dysregulated Lipid Metabolism and Insulin Resistance

Obese individuals present with increased expression and activity of SCD1 in skeletal muscle that relates to higher rates of IMTG synthesis. Acute and chronic bouts of endurance exercise show a similar phenotype to this in regards to levels of SCD1 (Amati et al. 2011; Dobrzyn et al. 2010; Dubé et al. 2008; Schenk and Horowitz 2007). These alterations appear to be dependent on the muscle-fiber type as these changes have been observed in oxidative, but not glycolytic muscles (Dobrzyn et al. 2010). This follows suit with data showing that when SCD1 is overexpressed in a CHO cell line there is a significant increase in TG synthesis (Listenberger et al. 2003). Over-expression of SCD1 in the L6 muscle cell line exhibits beneficial metabolic effects as they relate to glucose clearance. When treated with palmitate, the cells over-expressing SCD1 showed improvements in triacylglycerol synthesis, which attenuated ceramide and diacylglycerol synthesis that protected against insulin resistance (Pinnamaneni et al. 2006). These opposing data suggest that changes in SCD1 expression can have an important role in glucose clearance, but then it must also be tightly regulated to avoid metabolic consequences. There have also been mouse models supporting this hypothesis (Dobrzyn et al. 2005, 2010). In terms of causation, it is known that SREBP-1 may be central to the up-regulation of SCD1 under exercised conditions (Hodson and Fielding 2013; Sekiya et al. 2003). Data exists showing increases in this transcription factor under exercise conditions with both mice and humans that is associated with increases in SCD1 and IMTG production (Bergman et al. 2010; Ikeda et al. 2002). Thus it is likely that SREBP1c is functioning in both the deleterious and protective roles for SCD1. There is also data suggesting that FAS may have roles in regulating SCD1 under these conditions. FAS synthesizes palmitic acid de novo, a substrate for SCD1 (Conraads et al. 2002; Mooren et al. 2004). It has been shown that increases in the substrate for SCD1 leads to up-regulation of SCD1 protein levels (Thorn et al. 2010).

In terms of a protective mechanism(s), data suggests a role for SCD1 in alleviating the stressful conditions caused by the overabundance of FA. Exercise studies have shown a role for SCD1 in enhancing the partitioning of FA into IMTG, as well as an increase in the movement of FA into the mitochondria for beta-oxidation (Dobrzyn et al. 2010). Dobryzn et al. showed a significant increase in the expression of proteins involved in the movement of FA in to the mitochondria in mice in response to an endurance exercise protocol. Work by Schenk and Horowitz has demonstrated that an acute bout of exercise (1 h running on a treadmill) prior to being challenged with a lipid infusion was sufficient to increase TAG synthesis. These changes were characterized by increased SCD1 expression and concomitantly reduced DAG and ceramide production (Bergman et al. 2010; Bruce et al. 2006; Dubé et al. 2008). Certain studies have shown increases in DAG with exercise protocols whilst others have shown decreases (Dobrzyn et al. 2010). Schenk et al. was able to show significant increases in DAG content of soleus muscles after a 12-week treadmill running program in mice. Bioactive sphingolipids, such as ceramides, are affected by exercise regimens. Decreases in ceramide content in skeletal muscle have been shown, both in mice and in humans (Amati et al. 2011; Bruce et al. 2006; Dobrzyn et al. 2004b). These changes in lipid intermediates relate to changes seen in the local environment of skeletal muscle under exercised conditions. Adaptation to exercise by skeletal muscle entails both increasing lipogenic activity and oxidative capacity. The increased lipogenic activity serves the purpose of providing ample IMTG to provide a local fuel source of fatty acids to be oxidized in the mitochondria during exercise as well as during the postexercise recovery period. The increased oxidative capacity allows the muscle tissue to predominantly rely on fatty acids as an energy source, as this is the most efficient fuel for ATP production. It has been discussed previously that SCD1 KO mice show increases in AMPK phosphorylation, leading to increases in beta-oxidation. Along these same lines, there are increases in the ability and capacity for oxidizing fatty acids during and after endurance exercise; this appears to be through an AMPK-dependent mechanism (Bruce et al. 2006; Dobrzyn et al. 2010; Russell et al. 2003). Various exercise models using both animals and humans have shown increases in AMPK activity that correlate with increases in FA oxidation (Dobrzyn et al. 2010; Dubé et al. 2008; Schenk and Horowitz 2007). These increases in FA oxidation are shown along with increases in the activity of SCD1. Being that SCD1 is playing a prominent role in the shuttling of monounsaturated FA into TG synthesis (Paton and Ntambi 2009), and these TG are being hydrolyzed to allow influx of FA into the mitochondria, it is possible that SCD1 is playing a role in controlling FA oxidation, particularly the increases seen with endurance exercise.

This data also appears to be supported in humans. Multiple studies in human subjects have shown increases in IMTG that track with increases in SCD1 (Amati et al. 2011; Bergman et al. 2010). Bergman et al. showed that endurance-trained male cyclists have greater skeletal muscle IMTG and SCD1 mRNA expression and protein content than controls. This would suggest that high muscle SCD1 activity is an advantage for these individuals. In addition, Amati et al. showed increases in SCD1 content that was related to increases in IMTG in endurance-trained athletes. These individuals exhibited decreases in both ceramide and diacylglycerol content compared to sedentary counterparts. These data suggest a protective mechanism for SCD1 and a role for SCD1 in mediating the beneficial metabolic changes seen with exercise regimens.

The changes in FA partitioning towards oxidation and away from storage shown with the above models have also shown improvements in glucose tolerance (Amati et al. 2011; Bergman et al. 2010). Whereas increases in IMTG are correlated with insulin resistance in diseased states, in exercised states it has been shown that increases in IMTG are related to improvements in many metabolic characteristics including insulin sensitivity. This implies a certain paradox in which increases in IMTG can either be beneficial or harmful. Supporting this is research that shows individuals who are endurance trained have increased IMTG and increased mitochondria as well as increased insulin sensitivity compared to sedentary individuals. Initially, it was suspected that these changes in insulin sensitivity due to IMTG increases with endurance exercise where related to the removal of harmful lipid intermediates. However, more recent research has shown this may not be true as sedentary and exercised individuals show similar DAG content as well as reduced ceramide content in skeletal muscle (Amati et al. 2011). It is thought that neutral lipid synthesis, instead of concentration of lipid intermediate, may be causing the improvements in insulin sensitivity; that is enhanced clearance of intermediates into synthetic pathways. Research in this area has also shown a role for increases in FA partitioning into IMTG in reducing the inflammatory response seen in insulin-resistant situations (Schenk and Horowitz 2007). This is due to both the removal of pro-inflammatory FFA as they are partitioned into IMTG as well as the removal of lipid intermediates that can lead to activation of inflammatory responses (Schenk and Horowitz 2007).

Conclusions/Future Directions

At this time, knowledge regarding the exact role for SCD1 in mediating metabolic disease in skeletal muscle is lacking. It has been shown that decreases in the expression and activity of SCD1 track with improvements in various metabolic characteristics, including obesity/lipid partitioning as well as insulin resistance and inflammation. While obese and diabetic states both show increases in IMTG and concomitant increases in SCD1, the skeletal muscle of trained individuals also show similar effects. This begs the question, can SCD1 activity be both helpful and harmful.

In skeletal muscle, there appears to be a role for SCD1 in controlling the phosphorylation state of AMPK and subsequently for regulating FA oxidation. However, there are similar effects on AMPK when SCD1 is increased in exercising individuals. Either way, SCD1 appears to have direct affects on the insulin-signaling machinery through its affect of AMPK and the exact mechanism warrants further investigation. There also appears to be a role for SCD1 in controlling the expression of PTP-1b and its role in down-regulation of the insulin receptor and its substrates. PTP-1b appears to decrease with suppression of SCD1 activity and could be causative of the improvements shown in these models. This also warrants further investigation. Thirdly, a mechanism for SCD1 in either causing or suppressing an inflammatory response has been shown. This will become an important area of research as inflammatory processes are involved in many of the chronic diseases related to skeletal muscle metabolism.

Data demonstrates that there is a role for SCD1 in the modulation of skeletal muscle metabolism. However, there is disagreement whether this role is as a protective molecule preventing many of the aberrations caused by dysfunctional lipid metabolism, or whether SCD1 is a contributing factor whose malfunctioning activity is causing disease. It appears as though SCD1 may function on both sides of the argument, and this may be related to the particular metabolic environment by which SCD1 is being regulated. Research using skeletal muscle specific KO would allow for a better understanding of the molecule and its influences on the development and progression of metabolic disease. Of particular interest would be how these mice would adapt to endurance exercise. It is possible that these animals would have a decreased oxidative capacity and therefore a decreased ability to perform aerobic exercise. However, it is also possible that these animals would exhibit compensatory mechanisms that allow them to maintain correct handling and oxidation of FAs. Another research question involving a skeletal muscle SCD1 KO would be how these animals would adapt to a HF diet. It is possible that these animals would exhibit a phenotype similar to SCD1 KO in other tissues such as skin and liver in regards to insulin sensitivity given skeletal muscle's importance in whole-body glucose clearance and thermogenesis.

References

Agatha G, Hafer R, Zintl F (2001) Fatty acid composition of lymphocyte membrane phospholipids in children with acute leukemia. Cancer Lett 173:139–144

Aguilera CM, Gil-Campos M, Canete R, Gil A (2008) Alterations in plasma and tissue lipids associated with obesity and metabolic syndrome. Clin Sci (Lond) 114:183–193

Amati F, Dubé JJ, Alvarez-Carnero E, Edreira MM, Chomentowski P, Coen PM, Switzer GE, Bickel PE, Stefanovic-Racic M, Toledo FGS, Goodpaster BH (2011) Skeletal muscle triglycerides, diacylglycerols, and ceramides in insulin resistance: another paradox in endurance-trained athletes? Diabetes 60:2588–2597

Attie AD, Krauss RM, Gray-Keller MP, Brownlie A, Miyazaki M, Kastelein JJ, Lusis AJ, Stalenhoef AFH, Stoehr JP, Hayden MR, Ntambi JM (2002) Relationship between stearoyl-CoA

desaturase activity and plasma triglycerides in human and mouse hypertriglyceridemia. J Lipid Res 43:1899–1907
Baron AD, Brechtel G, Wallace P, Edelman SV (1988) Rates and tissue sites of non-insulin- and insulin-mediated glucose uptake in humans. Am J Physiol 255:E769–E774
Bergman BC, Perreault L, Hunerdosse DM, Koehler MC, Samek AM, Eckel RH (2010) Increased intramuscular lipid synthesis and low saturation relate to insulin sensitivity in endurance-trained athletes. J Appl Physiol 108:1134–1141
Bilan PJ, Samokhvalov V, Koshkina A, Schertzer JD, Samaan MC, Klip A (2009) Direct and macrophage-mediated actions of fatty acids causing insulin resistance in muscle cells. Arch Physiol Biochem 115:176–190
Black S, Kushner I, Samols D (2004) C-reactive protein. J Biol Chem 279:48487–48490
Blázquez C, Geelen MJH, Velasco G, Guzmán M (2001) The AMP-activated protein kinase prevents ceramide synthesis de novo and apoptosis in astrocytes. FEBS Lett 489:149–153
Boden G, She P, Mozzoli M, Cheung P, Gumireddy K, Reddy P, Xiang X, Luo Z, Ruderman N (2005) Free fatty acids produce insulin resistance and activate the proinflammatory nuclear factor-kappaB pathway in rat liver. Diabetes 54:3458–3465
Brenner RR (2003) Hormonal modulation of Δ6 and Δ5 desaturases: case of diabetes. Prostaglandins Leukot Essent Fatty Acids 68:151–162
Brown MS, Goldstein JL (1998) Sterol regulatory element binding proteins (SREBPs): controllers of lipid synthesis and cellular uptake. Nutr Rev 56:S1–S3; discussion S54–75
Brown JM, Chung S, Sawyer JK, Degirolamo C, Alger HM, Nguyen T, Zhu X, Duong M-N, Wibley AL, Shah R (2008) Inhibition of stearoyl-coenzyme A desaturase 1 dissociates insulin resistance and obesity from atherosclerosis. Circulation 118:1467–1475
Bruce CR, Thrush AB, Mertz VA, Bezaire V, Chabowski A, Heigenhauser GJF, Dyck DJ (2006) Endurance training in obese humans improves glucose tolerance and mitochondrial fatty acid oxidation and alters muscle lipid content. Am J Physiol Endocrinol Metab 291:E99–E107
Conraads VM, Beckers P, Bosmans J, De Clerck LS, Stevens WJ, Vrints CJ, Brutsaert DL (2002) Combined endurance/resistance training reduces plasma TNF-alpha receptor levels in patients with chronic heart failure and coronary artery disease. Eur Heart J 23:1854–1860
Dobrzyn A, Dobrzyn P (2006) Stearoyl-CoA desaturase—a new player in skeletal muscle metabolism regulation. J Physiol Pharmacol 57:31–42
Dobrzyn P, Dobrzyn A, Miyazaki M, Cohen P, Asilmaz E, Hardie DG, Friedman JM, Ntambi JM (2004) Stearoyl-CoA desaturase 1 deficiency increases fatty acid oxidation by activating AMP-activated protein kinase in liver. Proc Natl Acad Sci U S A 101:6409–6414
Dobrzyn A, Dobrzyn P, Lee S-H, Miyazaki M, Cohen P, Asilmaz E, Hardie DG, Friedman JM, Ntambi JM (2005) Stearoyl-CoA desaturase-1 deficiency reduces ceramide synthesis by down-regulating serine palmitoyltransferase and increasing β-oxidation in skeletal muscle. Am J Physiol Endocrinol Metab 288:E599–E607
Dobrzyn P, Pyrkowska A, Jazurek M, Szymanski K, Langfort J, Dobrzyn A (2010) Endurance training-induced accumulation of muscle triglycerides is coupled to upregulation of stearoyl-CoA desaturase 1. J Appl Physiol 109:1653–1661
Dubé JJ, Amati F, Stefanovic-Racic M, Toledo FGS, Sauers SE, Goodpaster BH (2008) Exercise-induced alterations in intramyocellular lipids and insulin resistance: the athlete's paradox revisited. Am J Physiol Endocrinol Metab 294:E882–E888
Dubé JJ, Amati F, Toledo FGS, Stefanovic-Racic M, Rossi A, Coen P, Goodpaster BH (2011) Effects of weight loss and exercise on insulin resistance, and intramyocellular triacylglycerol, diacylglycerol and ceramide. Diabetologia 54:1147–1156
Eizirik DL, Cardozo AK, Cnop M (2008) The role for endoplasmic reticulum stress in diabetes mellitus. Endocr Rev 29:42–61
Elchebly M, Payette P, Michaliszyn E, Cromlish W, Collins S, Loy AL, Normandin D, Cheng A, Himms-Hagen J, Chan CC, Ramachandran C, Gresser MJ, Tremblay ML, Kennedy BP (1999) Increased insulin sensitivity and obesity resistance in mice lacking the protein tyrosine phosphatase-1B gene. Science 283:1544–1548
Fessler MB, Rudel LL, Brown JM (2009) Toll-like receptor signaling links dietary fatty acids to the metabolic syndrome. Curr Opin Lipidol 20:379–385

Flowers MT, Ntambi JM (2009) Stearoyl-CoA desaturase and its relation to high-carbohydrate diets and obesity. Biochim Biophys Acta 1791:85–91

Flowers MT, Keller MP, Choi Y, Lan H, Kendziorski C, Ntambi JM, Attie AD (2008) Liver gene expression analysis reveals endoplasmic reticulum stress and metabolic dysfunction in SCD1-deficient mice fed a very low-fat diet. Physiol Genomics 33:361–372

Flowers MT, Ade L, Strable MS, Ntambi JM (2012) Combined deletion of SCD1 from adipose tissue and liver does not protect mice from obesity. J Lipid Res 53:1646–1653

Frangioudakis G, Garrard J, Raddatz K, Nadler JL, Mitchell TW, Schmitz-Peiffer C (2010) Saturated- and n-6 polyunsaturated-fat diets each induce ceramide accumulation in mouse skeletal muscle: reversal and improvement of glucose tolerance by lipid metabolism inhibitors. Endocrinology 151:4187–4196

Geraldes P, King GL (2010) Activation of protein kinase C isoforms and its impact on diabetic complications. Circ Res 106:1319–1331

Grimble RF, Tappia PS (1998) Modulation of pro-inflammatory cytokine biology by unsaturated fatty acids. Z Ernahrungswiss 37(Suppl 1):57–65

Hardie DG, Pan DA (2002) Regulation of fatty acid synthesis and oxidation by the AMP-activated protein kinase. Biochem Soc Trans 30:1064–1070

Hasty AH, Shimano H, Yahagi N, Amemiya-Kudo M, Perrey S, Yoshikawa T, Osuga J-i, Okazaki H, Tamura Y, Iizuka Y, Shionoiri F, Ohashi K, Harada K, Gotoda T, Nagai R, Ishibashi S, Yamada N (2000) Sterol regulatory element-binding protein-1 is regulated by glucose at the transcriptional level. J Biol Chem 275:31069–31077

Hayashi T, Hirshman MF, Kurth EJ, Winder WW, Goodyear LJ (1998) Evidence for 5′AMP-activated protein kinase mediation of the effect of muscle contraction on glucose transport. Diabetes 47:1369–1373

Hevener AL, Olefsky JM, Reichart D, Nguyen MT, Bandyopadyhay G, Leung HY, Watt MJ, Benner C, Febbraio MA, Nguyen AK, Folian B, Subramaniam S, Gonzalez FJ, Glass CK, Ricote M (2007) Macrophage PPAR gamma is required for normal skeletal muscle and hepatic insulin sensitivity and full antidiabetic effects of thiazolidinediones. J Clin Invest 117:1658–1669

Hodson L, Fielding BA (2013) Stearoyl-CoA desaturase: rogue or innocent bystander? Prog Lipid Res 52:15–42

Holland WL, Bikman BT, Wang LP, Yuguang G, Sargent KM, Bulchand S, Knotts TA, Shui G, Clegg DJ, Wenk MR, Pagliassotti MJ, Scherer PE, Summers SA (2011) Lipid-induced insulin resistance mediated by the proinflammatory receptor TLR4 requires saturated fatty acid-induced ceramide biosynthesis in mice. J Clin Invest 121:1858–1870

Hotamisligil GS (2006) Inflammation and metabolic disorders. Nature 444:860–867

Houdali B, Wahl HG, Kresi M, Nguyen V, Haap M, Machicao F, Ammon HP, Renn W, Schleicher ED, Haring HU (2003) Glucose oversupply increases Delta9-desaturase expression and its metabolites in rat skeletal muscle. Diabetologia 46:203–212

Hulver MW, Berggren JR, Carper MJ, Miyazaki M, Ntambi JM, Hoffman EP, Thyfault JP, Stevens R, Dohm GL, Houmard JA, Muoio DM (2005) Elevated stearoyl-CoA desaturase-1 expression in skeletal muscle contributes to abnormal fatty acid partitioning in obese humans. Cell Metab 2:251–261

Ikeda S, Miyazaki H, Nakatani T, Kai Y, Kamei Y, Miura S, Tsuboyama-Kasaoka N, Ezaki O (2002) Up-regulation of SREBP-1c and lipogenic genes in skeletal muscles after exercise training. Biochem Biophys Res Commun 296:395–400

Kahn BB, Flier JS (2000) Obesity and insulin resistance. J Clin Invest 106:473–481

Kelley DE, Goodpaster BH (2001) Skeletal muscle triglyceride: an aspect of regional adiposity and insulin resistance. Diabetes Care 24:933–941

Kim JJ, Sears DD (2010) TLR4 and insulin resistance. Gastroenterol Res Pract 2010

Kim F, Pham M, Luttrell I, Bannerman DD, Tupper J, Thaler J, Hawn TR, Raines EW, Schwartz MW (2007) Toll-like receptor-4 mediates vascular inflammation and insulin resistance in diet-induced obesity. Circ Res 100:1589–1596

Kim E, Lee JH, Ntambi JM, Hyun CK (2011) Inhibition of stearoyl-CoA desaturase1 activates AMPK and exhibits beneficial lipid metabolic effects in vitro. Eur J Pharmacol 672:38–44

Könner AC, Brüning JC (2011) Toll-like receptors: linking inflammation to metabolism. Trends Endocrinol Metab 22:16–23

Lamping KG, Nuno DW, Coppey LJ, Holmes AJ, Hu S, Oltman CL, Norris AW, Yorek MA (2013) Modification of high saturated fat diet with n-3 polyunsaturated fat improves glucose intolerance and vascular dysfunction. Diabetes Obes Metab 15:144–152

Lee CH, Olson P, Evans RM (2003) Minireview: lipid metabolism, metabolic diseases, and peroxisome proliferator-activated receptors. Endocrinology 144:2201–2207

Lee JS, Pinnamaneni SK, Eo SJ, Cho IH, Pyo JH, Kim CK, Sinclair AJ, Febbraio MA, Watt MJ (2006) Saturated, but not n-6 polyunsaturated, fatty acids induce insulin resistance: role of intramuscular accumulation of lipid metabolites. J Appl Physiol 100:1467–1474

Listenberger LL, Han X, Lewis SE, Cases S, Farese RV, Ory DS, Schaffer JE (2003) Triglyceride accumulation protects against fatty acid-induced lipotoxicity. Proc Natl Acad Sci U S A 100:3077–3082

Liu L, Zhang Y, Chen N, Shi X, Tsang B, Yu Y-H (2007) Upregulation of myocellular DGAT1 augments triglyceride synthesis in skeletal muscle and protects against fat-induced insulin resistance. J Clin Invest 117:1679–1689

Liu X, Strable MS, Ntambi JM (2011) Stearoyl CoA desaturase 1: role in cellular inflammation and stress. Adv Nutr 2:15–22

MacDonald ML, van Eck M, Hildebrand RB, Wong BW, Bissada N, Ruddle P, Kontush A, Hussein H, Pouladi MA, Chapman MJ, Fievet C, van Berkel TJ, Staels B, McManus BM, Hayden MR (2009) Despite antiatherogenic metabolic characteristics, SCD1-deficient mice have increased inflammation and atherosclerosis. Arterioscler Thromb Vasc Biol 29:341–347

Matsuzaka T, Shimano H, Yahagi N, Amemiya-Kudo M, Okazaki H, Tamura Y, Iizuka Y, Ohashi K, Tomita S, Sekiya M, Hasty A, Nakagawa Y, Sone H, Toyoshima H, Ishibashi S, Osuga J-i, Yamada N (2004) Insulin-independent induction of sterol regulatory element-binding protein-1c expression in the livers of streptozotocin-treated mice. Diabetes 53:560–569

McGarry JD (2002) Banting lecture 2001: dysregulation of fatty acid metabolism in the etiology of type 2 diabetes. Diabetes 51:7–18

Menaldino DS, Bushnev A, Sun A, Liotta DC, Symolon H, Desai K, Dillehay DL, Peng Q, Wang E, Allegood J, Trotman-Pruett S, Sullards MC, Merrill AH Jr (2003) Sphingoid bases and de novo ceramide synthesis: enzymes involved, pharmacology and mechanisms of action. Pharmacol Res 47:373–381

Miyazaki M, Ntambi JM (2003) Role of stearoyl-coenzyme A desaturase in lipid metabolism. Prostaglandins Leukot Essent Fatty Acids 68:113–121

Miyazaki M, Gomez FE, Ntambi JM (2002) Lack of stearoyl-CoA desaturase-1 function induces a palmitoyl-CoA Delta6 desaturase and represses the stearoyl-CoA desaturase-3 gene in the preputial glands of the mouse. J Lipid Res 43:2146–2154

Miyazaki M, Dobrzyn A, Man WC, Chu K, Sampath H, Kim H-J, Ntambi JM (2004) Stearoyl-CoA desaturase 1 gene expression is necessary for fructose-mediated induction of lipogenic gene expression by sterol regulatory element-binding protein-1c-dependent and -independent mechanisms. J Biol Chem 279:25164–25171

Miyazaki M, Sampath H, Liu X, Flowers MT, Chu K, Dobrzyn A, Ntambi JM (2009) Stearoyl-CoA desaturase-1 deficiency attenuates obesity and insulin resistance in leptin-resistant obese mice. Biochem Biophys Res Commun 380:818–822

Mooren FC, Lechtermann A, Volker K (2004) Exercise-induced apoptosis of lymphocytes depends on training status. Med Sci Sports Exerc 36:1476–1483

Ntambi JM (1999) Regulation of stearoyl-CoA desaturase by polyunsaturated fatty acids and cholesterol. J Lipid Res 40:1549–1558

Ntambi JM, Miyazaki M (2004) Regulation of stearoyl-CoA desaturases and role in metabolism. Prog Lipid Res 43:91–104

Ntambi JM, Miyazaki M, Stoehr JP, Lan H, Kendziorski CM, Yandell BS, Song Y, Cohen P, Friedman JM, Attie AD (2002) Loss of stearoyl-CoA desaturase-1 function protects mice against adiposity. Proc Natl Acad Sci U S A 99:11482–11486

Odegaard JI, Ricardo-Gonzalez RR, Goforth MH, Morel CR, Subramanian V, Mukundan L, Red Eagle A, Vats D, Brombacher F, Ferrante AW, Chawla A (2007) Macrophage-specific PPARgamma controls alternative activation and improves insulin resistance. Nature 447:1116–1120

Ortinau LC, Nickelson KJ, Stromsdorfer KL, Naik CY, Pickering RT, Haynes RA, Fritsche KL, Perfield JW II (2013) Sterculic oil, a natural inhibitor of SCD1, improves the metabolic state of obese OLETF rats. Obesity 21(2):344–352

Paton CM, Ntambi JM (2009) Biochemical and physiological function of stearoyl-CoA desaturase. Am J Physiol Endocrinol Metab 297:E28–E37

Perry DK (2002) Serine palmitoyltransferase: role in apoptotic de novo ceramide synthesis and other stress responses. Biochim Biophys Acta 1585:146–152

Peter A, Weigert C, Staiger H, Machicao F, Schick F, Machann J, Stefan N, Thamer C, Haring HU, Schleicher E (2009) Individual stearoyl-coa desaturase 1 expression modulates endoplasmic reticulum stress and inflammation in human myotubes and is associated with skeletal muscle lipid storage and insulin sensitivity in vivo. Diabetes 58:1757–1765

Petersen KF, Shulman GI (2002) Cellular mechanism of insulin resistance in skeletal muscle. J R Soc Med 95(Suppl 42):8–13

Pinnamaneni SK, Southgate RJ, Febbraio MA, Watt MJ (2006) Stearoyl CoA desaturase 1 is elevated in obesity but protects against fatty acid-induced skeletal muscle insulin resistance in vitro. Diabetologia 49:3027–3037

Rahman SM, Dobrzyn A, Dobrzyn P, Lee SH, Miyazaki M, Ntambi JM (2003) Stearoyl-CoA desaturase 1 deficiency elevates insulin-signaling components and down-regulates protein-tyrosine phosphatase 1B in muscle. Proc Natl Acad Sci U S A 100:11110–11115

Rahman SM, Dobrzyn A, Lee SH, Dobrzyn P, Miyazaki M, Ntambi JM (2005) Stearoyl-CoA desaturase 1 deficiency increases insulin signaling and glycogen accumulation in brown adipose tissue. Am J Physiol Endocrinol Metab 288:E381–387

Roden M (2004) How free fatty acids inhibit glucose utilization in human skeletal muscle. News Physiol Sci 19:92–96

Roden M, Price TB, Perseghin G, Petersen KF, Rothman DL, Cline GW, Shulman GI (1996) Mechanism of free fatty acid-induced insulin resistance in humans. J Clin Invest 97:2859–2865

Russell AP, Feilchenfeldt J, Schreiber S, Praz M, Crettenand A, Gobelet C, Meier CA, Bell DR, Kralli A, Giacobino J-P, Dériaz O (2003) Endurance training in humans leads to fiber type-specific increases in levels of peroxisome proliferator-activated receptor-γ coactivator-1 and peroxisome proliferator-activated receptor-α in skeletal muscle. Diabetes 52:2874–2881

Sampath H, Ntambi JM (2011) The role of stearoyl-CoA desaturase in obesity, insulin resistance, and inflammation. Ann N Y Acad Sci 1243:47–53

Savage DB, Petersen KF, Shulman GI (2007) Disordered lipid metabolism and the pathogenesis of insulin resistance. Physiol Rev 87:507–520

Schäffler A, Müller-Ladner U, Schölmerich J, Büchler C (2006) Role of adipose tissue as an inflammatory organ in human diseases. Endocr Rev 27:449–467

Schenk S, Horowitz JF (2007) Acute exercise increases triglyceride synthesis in skeletal muscle and prevents fatty acid-induced insulin resistance. J Clin Invest 117:1690–1698

Scheuermann-Freestone M, Madsen PL, Manners D, Blamire AM, Buckingham RE, Styles P, Radda GK, Neubauer S, Clarke K (2003) Abnormal cardiac and skeletal muscle energy metabolism in patients with type 2 diabetes. Circulation 107:3040–3046

Sekiya M, Yahagi N, Matsuzaka T, Najima Y, Nakakuki M, Nagai R, Ishibashi S, Osuga J, Yamada N, Shimano H (2003) Polyunsaturated fatty acids ameliorate hepatic steatosis in obese mice by SREBP-1 suppression. Hepatology 38:1529–1539

Shi H, Kokoeva MV, Inouye K, Tzameli I, Yin H, Flier JS (2006) TLR4 links innate immunity and fatty acid-induced insulin resistance. J Clin Invest 116:3015–3025

Shimabukuro M, Zhou Y-T, Levi M, Unger RH (1998) Fatty acid-induced β cell apoptosis: a link between obesity and diabetes. Proc Natl Acad Sci U S A 95:2498–2502

Shoelson SE, Lee J, Goldfine AB (2006) Inflammation and insulin resistance. J Clin Invest 116:1793–1801

Staiger K, Staiger H, Weigert C, Haas C, Haring HU, Kellerer M (2006) Saturated, but not unsaturated, fatty acids induce apoptosis of human coronary artery endothelial cells via nuclear factor-kappaB activation. Diabetes 55:3121–3126

Storlien L, Baur L, Kriketos A, Pan D, Cooney G, Jenkins A, Calvert G, Campbell L (1996) Dietary fats and insulin action. Diabetologia 39:621–631

Stryjecki C, Roke K, Clarke S, Nielsen D, Badawi A, El-Sohemy A, Ma DWL, Mutch DM (2012) Enzymatic activity and genetic variation in SCD1 modulate the relationship between fatty acids and inflammation. Mol Genet Metab 105:421–427

Suganami T, Tanimoto-Koyama K, Nishida J, Itoh M, Yuan X, Mizuarai S, Kotani H, Yamaoka S, Miyake K, Aoe S, Kamei Y, Ogawa Y (2007) Role of the Toll-like receptor 4/NF-kappaB pathway in saturated fatty acid-induced inflammatory changes in the interaction between adipocytes and macrophages. Arterioscler Thromb Vasc Biol 27:84–91

Talukdar S, Olefsky JM, Osborn O (2011) Targeting GPR120 and other fatty acid-sensing GPCRs ameliorates insulin resistance and inflammatory diseases. Trends Pharmacol Sci 32:543–550

Thorn K, Hovsepyan M, Bergsten P (2010) Reduced levels of SCD1 accentuate palmitate-induced stress in insulin-producing beta-cells. Lipids Health Dis 9:108

Tomas E, Tsao T-S, Saha AK, Murrey HE, Zhang C, Itani SI, Lodish HF, Ruderman NB (2002) Enhanced muscle fat oxidation and glucose transport by ACRP30 globular domain: acetyl-CoA carboxylase inhibition and AMP-activated protein kinase activation. Proc Natl Acad Sci U S A 99:16309–16313

Turban S, Hajduch E (2011) Protein kinase C isoforms: mediators of reactive lipid metabolites in the development of insulin resistance. FEBS Lett 585:269–274

Vassiliou EK, Gonzalez A, Garcia C, Tadros JH, Chakraborty G, Toney JH (2009) Oleic acid and peanut oil high in oleic acid reverse the inhibitory effect of insulin production of the inflammatory cytokine TNF-alpha both in vitro and in vivo systems. Lipids Health Dis 8:25

Virtue S, Vidal-Puig A (2010) Adipose tissue expandability, lipotoxicity and the Metabolic Syndrome—an allostatic perspective. Biochim Biophys Acta 1801:338–349

Voss MD, Beha A, Tennagels N, Tschank G, Herling AW, Quint M, Gerl M, Metz-Weidmann C, Haun G, Korn M (2005) Gene expression profiling in skeletal muscle of Zucker diabetic fatty rats: implications for a role of stearoyl-CoA desaturase 1 in insulin resistance. Diabetologia 48:2622–2630

Warensjo E, Rosell M, Hellenius ML, Vessby B, De Faire U, Riserus U (2009) Associations between estimated fatty acid desaturase activities in serum lipids and adipose tissue in humans: links to obesity and insulin resistance. Lipids Health Dis 8:37

Weigert C, Brodbeck K, Staiger H, Kausch C, Machicao F, Haring HU, Schleicher ED (2004) Palmitate, but not unsaturated fatty acids, induces the expression of interleukin-6 in human myotubes through proteasome-dependent activation of nuclear factor-kappaB. J Biol Chem 279:23942–23952

Wellen KE, Hotamisligil GS (2005) Inflammation, stress, and diabetes. J Clin Invest 115:1111–1119

Yoon MJ, Lee GY, Chung J-J, Ahn YH, Hong SH, Kim JB (2006) Adiponectin increases fatty acid oxidation in skeletal muscle cells by sequential activation of AMP-activated protein kinase, p38 mitogen-activated protein kinase, and peroxisome proliferator-activated receptor α. Diabetes 55:2562–2570

Chapter 10
The Cellular Function of Stearoyl-CoA Desaturase-2 in Development and Differentiation

Jennifer L. Cantley, Lucas M. O'Neill, James M. Ntambi, and Michael P. Czech

Introduction

Stearoyl-CoA desaturase-2 (SCD2) is a key member of the murine Δ9-desaturase family and is documented to be the major isoform required for the early development of mouse (Miyazaki et al. 2005). Despite this, little is known about its physiological roles. Historically, SCD2 was labeled as "brain SCD," because at the time, SCD2 was thought to be the only SCD isoform expressed in the brain (Kaestner et al. 1989; DeWillie and Farmer 1992). Despite being highly expressed in the brain, the role of SCD2 in this organ has been largely unexplored. Today, we know SCD2 to be expressed ubiquitously and is the predominate isoform in tissues such as B lymphocytes, epidermis, lung, pancreas, spleen, testis, and ovaries (Sampath and Ntambi 2011a, b; Swick and Lane 1992; Saether et al. 2003). We are also beginning to grasp the temporal expression pattern of SCD2 and its role in adipocytes.

SCD2 is expressed in adult mice, but it is more highly expressed and critical during embryonic development and in newborn mice prior to weaning. Complete loss of SCD2 is embryonic lethal in the C57Bl6 genetic background (B6) (Miyazaki et al. 2005). In the 129SvEv genetic background (129), complete loss of SCD2 is

Jennifer L. Cantley and Lucas M. O'Neill contributed equally to this work.

J.L. Cantley
Departments of Internal Medicine and Cellular and Molecular Physiology,
Yale School of Medicine, New Haven, CT 06510, USA

L.M. O'Neill
Department of Biochemistry, University of Wisconsin-Madison, Madison, WI 53705, USA

J.M. Ntambi, Ph.D.
Departments of Biochemistry and Nutritional Sciences, University of Wisconsin-Madison,
Madison, WI 53705, USA

M.P. Czech (✉)
Program in Molecular Medicine, University of Massachusetts Medical School,
Suite 100, 373 Plantation Street, Worcester, MA 01605, USA
e-mail: Michael.Czech@umassmed.edu

J.M. Ntambi (ed.), *Stearoyl-CoA Desaturase Genes in Lipid Metabolism*, 119
DOI 10.1007/978-1-4614-7969-7_10, © Springer Science+Business Media New York 2013

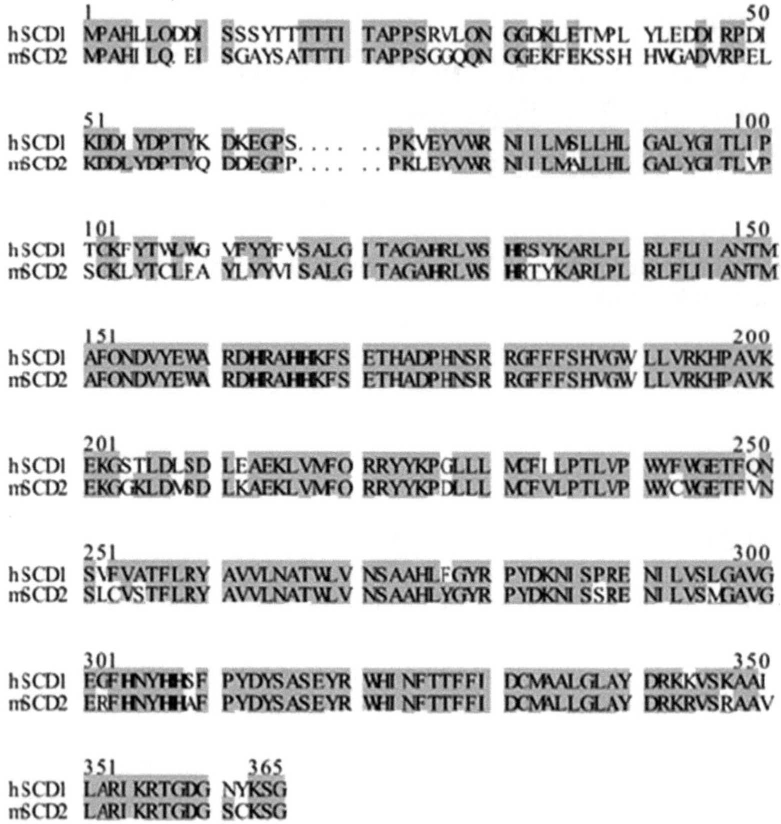

Fig. 10.1 Comparison of the amino acid sequences of the mouse SCD2 (mSDC2) and human SDC1 (hSDC1). There is very high sequence similarity between these mouse and human proteins, with highest variability in the N-terminus region. Both the N-terminus and the C-terminus regions of these proteins are thought to be cytoplasmic, while four putative transmembrane regions of these proteins are thought to cross the endoplasmic reticulum membrane. The variable N-terminus region is far removed from the more central segments containing the histidine regions that form the core of the catalytic domain, suggesting they may not be involved in substrate recognition (Zhang et al. 2005)

~70-100 % lethal at birth with few survivors (Miyazaki et al. 2005). In cultured 3T3L1 adipocytes, SCD2 is the only SCD isoform required for adipocyte differentiation (Christianson et al. 2007). The mechanism for this is currently unknown, but it is thought that SCD2 regulates the expression of PPARγ, a key determinant of the adipogenic process. In fact, knockdown of SCD2 is just as powerful at inhibiting adipocyte differentiation as silencing PPARγ itself (Christianson et al. 2007).

There are four known SCD isoforms in mouse (SCD1–4) and two in human (hSCD1 and hSCD5), (Beiragi et al. 2003; Man et al. 2005; Ntambi and Miyazaki 2004; Zhang et al. 1999) but only human SCD1 is considered an orthologue of the mouse SCD isoforms (Evans 2008; Wang et al. 2005). Both mouse SCD1 and SCD2 display approximately 80 % sequence identity to human SCD1 (Fig. 10.1), utilize the same substrates, and share common promoter elements (Fig.10.2) and tissue

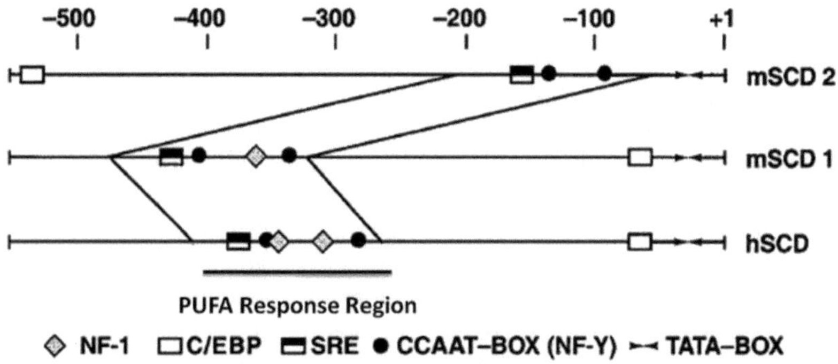

Fig. 10.2 Location of transcription factor recognition sites in the promoter regions of mouse SCD1 (mSCD1), mouse SCD2 (mSCD2), and human SCD1 (hSCD1). The differences in recognition sites and their locations between mSCD1 and mSCD2 likely explain in part their distinct tissue expression profiles, for example that mSCD2 is more highly expressed in embryonic and newborn liver than mSCD1. *NF* nuclear factor, *C/EBP* CCAAT enhancer binding protein, *SRE* serum response factor (Bene et al. 2001)

expression profiles (Hodson and Fielding 2012; Waters et al. 1997; Zhang et al. 2005, 1999). Similar to mouse SCD1 and SCD2, human SCD1 is also ubiquitously expressed with the highest expression in the adipose tissue (Hodson and Fielding 2012; Roongta et al. 2011; Zhang et al. 1999). Human SCD5 is highly expressed in the brain and pancreas, similar to mouse SCD2, despite not sharing functional homology (Wang et al. 2005). Functionally, human SCD1 may encompass the characteristics of both SCD1 and SCD2 in one enzyme: while mouse SCD1 and SCD2 appear to have redundant functions in most tissues, functional differences for these isoforms do appear to exist in adipose tissue, which may be fulfilled solely by SCD1 in humans (Christianson et al. 2007).

The study of SCD2 opens the door to discovering new lipid regulated developmental signaling pathways along with exploring new roles for monounsaturated fatty acids (MUFAs). The legacy of SCD2 is still developing and many questions regarding the physiological roles of this enzyme remain unanswered. Undoubtedly, SCD2 plays a unique and essential role. However, more research is warranted to determine the exact pathways that SCD2 affects and how to use this information in relation to human development.

This chapter will be a cumulative review about SCD2 that will include its regulation and its known physiological roles in different tissues.

Discovery of SCD2

During the initial identification of SCD1, southern blots were performed on 3T3-L1 genomic DNA using a variety of probes against multiple regions of the SCD gene. One probe in particular, containing the entire coding region of SCD1, was hybridized to 3T3-L1 genomic DNA that was collected 5 days after the addition of a differentiation cocktail consisting of methylisobutylxanthine, dexamethasone, and insulin, and digested with EcoR1. Surprisingly, in addition to the three EcoR1 fragments that comprised the complete gene of SCD1, this probe hybridized to a 14- and 2.2-kb genomic DNA EcoRI fragment (Kaestner et al. 1989). The additional hybridizations could be explained by two possibilities; the 14- and 2.2-kb fragments could be a part of an SCD pseudogene, or secondly, the fragments could be representative of another SCD isoform (Ntambi et al. 1988).

It is at this point, in 1986, that Klaus Kaestner, a graduate student working with James Ntambi in M. Daniel Lane's laboratory at The Johns Hopkins School of Medicine, decided to further explore these unidentified fragments for the possibility that they represented a second SCD gene. Kaestner began his investigation with a clone, designated λE14, which was created during the initial cloning of SCD1 (Kaestner et al. 1989). It contained five exons that were very similar to those in the SCD1 gene, but they were not identical. In order to determine if this clone was a second SCD, Kaestner set out to find its corresponding cDNA. Using a probe containing part of the first exon (nucleotides 205–324), generated by the digestion of λE14 with *PstI*, he screened a mouse 3T3-L1 cDNA library. The longest clone purified from the cDNA library was 2.7kb fragment called λ32 (Kaestner et al. 1989).

The last 200 bp of λ32 was then used to probe the same mouse 3T3-L1 cDNA library, and from this a clone λ53, containing a sequence containing a poly(A)-track, was isolated (Kaestner et al. 1989). Together λ32 and λ53 represent the complete mRNA of 5 kb. This was subcloned and sequenced using dideoxy chain termination. The sequence encoded for a protein that was >87 % similar to SCD1 and was named SCD2 (Kaestner et al. 1989).

SCD2 *Regulation*

Unlike *Scd1*, which is located on the negative strand of chromosome 19, *Scd2* and the other two SCD family members, *Scd3* and *Scd4*, are found on the positive strand of chromosome 19 (Man et al. 2005). *Scd2* is ~15 kb in length and has six exons, five introns, and a uniquely large 3′ UTR that is indicative of the SCD family (Kaestner et al. 1989). Most often, genes that have a long 3′ UTR, including the other SCD isoforms, have multiple polyadenylation signals, so it is surprising that SCD2 has only one such signal. To date, the function of this long 3′ UTR remains unknown.

Many of the regulatory components of *Scd2* were discovered by the differential tissue expression patterns of SCD2 and SCD1 in response to diet. SCD1 and SCD2 are both induced by diets low in unsaturated fatty acids. However, they are induced in different tissues. For example, mice fed a low-fat diet induce SCD1, but not SCD2, expression in the liver; in contrast, this diet induces SCD2 expression in the kidney to a much higher extent than SCD1 (Kaestner et al. 1989). The differences in tissue expression and dietary induction were thought to be differences in the control of gene expression. Kaestner recognized that the promoters of SCD2 and SCD1 are significantly different and that these differences may have an effect on the expression and induction differences between the two isoforms (Kaestner et al. 1989). A detailed examination of the promoters would come in the years that followed his initial work, but at the time, Kaestner was able to determine that SCD2 has two conserved CCAAT boxes found in SCD1 and that SCD2 has a missing cAMP core regulatory sequence that is present in the promoter of SCD1 (Fig. 10.2) (Kaestner et al. 1989).

During the differentiation of 3T3L1 preadipocytes into adipocytes, genes involved in fatty acid uptake, synthesis, and storage are activated (Christianson et al. 2007; Hodson and Fielding 2012). In 1992, Andrew Swick and M. Daniel Lane discovered that SCD2 contains a preadipocyte repressor element (PRE) site in the promoter (between –435 and –410) that is targeted by a ~58 kDa PRE binding protein causing transcription repression (Swick and Lane 1992).

In 1998, David Tabor and Peter Edwards determined that SCD2 is regulated by the nuclear levels of the pro-lipogenic transcription factor, sterol regulatory element binding protein (SREBP) (Tabor et al. 1999). They also identified a novel sterol regulatory element (SRE) located within the proximal promoter of SCD2 that does not comprise of the usual C/TCAC/T repeats that are separated by a single nucleotide (Fig. 10.2) (Tabor et al. 1999). Additionally, this group found that either a mutation in the SRE or the CCAAT boxes could attenuate SCD2 expression in response to sterols (Tabor et al. 1999). This demonstrated that that sterol-induced transcription is both NF-Y- and SREBP-dependent.

SCD2 Function in the Skin

SCD2 appears to be critical for lipid synthesis during neonatal development, since the SCD2 null mice have a severe skin permeability barrier defect and decreased triglyceride and FFA (free fatty acid) levels in the liver, skin, and plasma. This lack in lipid synthesis appears to be critical for survival because while the SCD2 null mice are born with the expected frequency, >70 % of them die within 24 h after birth when the genetic background is 129SvEv and 100 % die during embryonic development when the genetic background is C57Bl6. Most likely the lethality displayed in the 129SvEv genetic backgound is due to severe dehydration from the skin permeability dysfunction.

The SCD2 null mice that do survive are significantly smaller than the wild-type mice and have reduced body weights. The newborn SCD2 null mice have dry, cracked skin due to the skin permeability barrier defect, with a tightly packed and thickened epidermal stratum corneum. Examination of the lamellar membrane system in the skin reveals there are normal numbers of lamellar bodies in the knockout mice, but a decrease in their internal contents and a delay in lamellar membrane formation. While the epidermis has increased saturated FFA content, it has significantly decreased levels of cholesterol esters, triglyceride, acylceramide, and glucosylceramide, which are the primary constituents of lamellar bodies. Therefore, there appears to be decreased lipid delivery to lamellar granules, leading to deficient deposition of lipids in the stratum corneum interstices.

Additionally, while the total content of linoleic acid is not altered, there is a 80 % decrease of linoleic acid in the acylceramide fraction and an increase of this fatty acid in the phospholipid fraction. Since acylceramide is a major lipid involved in epidermal barrier function, decreasing its content or altering its composition will affect membrane barrier integrity. The epidermal lipid composition also showed the expected decrease in palmitoleate and oleate levels in total lipids of SCD2 null mice, with corresponding increases in saturated fatty acid levels. These differences occurred despite a several fold increase in SCD1 expression in both the epidermis and dermis of the SCD2 null mice, presumably a response to the loss of SCD2. Depletion of SCD2 also decreased the expression of keratinocyte differentiation genes transglutaminase-1 and involucrin, indicating that SCD2 products may play a role in gene expression programming during development. Nonetheless two genes that function in modulating the skin permeability barrier, glucosidase beta and diacylglycerol acyltransferase-2, were not altered in their mRNA levels. Together, these studies indicate that SCD2 is crucial for proper epidermal lipid synthesis, lipid channeling into different epidermal cellular compartments, and skin permeability barrier formation in neonate mice.

SCD2 Function in the Liver

While SCD1 regulates hepatic triglyceride synthesis in adult mice, SCD2 controls hepatic lipid metabolism in embryonic and newborn mice. This concept is consistent with the finding that hepatic total SCD catalytic activity is decreased by 70 % in newborn SCD2 null mice compared to wild-type mice, but not different between adult SCD2 null mice and age-matched controls. In the absence of SCD2, there is a near 50 % reduction in hepatic and plasma triglyceride levels and a 27 % reduction in total hepatic FFA in newborn mice. The decrease in triglyceride content appears to be due to decreased triglyceride synthesis, since ^{3}H-glycerol incorporation after IP injection is 54 % reduced in the triglyceride fraction[11]. However, there are no differences in the plasma glucose concentration, hepatic glycogen content, or hepatic phospholipid and cholesterol ester content. Also, there are no differences in the hepatic phospholipid synthesis rate or in the phospholipid composition as seen in the epidermis of the knockout mice. In the mice that survive past weaning, the defects in lipid synthesis are normalized due to increased SCD1 expression. Therefore, SCD2 is required for appropriate FFA, triglyceride, and VLDL production in the liver of embryos and neonates, but not in adult mice.

SCD2 Function in the Adipose Tissue

In adult mice, SCD1 is highly expressed in white adipose tissue depots and regulates MUFA production (Miyazaki et al. 2001; Ntambi and Miyazaki 2004). However, SCD2 is also expressed in adipose tissue, despite the abundant expression of SCD1. Evidence suggests that SCD1 and SCD2 may have disparate cellular functions in the white adipose tissue, as opposed to the liver and skin, where SCD1 appears to compensate for SCD2 function. It was reported that in mice fed a high-fat diet there is a large fold induction of SCD2 mRNA expression, much larger than effects on SCD1. In fact, out of 26 lipid metabolizing enzymes examined, SCD2 mRNA expression increased in per cent more than any other enzyme. One possible explanation for this observation is that SCD2 may be required to metabolize particular fatty acid species under high-fat feeding conditions. However, since increasing obesity on a high-fat diet also promotes adipogenesis, the process whereby preadipocyte fibroblasts differentiate into fully mature adipocytes, a role for SCD2 in controlling adipocyte development in the mouse is also possible.

Consistent with this latter hypothesis, studies in 3T3-L1 adipocytes indicate that significant levels of SCD2, but not of SCD1, are required for adipogenesis and for the maintenance of the adipocyte phenotype in fully differentiated adipocytes. The process of adipogenesis occurs through a transcriptional cascade beginning with the rapid and transient expression of C/EBPβ and C/EBPδ. These transcriptional factors then induce the expression of the ligand-activated receptor, PPARγ, and C/EBPα, which are critical for the completion of adipogenesis as well as the maintenance of adipocyte gene expression in fully differentiated cells (Ishida-Yamamoto and Kishibe 2011; Tang and Lane 2012). Recent experiments showed that siRNA-mediated depletion of SCD2 prevented PPARγ mRNA induction during adipogenesis and blocked the conversion of fibroblasts to adipocytes as indicated by oil red O staining of neutral lipid accumulation. Remarkably, silencing SCD2 was as powerful at inhibiting adipocyte differentiation as silencing PPARγ in this cultured preadipocyte system. In line with this interpretation, GLUT4 expression in 3T3-L1 cells was inhibited by about 90 % when either SCD2 or PPARγ were depleted with siRNA prior to the induction of differentiation. PPARγ protein levels were also reduced by about 90 % upon siRNA-mediated silencing of SCD2 under these same conditions. This defective adipogenesis observed upon SCD2 depletion was not restored by the addition of the PPARγ-specific ligand, rosiglitazone, suggesting SCD2 is not controlling the production of a PPARγ ligand.

In fully differentiated adipocytes, SCD2 regulates PPARγ protein expression, rather than transcript levels, which are unchanged after SCD2 depletion. It appears that PPARγ translation is inhibited in the absence of SCD2, while degradation is not affected. This downregulation of PPARγ translation and protein expression then leads to decreased expression of PPARγ target genes and loss of adipocyte function (Fig. 10.3). However, the effect of SCD2 depletion on protein synthesis is not specific to PPARγ since it appears SCD2 is required to maintain general protein synthesis. In fact, [^{35}S] methionine/cysteine labeling shows a small 15 % decrease in newly synthesized protein in 3T3-L1 adipocytes depleted of SCD2 by siRNA silencing. However, unlike PPARγ, many proteins actually decrease at the transcript level, while others increase. Thus, in the mature adipocytes, lack of SCD2 decreases PPARγ protein, which in turn decreases expression of downstream genes important for maintaining the adipocyte phenotype, such as SCD1, fatty acid synthase, and acetyl CoA carboxylase (Fig. 10.3).

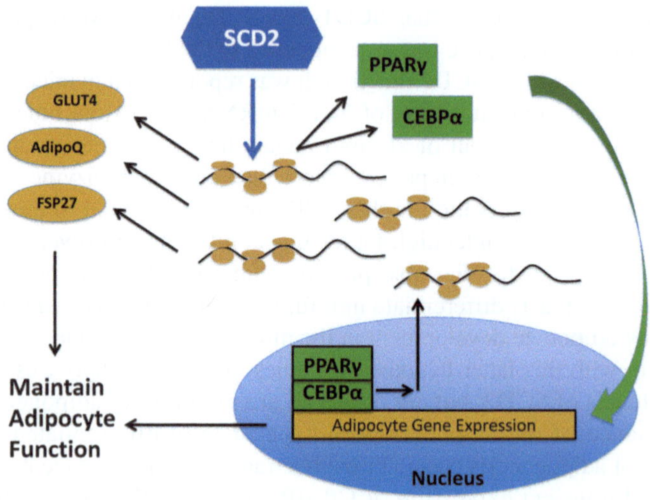

Fig. 10.3 Hypothetical model whereby mSCD2 may function directly or indirectly to selectively stimulate protein translation in cultured adipocytes. Such an action of mSCD2 protein or products that it produces increase PPARγ and CEBPα protein-levels (among other proteins), that in turn function to promote expression of genes [e.g., GLUT4 glucose transporter, Fat-specific protein (FSP) 27, Adiponectin (AdipoQ)] required for adipocyte differentiation. See text and Christianson et al. (2007)

The mechanism by which SCD2 regulates PPARγ mRNA expression in preadipocytes and protein translation in fully differentiated adipocytes remains a mystery. It may be through the production of an SCD2-specific unsaturated fatty acid or the proper shuttling of a fatty acid, as seen in the SCD2 knockout mouse, that regulates the translational machinery. Another possibility is that SCD2 is necessary for a protein–protein interaction that regulates translation. Regardless of the mechanism, these studies have described a distinct function for SCD2, apparently divergent from SCD1, in 3T3-L1 adipocytes. This may also be the case in other mouse tissues, explaining why these two isoforms are often simultaneously expressed. Furthermore, it is possible that human SCD1 also controls adipogenesis and adipocyte function, a hypothesis that deserves testing in future experiments.

SCD2 Expression in Immune Cells

Stearic acid potently inhibits the growth and proliferation of T lymphocytes, whereas B lymphocytes display a high degree of immunity (Pourbohloul et al. 1985; Pourbohloul and Buttke 1990). The reason for the differences in T lymphocyte and B lymphocyte susceptibility can be attributed to their Δ9-desaturase activity. B lymphocytes exclusively express SCD2 to a high level, and in contrast T lymphocytes do not have any detectable SCD transcript (Tebby and Buttke 1992).

Accordingly, when B and T lymphocytes are treated with stearate in vitro, the B cells are able to desaturate up to 25 % of the stearate into oleate, whereas the T cells are not able to desaturate stearate (Buttke et al. 1989). The inability of T cells to desaturate stearate causes the T cell to display 2–4 times more stearate in distearoylphosphatidylcholine than B cells. This level is equivalent to the amount of distearoylphosphatidylcholine that causes hemolysis of erythrocytes (Kuypers et al. 1984). Our knowledge of the role SCD2 in mouse immunology is limited. However, changes in the levels of dietary fat can profoundly alter immune responses in vivo, and it has been suggested that T cells are the predominate sensitive immune cell population (Buttke et al. 1989).

SCD2 Function in Reproduction

Functional testes and ovaries are in part dependent on the proper balance of fatty acids. This balance is critical for maintaining membrane fluidity and motility of sperm, and for membrane fluidity and the proper lipid reserve in the oocytes (Moreau et al. 2005; Saether et al. 2003). Although these alone can be attributed to the abundance and distribution of 22:6(*n*-3), functional sperm also require a balanced composition of *n*-3, *n*-6, and *n*-9 fatty acids. SCD2 is the predominate Δ9 desaturase expressed in the testes and ovaries and is responsible for the Δ9-desaturase activity in these organs (Moreau et al. 2005; Saether et al. 2003).

The roles of palmitoleate and oleate in the testis are not clear. However, the expression pattern of SCD1 and SCD2 within the testis has been established. SCD1 and SCD2 are both highly expressed in the epididymis, whereas only SCD2 is predominate in the testis, germ cells, and Sertoli cells (Saether et al. 2003). Sertoli cells are also known as "nurse cells" for their function of providing nutrients, such as PUFAs, to developing sperm cells. They are thought to be the key provider of PUFAs to the germ cells (Saether et al. 2003). Since SCD2 is highly expressed in the Sertoli cells, it is possible that in addition to providing PUFAs, the Sertoli cells also provide the germ cells with MUFAs (Saether et al. 2003).

The regulation of SCD1 and SCD2 within the testis is very interesting. SCD2 expression is up-regulated by insulin and dexamethasone while SCD1 is down-regulated (Saether et al. 2003). This is surprising, given the opposite effect these hormones have on SCD1 in the liver, where SCD1 is up-regulated by insulin and dexamethasone and SCD2 is down-regulated. One such explanation for this difference is that the regulation may reflect the roles that the testis and liver play in fatty acid metabolism. The primary role of SCD1 in the liver is to provide MUFAs for export, whereas the primary role of SCD2 in the testis is to replenish lost MUFAs (Saether et al. 2003).

Another key regulator of desaturase activity in the testis is follical stimulating hormone (FSH). FSH stimulates both SCD2 and SCD1 to very high levels in the Sertoli cells (Saether et al. 2003). It is reported that in the developing testis, FSH enhances the desaturase expression to levels needed for proper spermatogenesis (Saether et al. 2003).

Female oocytes of a variety of vertebrates have a high abundance of MUFAs, mainly palmitoleate and oleate, in the form of a lipid reserve that is critical for reproduction (Moreau et al. 2005). Much of the work analyzing the expression and regulation of SCD2 in the ovaries was by Céline Moreau and Joëlle Dupont (Moreau et al. 2005). In agreement with the work done on rat testis, by the Haugen group, the level of *Scd2* mRNA was high and predominate in whole ovary compared to *Scd1*, while the opposite is true for liver (Moreau et al. 2005; Saether et al. 2003). Both SCD1 and SCD2 are expressed in the granulosa cells of large follicles, corpus luteum, and cumulus oophorus of the ovary, but SCD2 is predominate in all of these tissues (Moreau et al. 2005). The cumulus cells provide nutrition for the oocyte and given expression of SCD2 is absent in the oocyte (Moreau et al. 2005); it can be speculated that SCD2 is the main depositor of MUFAs within that tissue. Also, SCD2 is highly expressed within the granulosa cell of the cumulus and is virtually absent from the oocytes (Moreau et al. 2005), which supports the hypothesis that lipids are transported into the oocytes from the cumulus. In the ovary *SCD2* is increased during follicular development and is positively regulated by pregnancy hormones that affect follicular development such as eCG and hCG as well as the hormones FSH and IGF1 (Moreau et al. 2005). Using rat granulosa cells to investigate the molecular mechanisms involving IGF1- and FSH-induced SCD2 expression, it was determined that the MAPK3/MAP1 and AKT signaling are involved in inducing expression, respectively (Moreau et al. 2005). This is in agreement with evidence suggesting that MAP3K/MAP1 and AKT pathways are involved in granulosa cell proliferation and steroidgenesis (Moreau et al. 2005).

Human Relevance

The two known human SCD isoforms are hSCD1 and hSCD5. hSCD1 shares high amino acid sequence similarity to SCD2. However hSCD5, also shares a high amino acid sequence similarity to SCD2, but unlike hSCD1, hSCD5 is expressed in a tissue specific manner similar to SCD2 (Beiragi et al. 2003; Dobrzyn and Ntambi 2005; Miyazaki et al. 2006; Wang et al. 2005). In addition, it was discovered that a pericentric inversion of hSCD5 was found in two generations of a family with cheiloschisis or cleft lip (Beiragi et al. 2003). Although defects in SCD2 and hSCD5 produce different tissue morphologies (Beiragi et al. 2003; Miyazaki et al. 2005), there is a strong connection between MUFA synthesis and tissue development.

Conclusions

In recent years it has become evident that SCD2 and the other isoforms of SCD are playing distinct and nonredundant roles. In this chapter we examined the known function of SCD2, both tissue specifically and temporally. Future research on SCD2 will be exciting as there are many unanswered questions about how exactly MUFAs influence early development, tissue morphogenesis, signaling pathways, and adipocyte differentiation.

Acknowledgments The authors thank members of the Czech and Ntambi laboratory for stimulating discussions on the topics addressed in this chapter. The work in the Czech laboratory related to the issues discussed in this review was funded by grants DK085753 and DK030898 from the National Institutes of Health and from the international Research Alliance at the Novo Nordisk Foundation Center for Metabolic Research. The work in the Ntambi lab related to this review was funded by NIH grant NIDDK-RO162388, AAHA Grant-in aid Wisconsin Affiliate, USDA (Hatch) grant #142–4306, and in part by funds from Xenon Genetics, Inc. Lucas M. O'Neill is a fellow supported by NIH grant 5T32GM007215-37 and SciMed GRS.

References

Beiragi S, Zhou M, Talmadge CB, Went-Sumegi N, Davis JR, Huang D, Saal H, Seemayer T, Sumegi J (2003) Identification and characterization of a novel gene disrupted by a pericentric inversion inv(4)(p13.lq21.1) in a family with cleft lip. Gene 309:11–21

Bene H, Lasky D, Ntambi JM (2001) Cloning and Characterization of the Human Stearoyl CoA Desaturase Gene Promoter: Transcriptional Activation by Sterol Regulatory Element Binding Protein and Repression by Polyunsaturated Fatty Acids and Cholesterol. Biochem Biophys Res Commun 284:1194–1198

Buttke T, Van Cleave S, Steelman L, McCubrey J (1989) Absence of unsaturated fatty acid synthesis in murine T lymphocytes. Proc Natl Acad Sci U S A 86:6133–6137

Christianson JL, Nicoloro S, Straubhaar J, Czech MP (2007) Stearoyl-CoA desaturase 2 is required for peroxisome proliferator-activated receptor γ expression and adipogenesis in cultured 3T3–L1 cells. J Biol Chem 283(5):2906–2916

DeWillie JW, Farmer SJ (1992) Quaking phenotype influences brain lipid-related mRNA levels. Neurosci Lett 141(12):195–198

Dobrzyn A, Ntambi JM (2005) Stearoyl-CoA desaturase as a new drug target for obesity treatment. Obes Rev 6(2):169–174

Evans H (2008) Ancient and modern duplication events and the evolution of stearoyl-CoA desaturases in teleost fishes. Physiol Genomics 35:18–29

Hodson L, Fielding BA (2012) Stearoyl-CoA desaturase: rogue or innocent bystander? Prog Lipid Res 52:15–42

Ishida-Yamamoto A, Kishibe M (2011) Involvement of corneodesmosome degradation and lamellar granule transportation in the desquamation process. Med Mol Morphol 44:1–6

Kaestner KH, Ntambi JM, Kelly TJ, Lane MD (1989) Differentiation-induced gene expression in 3T3-L1 preadipocytes: a second differentially-expressed gene encoding stearoyl-CoA desaturase. J Biol Chem 264:14755–14761

Kuypers FA, Roelofsen B, Berendsen W, Op den Kamp JA, van Deenen LL (1984) Shape changes in human erythrocytes induced by replacement of the native phosphatidylcholine with species containing various fatty acids. J cell Biol 99(6):2260– 2267

Man WC, Miyazaki M, Chu K, Ntambi JM (2005) Membrane topology of mouse stearoyl-CoA desaturase 1. J Biol Chem 281(2):1251–1260

Miyazaki M, Kim YC, Ntambi JM (2001) A lipogenic diet in mice with a disruption of the stearoyl-CoA desaturase 1 gene reveals a stringent requirement of endogenous monounsaturated fatty acids for triglyceride synthesis. J Lipid Res 42:1018–1024

Miyazaki M, Dobrzyn A, Elias P, Ntambi J (2005) Stearoyl-CoA desaturase-2 gene expression is required for lipid synthesis during early skin and liver development. Proc Natl Acad Sci USA 102(35):12501–12506

Miyazaki M, Bruggink SM, Ntambi JM (2006) Identification of mouse palmitoyl-coenzyme A Δ9-desaturase. J Lipid Res 47:700–704

Moreau C, Froment P, Tosca L, Moreau V, Dupont J (2005) Expression and regulation of the SCD2 desaturase in the rat ovary. Biol Reprod 74:75–87

Ntambi JM, Miyazaki M (2004) Regulation of stearoyl-CoA desaturases and role in metabolism. Prog Lipid Res 4:91–104

Ntambi JM, Buhrow S, Kaestner KH, Christy RJ, Sibley E, Kelly TJ, Lane MD (1988) Differentiation-induced gene expression in 3T3-L1 preadipocytes: characterization of a differentially-expressed gene encoding stearoyl-CoA desaturase. J Biol Chem 3:17291–17300

Pourbohloul S, Mallett GS, Buttke TM (1985) Inhibition of lymphocyte proliferation by free fatty acids. III. Modulation of thymus-dependent immune responses. Immun. 56(4):659–666

Pourbohloul SC, Buttke TM (1990) Identification of T-helper cells as the target of stearic acid-inhibition in primary antibody responses in vitro. Int J Immunopharmcol 12(7):799–807

Roongta UV, Pabalan JG, Wang X, Ryseck RP, Fargnoli J, Henely BJ, Yang WP, Zhu J, Madireddi MT, Lawrence RM, Wong TW, Rupnow BA (2011) Cancer cell dependence on unsaturated fatty acids implicates stearoyl-CoA desaturase as a target for cancer therapy. Mol Cancer Res 9:1551–1561

Saether T, Tran TN, Rootwelt H, Christophersen BO, Haugen TB (2003) Expression and regulation of Δ^5-desaturase, Δ^6-desaturase, stearoyl-coenzyme A (CoA) desaturase 1, and stearoyl-CoA desaturase 2 in rat testis. Biol Reprod 69(1):117–124

Sampath H, Ntambi JM (2011a) The role of stearoyl-CoA desaturase in obesity, insulin resistance, and inflammation. Ann N Y Acad Sci 1243:47–53

Sampath H, Ntambi JM (2011b) The role of fatty acid desaturases in epidermal metabolism. Dermatoendocrinol 3(2):62–64

Swick AG, Lane MD (1992) Identification of a transcriptional repressor down-regulated during preadipocyte differentiation. Proc Natl Acad Sci U S A 89:7895–7599

Tabor DE, Kim JB, Spiegelman BM, Edwards PA (1999) Identification of conserved *cis*-elements and transcription factors required for sterol-regulated transcription of stearoyl-CoA desaturase 1 and 2. J Biol Chem 274(29):20603–20610

Tang QQ, Lane MD (2012) Adipogenesis: from stem cell to adipocyte. Annu Rev Biochem 81:715–736

Tebby PW, Buttke TM (1992) Stearoyl-CoA desaturase gene expression in lymphocytes. Biochem Biophys Res Commun 186(1):531–536

Wang J, Yu L, Schmidt RE, Su C, Huang X, Gould K, Cao G (2005) Characterization of HSCD5, a novel human stearoyl-CoA desaturase unique to primates. Biochem Biophys Res Commun 332:735–742

Waters KM, Miller CW, Ntambi JM (1997) Localization of a polyunsaturated fatty acid response region in stearoyl-CoA desaturase gene 1. Biochim Biophys Acta 1349:33–42

Zhang L, Ge L, Parimoo S, Stenn K, Prouty SM (1999) Human stearoyl-CoA desaturase: alternative transcripts generated from a single gene by usage of tandem polyadenylation sites. Biochem J 340(Pt 1):255–264

Zhang S, Yang Y, Shi Y (2005) Characterization of human SCD2, an oligomeric desaturase with improved stability and enzyme activity by cross-linking in intact cells. Biochem J 388:135–142

Chapter 11
Stearoyl-CoA Desaturase Isoforms 3 and 4: Avenues for Tissue-Specific Δ9 Desaturase Activity

Laura M. Bond and James M. Ntambi

Introduction

Murine stearoyl-CoA desaturase (SCD) isoform 1 was discovered in 1985 by James Ntambi while studying adipocyte differentiation in 3T3-L1 cells (Ntambi et al. 1988). SCD is a critical enzyme in the de novo lipogenesis pathway and catalyzes the Δ9 desaturation of saturated fatty acids to yield monounsaturated fatty acids (MUFAs). As illustrated in the previous chapters of this book, tremendous progress has been made in elucidating the regulation and function of SCD isoform 1. However, in contrast to SCD1, additional isoforms of SCD have been relatively unexplored. Suggestion of multiple SCD isoforms in mice arose just one year after the discovery of *Scd1*, when southern blotting of digested genomic DNA against probes made against *Scd1* revealed DNA bands that did not correspond to the single known *Scd* isoform (Kaestner et al. 1989). After further studies, *Scd2* was isolated and characterized, but a potential third isoform was not investigated until 2001 (Kaestner et al. 1989; Zheng et al. 2001).

L.M. Bond
Department of Biochemistry, University of Wisconsin-Madison, Madison, WI 53706, USA

J.M. Ntambi, Ph.D. (✉)
Departments of Biochemistry and Nutritional Sciences, University of Wisconsin-Madison, Madison, WI 53706, USA
e-mail: ntambi@biochem.wisc.edu

J.M. Ntambi (ed.), *Stearoyl-CoA Desaturase Genes in Lipid Metabolism*,
DOI 10.1007/978-1-4614-7969-7_11,

SCD3

Over a decade after the initial discovery of *Scd1*, *Scd3* was reported by scientists at Johnson & Johnson who were studying lipid metabolism in skin to aid developments in pharmaceutics (Zheng et al. 2001). The group was interested in the unique phenotypes of the asebia mouse and was trying to determine the exact location of the mutation that causes alopecia and defects in sebaceous glands (Sundberg et al. 2000). They mapped the asebia mutation to *Scd1*, but in the process of characterizing this mutation, Zheng et al. 2001 identified a previously unreported isoform, which they named *Scd3*.

Tissue Expression

Scd3 is unique to mice and has not been identified in humans, which express h*SCD1* and h*SCD5* (Wang et al. 2005). Zheng and colleagues established with Northern blotting that *Scd3* is expressed in differentiated sebocytes (Zheng et al. 2001). *Scd1*, on the other hand, is expressed in undifferentiated sebocytes, and *Scd2* expression is restricted to the hair follicle (Paton and Ntambi 2009).

The preliminary tissue panel by Zheng et al. surveyed several tissues for expression of *Scd3* by hybridizing tissue blots containing mRNA from 12 tissues with *Scd3* specific probes. Their results suggested that *Scd3* is exclusively expressed in the skin. However, later studies in the mouse harderian gland and preputial gland revealed the expression of *Scd3* in these esoteric tissues (Miyazaki et al. 2001, 2002). Both of these tissues are highly lipogenic; the harderian gland is located within the orbit of the eyeball and is proposed to have a myriad of functions, such as the secretion of hormones and neutral lipids for lubrication (Payne 1994). The preputial glands are anterior to the genitals and are important for reproduction (Thody et al. 1976). *Scd3* expression has also been detected in murine liver, but only in conditions of SREBP-1a overexpression (Horton et al. 2003).

Recent publications report the expression of *Scd3* in subcutaneous white adipose tissue (Bostrom et al. 2012; Yu et al. 2010). Notably, the Speigelman laboratory reported that irisin, a myokine that induces white adipose "beiging", induces expression of *Scd3* (Bostrom et al. 2012).

The detection of *Scd3* expression in white adipose tissue is not entirely surprising, as *Scd3* expression has been reported in 3T3-L1 cultured pre-adipocytes undergoing differentiation (Wang et al. 2008). In fact, *Scd3* exhibits a unique temporal expression pattern. *Scd1* and *Scd2* are expressed in differentiated adipocytes (Kaestner et al. 1989). In contrast, *Scd3* expression is transient, appearing around day 2 and vanishing by day 7 (Wang et al. 2008). Importantly, *Scd3* is not expressed in cultured mature adipocytes. The significance of *Scd3* expression in adipocyte differentiation has not been explored, but it may account for its detection in adipose tissue of adult mice. Alternatively, *Scd3* may be expressed in mature adipocytes in vivo, but cultured mature adipocytes derived from fibroblasts do not express *Scd3* once differentiated, illustrating the limitations of in vitro models.

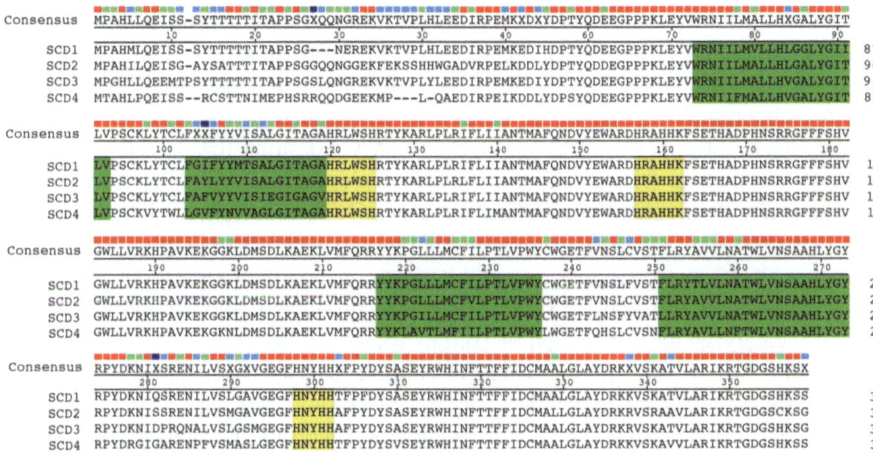

Fig. 11.1 Conservation analysis of murine SCD protein sequences. All murine SCD isoforms contain four transmembrane regions (*green*) and three cytosolic histidine motifs (*yellow*). These histidine residues coordinate elemental iron and are essential for catalytic activity (Shanklin et al. 1994). The histidine motifs are completely conserved in all four isoforms. Degree of conservation is depicted as red (high), green (medium), and blue (low) in the color bar. The consensus sequence is shown below the color bar. Sequence alignment was completed using DNAStar MegAlign ClustalW analysis

Table 11.1 Murine SCD isoforms share strong protein sequence identity

	SCD1	SCD2	SCD3	SCD4
SCD1	–	86	88	79
SCD2	86	–	85	79
SCD3	88	85	–	82
SCD4	79	79	82	–

Percent sequence identity was determined using NCBI BLAST

Protein Structure and Substrate Specificity

SCD3 contains four transmembrane domains and the three histidine motifs that are essential for catalytic activity (Fig. 11.1) (Shanklin et al. 1994). SCD3 shares 88 % and 84 % protein sequence identity with SCD1 and SCD2, respectively (Table 11.1). According to the proposed topology of mouse SCD1, the N and C termini are oriented toward the cytosol (Man et al. 2006).

Despite the strong structural similarities between SCD isoforms 1, 2, and 3, SCD3 exhibits unique substrate specificity. SCD1 and SCD2 introduce a double bond between carbons 9 and 10 of both palmitate and stearate, whereas SCD3 preferentially desaturates palmitate. The first indication that SCD3 desaturates primarily 16:0 came from studies in the mouse harderian gland. SCD1$^{-/-}$ mice show a complete loss of desaturation toward stearoyl-CoA, but some desaturation toward palmitoyl-CoA remains (Miyazaki et al. 2001). This provides reason to suspect that a harderian gland-specific palmitoyl-CoA Δ9 desaturase must be catalyzing the synthesis of the 16:1(*n*-7) that remains. In 2006, Miyazaki et al. determined the substrate specificity of all four SCD isoforms (Miyazaki et al. 2006). A yeast strain that is deficient in the yeast Δ9 desaturase, OLE1, was transformed with plasmids encoding each mouse SCD gene and then treated with saturated fatty acyl-CoAs of specific lengths. Lipids were extracted and fatty acids were identified with gas chromatography. Yeast given the mouse SCD3 expression vector exhibited no conversion of 18:0 to 18:1(*n*-9), but rather desaturated 16:0 and 14:0. The fatty acid products demonstrated that SCD3 primarily catalyzes the conversion of 16:0 to 16:1(*n*-7). The authors also transfected HeLa cells, which have low basal Δ9 desaturase activity, with the SCD isoforms cloned into a mammalian expression vector. Gas chromatography revealed the identity of the fatty acid products and confirmed that SCD3 primarily catalyzes the conversion of 16:0 to 16:1(*n*-7). The unique substrate specificity of SCD3 is of particular interest in the metabolism field because 16:1(*n*-7) has been proposed to be a lipid-derived hormone, or lipokine, that elicits beneficial metabolic effects. In 2008, this lipokine was reported to improve insulin sensitivity and prevent hepatic steatosis (Cao et al. 2008). However, the effects of 16:1(*n*-7) are controversial and subsequent studies question the original results (Guo et al. 2012; Hodson and Karpe 2013). In vivo and in vitro studies modulating SCD3 activity may elucidate the specific effects of this fatty acid and its participation in tissue cross-talk.

Gene Structure and Regulation

Like other SCD isoforms, *Scd3* maps to chromosome 19 D2 (Fig. 11.2a), has six exons and five introns, and encodes a transcript ~4.9 kb in length (Fig. 11.2b) (Zheng et al. 2001). Scd3 has distinct 5' and 3' noncoding regions and intronic sequences that are significantly longer than those of *Scd1*, *2*, and *4* (Fig. 11.2b).

DNA sequence analysis of the *Scd3* promoter demonstrated that *Scd3* shares a few common regulatory elements with *Scd1* and *Scd2* but lacks several distinct transcription factor binding sites. For example, *Scd3* promoter has two noncanonical TATA boxes, a feature shared with *Scd1*. However, *Scd3* lacks the PUFA response region, SREBP site, and the CHOP/CEBP α-heterodimer binding site, all three of which are present on the promoter of *Scd1* and *Scd2*. Other distinguishing features of *Scd3* include the presence of an estrogen receptor binding site, which is absent in the promoters for *Scd1* and *Scd2*.

Transcription factor binding sites related to sex hormones and determination were identified in the 5' upstream regulatory sequence of *Scd3*, which is consistent with the observation that *Scd3* expression in skin is higher in males than females (Zheng et al. 2001). *Scd3* contains sites for estrogen receptor and sex-determining Y gene product that could facilitate gene repression and activation, respectively, to yield increased expression in males. Consistent with these findings, subcutaneous injection of

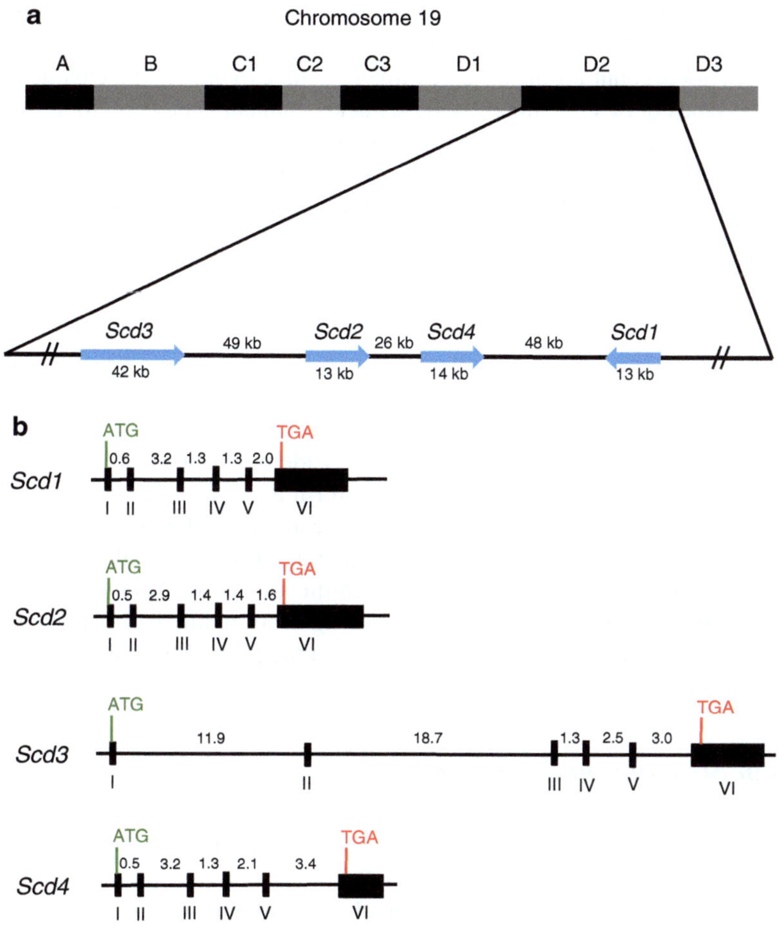

Fig. 11.2 (**a**) All murine SCD isoforms are present on chromosome 19 region D2. *Scd3*, *Scd2*, and *Scd4* are transcribed in the forward direction, while *Scd1* is transcribed in the reverse direction. (**b**) *Scd* isoforms 1-4 contain five introns and six exons. The majority of exon VI is noncoding due to a stop codon that is present towards the 5′ end of the exon

testosterone in mice has been shown to induce *Scd3* expression in the preputial gland; however, the specific transcription factor binding site has not yet been identified (Miyazaki et al. 2002).

Studies in SCD1$^{-/-}$ mice show that loss of SCD1 activity represses *Scd3* expression. The Ntambi lab showed that *Scd3* expression is significantly reduced in the skin, harderian gland, and preputial glands of mice that lack *Scd1* in these tissues (Miyazaki et al. 2001, 2002; Sampath et al. 2009). Zheng et al. also observed a strong reduction of *Scd3* expression in adult *Scd1*-mutant asebia mice, but detected *Scd3* expression in young asebia mice (Zheng et al. 2001). The authors propose that loss of *Scd3* expression may stem from destruction of the sebocytes, when SCD3 activity is not able to compensate for loss of SCD1 and rescue the sebaceous gland. While this hypothesis is consistent with the requirement for SCD1 activity in sebocyte differentiation, it likely does not account for the observed loss of *Scd3* expression in the harderian and preputial glands of SCD1$^{-/-}$ mice, as these glands maintain their core lipogenic properties and do not exhibit severe phenotypes in the absence of *Scd1*.

SCD products may directly or indirectly regulate transcription, but the mechanism underlying the regulation between isoforms is not well understood. Studies in SCD3$^{-/-}$ knockout mice may reveal if SCD3 activity regulates SCD1 or SCD2 gene expression and forms a positive or negative feedback loop.

SCD4

Discovered in 2003, SCD4 is the most recently identified SCD isoform (Miyazaki et al. 2003). Suggestion of a fourth SCD isoform arose from studies in the *ob/ob* mouse that revealed a paradoxical increase in cardiac MUFA levels despite unaltered expression of *Scd1* and *Scd2* (Miyazaki et al. 2003). The unknown source of Δ9 desaturation was probed and led to the identification of SCD isoform 4. Northern blot analysis of a standard tissue panel, including heart, brain, liver, skeletal muscle, and lung, demonstrated that *Scd4* is expressed in the heart (Miyazaki et al. 2003).

Protein Structure and Function

SCD4 shares 79 % protein sequence identity with SCD1 and SCD2 and 82 % sequence identity with SCD3 (Table 11.1). SCD4 contains the three histidine motifs required for catalytic activity, and hydropathy measurements suggest that all SCD isoforms have four transmembrane domains (Miyazaki et al. 2003) (Fig. 11.1). The conservation of the structural elements in SCD4 that are required by SCD1 suggests that SCD4 possesses similar Δ9 desaturase activity (Shanklin et al. 1994). To demonstrate the activity of SCD4 in vitro, mammalian cells with low endogenous desaturase activity were transfected with a plasmid containing *Scd4*. Microsomal fractions of these cells exhibited Δ9 desaturase activity. Analysis of fatty acid products demonstrated that SCD4 has substrate specificity similar to SCD1 and SCD2 and desaturates both stearate and palmitate (Miyazaki et al. 2003).

Gene Structure and Regulation

Scd4 maps to chromosome 19 region D2 and is located upstream of *Scd2* on the positive strand. *Scd4* is transcribed in the forward direction along with *Scd2* and *Scd3*, but in the reverse direction of *Scd1* (Fig. 11.2a). The predominant *Scd4* transcript spans 3.1 kb, which is shorter than the other *Scd* isoforms (Fig. 11.2b). Two smaller *Scd4* transcripts, 2.8 and 1.5 kb long, were also detected (Miyazaki et al. 2003). These minor transcripts may result from GAAA repeats in exon 6 that predispose the nascent transcript to premature polyadenylation. It is unknown if these truncated *Scd4* transcripts encode functional desaturase enzymes.

Gene regulation studies constitute the bulk of SCD4 characterization and provide a few clues to the function of this novel isoform. Miyazaki and colleagues selected several key regulatory elements known to modulate murine *Scd1* expression in liver and investigated their effect on *Scd4* expression in the heart (See Hodson and Fielding, 2013 for extensive review of *Scd1* regulation) (Miyazaki et al. 2003). They learned that *Scd4* shares some regulatory patterns with *Scd1* and *Scd2* in the heart but is also subject to unique control mechanisms.

Feeding mice a high-carbohydrate, low-fat diet upregulates both *Scd4* and *Scd1* expression in the heart (Miyazaki et al. 2003; Ntambi 1992). This suggests that *Scd4* is a target of the pro-lipogenic transcription factor, SREBP-1c, which drives *Scd1* expression, but the SREBP binding site, SRE, was not found on the *Scd4* promoter. This element may exist further upstream in an enhancer region not yet explored. Alternatively, the induction of *Scd4* expression after high-carbohydrate feeding could stem from gene activation by ChREBP, another pro-lipogenic transcription factor that was not well-defined at the time of the study (Yamashita et al. 2001). Dietary supplementation of fish oil, a rich source of polyunsaturated fatty acids (PUFAs), decreases *Scd1* and *Scd2* expression, but does not affect expression of *Scd4* (Miyazaki et al. 2003). This dichotomy is consistent with the presence of a PUFA response element on the promoters of *Scd1* and *Scd2*, but not on the *Scd4* promoter. LXR agonists activate expression of *Scd1*, *Scd2*, and *Scd4*, suggesting that all of the isoforms in the heart can be regulated by cholesterol (Miyazaki et al. 2003).

Leptin negatively regulates *Scd4* in the heart, as demonstrated in the *ob/ob* mouse studies that led to the discovery of SCD4 (Miyazaki et al. 2003). Leptin is a hormone secreted by adipose tissue that has garnered much attention for its anorectic properties, but its repression of lipogenic genes shows that this adipokine has additional features distinct from its function in satiety (Cohen et al. 2002; Friedman and Halaas 1998). *ob/ob* mice have increased cardiac SCD activity and, subsequently, increased MUFA and triglyceride content (Christoffersen et al. 2003; Miyazaki et al. 2003). These phenotypes normalize with intravenous treatment of leptin, and gene expression studies revealed that SCD4—not SCD1 or SCD2—mediates the fatty acid changes observed in the heart (Miyazaki et al. 2003). SCD4 most likely plays a predominant role in the pathogenesis of "fatty heart" that is exhibited in *ob/ob* mice. The exact mechanism by which leptin represses *Scd4* expression in the heart has not been determined, but transcriptional reporter assays may reveal the relevant binding region. While *Scd1* and *Scd2* expression are not regulated by leptin in the heart (Miyazaki et al. 2003), the transcriptional repression of *Scd1* and *Scd2* expression by this adipokine has been established in mouse liver (Cohen et al. 2002). Further studies will identify tissue-specific factors that mediate the differential regulation of these isoforms in various organs.

The tissue-specific expression pattern and regulation of *Scd4* expression helps validate the function of having several SCD isoforms. Regulation between the isoforms would further justify having multiple isoforms with high protein sequence identity. Upon the discovery of *Scd4*, Miyazaki and colleagues investigated potential compensatory mechanisms between the predominant heart *Scd* isoforms. $SCD1^{-/-}$ mice exhibit a 3-fold increase in *Scd4* expression (Miyazaki et al. 2003). This compensatory mechanism wherein *Scd4* acts in place of SCD1 explains why $SCD1^{-/-}$ mice have a less dramatic decrease in MUFAs in the heart than in the liver.

Summary

SCD isoforms 3 and 4 have not been fully explored. However, preliminary studies demonstrate that these novel isoforms have unique regulatory properties that manifest in restricted tissue expression. *Scd3* is expressed mainly in the skin, harderian gland, and preputial gland, and *Scd4* is expressed in the heart. Exploring SCD3 and SCD4 will help explain if 16:1(*n*-7) or 18:1(*n*-9) derived from a specific tissue, such as skin or heart, affects the function of these fatty acids as signaling molecules. Since SCD3 desaturates primarily palmitate to yield palmitoleate, a proposed lipokine, this isoform may be a tool for studying increased or decreased levels of this specific MUFA.

These unexplored SCD isoforms are now active sites of investigation. The recent development of SCD3 and SCD4 global knockout mouse models are current tools helping researchers shed light on the nuanced functions of these SCD isoforms. We expect the contents of this chapter to broaden and grow over the next decade.

Acknowledgements Our work has been supported by NIH grant NIDDK-RO162388 (J.M.N.), AHA Grant-in Aid Wisconsin Affiliate, and USDA (Hatch) grant # 142–4306. L.M.B. was supported by NIH National Research Service Award T32 GM07215.

References

Bostrom P, Wu J, Jedrychowski MP, Korde A, Ye L, Lo JC, Rasbach KA, Bostrom EA, Choi JH, Long JZ et al (2012) A PGC1-alpha-dependent myokine that drives brown-fat-like development of white fat and thermogenesis. Nature 481:463–468

Cao H, Gerhold K, Mayers JR, Wiest MM, Watkins SM, Hotamisligil GS (2008) Identification of a lipokine, a lipid hormone linking adipose tissue to systemic metabolism. Cell 134:933–944

Christoffersen C, Bollano E, Lindegaard ML, Bartels ED, Goetze JP, Andersen CB, Nielsen LB (2003) Cardiac lipid accumulation associated with diastolic dysfunction in obese mice. Endocrinology 144:3483–3490

Cohen P, Miyazaki M, Socci ND, Hagge-Greenberg A, Liedtke W, Soukas AA, Sharma R, Hudgins LC, Ntambi JM, Friedman JM (2002) Role for stearoyl-CoA desaturase-1 in leptin-mediated weight loss. Science 297:240–243

Friedman JM, Halaas JL (1998) Leptin and the regulation of body weight in mammals. Nature 395:763–770

Guo X, Li H, Xu H, Halim V, Zhang W, Wang H, Ong KT, Woo SL, Walzem RL, Mashek DG et al (2012) Palmitoleate induces hepatic steatosis but suppresses liver inflammatory response in mice. PLoS One 7:e39286

Hodson L, Karpe F (2013) Is there something special about palmitoleate? Curr Opin Clin Nutr Metab Care 16:225–231

Hodson L, Fielding BA (2013) Stearoyl-CoA desaturase: rogue or innocent bystander? Prog Lipid Res 52:15–42.

Horton JD, Shah NA, Warrington JA, Anderson NN, Park SW, Brown MS, Goldstein JL (2003) Combined analysis of oligonucleotide microarray data from transgenic and knockout mice identifies direct SREBP target genes. Proc Natl Acad Sci USA 100:12027–12032

Kaestner KH, Ntambi JM, Kelly TJ Jr, Lane MD (1989) Differentiation-induced gene expression in 3T3-L1 preadipocytes. A second differentially expressed gene encoding stearoyl-CoA desaturase. J Biol Chem 264:14755–14761

Man WC, Miyazaki M, Chu K, Ntambi JM (2006) Membrane topology of mouse stearoyl-CoA desaturase 1. J Biol Chem 281:1251–1260

Miyazaki M, Kim HJ, Man WC, Ntambi JM (2001) Oleoyl-CoA is the major de novo product of stearoyl-CoA desaturase 1 gene isoform and substrate for the biosynthesis of the harderian gland 1-alkyl-2,3-diacylglycerol. J Biol Chem 276:39455–39461

Miyazaki M, Gomez FE, Ntambi JM (2002) Lack of stearoyl-CoA desaturase-1 function induces a palmitoyl-CoA Delta6 desaturase and represses the stearoyl-CoA desaturase-3 gene in the preputial glands of the mouse. J Lipid Res 43:2146–2154

Miyazaki M, Jacobson MJ, Man WC, Cohen P, Asilmaz E, Friedman JM, Ntambi JM (2003) Identification and characterization of murine SCD4, a novel heart-specific stearoyl-CoA desaturase isoform regulated by leptin and dietary factors. J Biol Chem 278:33904–33911

Miyazaki M, Bruggink SM, Ntambi JM (2006) Identification of mouse palmitoyl-coenzyme A Delta9-desaturase. J Lipid Res 47:700–704

Ntambi JM (1992) Dietary regulation of stearoyl-CoA desaturase 1 gene expression in mouse liver. J Biol Chem 267:10925–10930

Ntambi JM, Buhrow SA, Kaestner KH, Christy RJ, Sibley E, Kelly TJ Jr, Lane MD (1988) Differentiation-induced gene expression in 3T3-L1 preadipocytes. Characterization of a differentially expressed gene encoding stearoyl-CoA desaturase. J Biol Chem 263:17291–17300

Paton CM, Ntambi JM (2009) Biochemical and physiological function of stearoyl-CoA desaturase. Am J Physiol Endocrinol Metab 297:E28–E37

Payne AP (1994) The harderian gland: a tercentennial review. J Anat 185(Pt 1):1–49

Sampath H, Flowers MT, Liu X, Paton CM, Sullivan R, Chu K, Zhao M, Ntambi JM (2009) Skin-specific deletion of stearoyl-CoA desaturase-1 alters skin lipid composition and protects mice from high fat diet-induced obesity. J Biol Chem 284:19961–19973

Shanklin J, Whittle E, Fox BG (1994) Eight histidine residues are catalytically essential in a membrane-associated iron enzyme, stearoyl-CoA desaturase, and are conserved in alkane hydroxylase and xylene monooxygenase. Biochemistry 33:12787–12794

Sundberg JP, Boggess D, Sundberg BA, Eilertsen K, Parimoo S, Filippi M, Stenn K (2000) Asebia-2J (Scd1(ab2J)): a new allele and a model for scarring alopecia. Am J Pathol 156:2067–2075

Thody AJ, Cooper MF, Bowden PE, Meddis D, Shuster S (1976) Effect of alpha-melanocyte-stimulating hormone and testosterone on cutaneous and modified sebaceous glands in the rat. J Endocrinol 71:279–288

Wang J, Yu L, Schmidt RE, Su C, Huang X, Gould K, Cao G (2005) Characterization of HSCD5, a novel human stearoyl-CoA desaturase unique to primates. Biochem Biophys Res Commun 332:735–742

Wang H, Qiang L, Farmer SR (2008) Identification of a domain within peroxisome proliferator-activated receptor gamma regulating expression of a group of genes containing fibroblast growth factor 21 that are selectively repressed by SIRT1 in adipocytes. Mol Cell Biol 28:188–200

Yamashita H, Takenoshita M, Sakurai M, Bruick RK, Henzel WJ, Shillinglaw W, Arnot D, Uyeda K (2001) A glucose-responsive transcription factor that regulates carbohydrate metabolism in the liver. Proc Natl Acad Sci USA 98:9116–9121

Yu CY, Mayba O, Lee JV, Tran J, Harris C, Speed TP, Wang JC (2010) Genome-wide analysis of glucocorticoid receptor binding regions in adipocytes reveal gene network involved in triglyceride homeostasis. PLoS One 5:e15188

Zheng Y, Prouty SM, Harmon A, Sundberg JP, Stenn KS, Parimoo S (2001) Scd3–a novel gene of the stearoyl-CoA desaturase family with restricted expression in skin. Genomics 71:182–191

Chapter 12
Functional Development of Stearoyl-CoA Desaturase Gene Expression in Livestock Species

Stephen B. Smith

Introduction

Foods high in oleic acid (18:1*n*-9), including oleic acid-enriched beef, reduce risk factors for cardiovascular disease (Kris-Etherton et al. 1999; Adams et al. 2010; Gilmore et al. 2011). Higher concentrations of oleic acid also are positively correlated with overall palatability of beef (Westerling and Hedrick 1979), whereas increased stearic acid (18:0) is the primary determinant of fat hardness (i.e., lipid melting point; Smith et al. 1998; Wood et al. 2004; Chung et al. 2006b). Therefore, increasing the monounsaturated fatty acid (MUFA) to saturated fatty acid (SFA) ratio would increase healthfulness, palatability, and fat softness of animal products. The conversion of SFA to MUFA by the fatty acid Δ^9 desaturase accounts for the majority of MUFA in muscle and adipose tissues of livestock species, i.e., the edible portions of beef, lamb, and pork carcasses (St John et al. 1991; Klingenberg et al. 1995; Lengli and Corl 2007; Jiang et al. 2008; Duckett et al. 2009). The Δ^9 desaturase is encoded by stearoyl-coenzyme A desaturase (SCD), which also is responsible for the conversion of *trans*-vaccenic acid to its corresponding CLA isomer, 18:2 *cis*-9, *trans*-11 CLA (Miyazaki and Ntambi 2003). Several studies have indicated that *SCD* gene expression and/or SCD activity is elevated in adipose tissues containing a high concentration of MUFA (Archibeque et al. 2005; Chung et al. 2007; Duckett et al. 2009; Brooks et al. 2011).

In many species, the concentration of oleic acid in adipose tissue reflects the relative proportion of oleic acid in the diet (St John et al. 1987; Go et al. 2012), but in ruminants such as beef cattle and sheep, dietary oleic acid is hydrogenated largely to stearic acid by ruminal microorganisms before reaching the small intestine (Ekeren et al. 1992). Thus, any accumulation of oleic acid in tissues of ruminants is

S.B. Smith (✉)
Department of Animal Science, Texas A&M University, 2471 TAMU,
College Station, TX 77843, USA
e-mail: sbsmith@tamu.edu

J.M. Ntambi (ed.), *Stearoyl-CoA Desaturase Genes in Lipid Metabolism*,
DOI 10.1007/978-1-4614-7969-7_12, © Springer Science+Business Media New York 2013

dependent on SCD catalytic activity. In this review we describe the genetic, developmental, and nutritional regulation of *SCD* gene expression and catalytic activity and fatty acid composition in tissues of livestock species.

SCD Tissue Distributions in Livestock Species

In laboratory rodents, *SCD* is expressed in both liver (Ntambi 1992; Waters and Ntambi 1994) and adipose tissue (Jones et al. 1996; Kang et al. 2004), although SCD activity is at least two orders of magnitude higher in mouse liver than in adipose tissue (Enser 1979). Bovine, ovine, and porcine adipose tissues exhibit substantially higher SCD catalytic activity than muscle, liver, or intestinal mucosa (Payne and Masters 1971; St John et al. 1991; Chang et al. 1992; Klingenberg et al. 1995; Archibeque et al. 2005) (Fig. 12.1). *SCD* gene expression is virtually undetectable in bovine liver (Cameron et al. 1994; Brooks et al. 2011), although *SCD* mRNA is readily detectable in porcine liver (Go et al. 2012). Chicken liver exhibits high SCD activity (Legrand and Lemarchal 1991; Legrand and Hermier 1992), although neither SCD activity nor *SCD* gene expression have been reported for

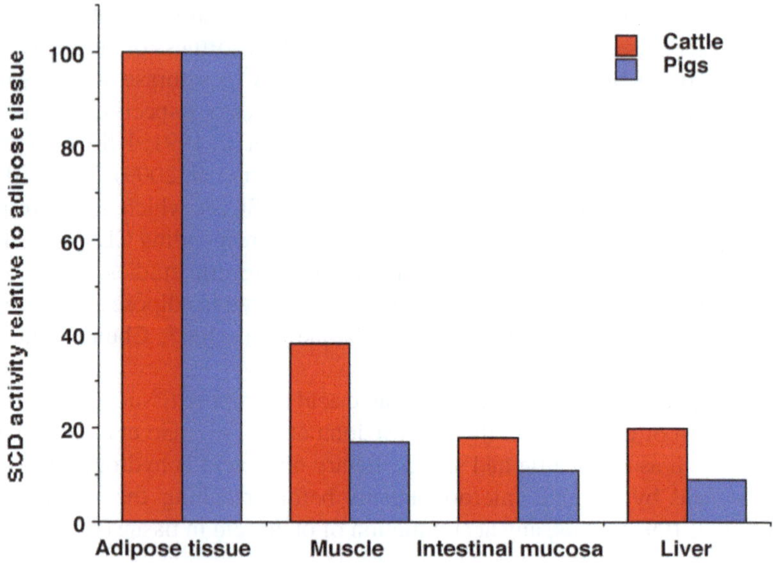

Fig. 12.1 Relative SCD catalytic activities in adipose tissue, longissimus muscle, duodenal intestinal mucosa, and liver of cattle and pigs. Data are expressed relative to adipose tissue activity, and are averages of values reported previously (St John et al. 1991; Chang et al. 1992; Cameron et al. 1994; Klingenberg et al. 1995; Page et al. 1997; Archibeque et al. 2005)

chicken adipose tissue. Consistent with the low hepatic SCD activity in pigs and cattle, the livers of these species are especially high in stearic acid and low in oleic acid (St John et al. 1991; Brooks et al. 2011; Go et al. 2012). Whereas liver lipids of mice contain approximately 12 % stearic acid (Kang et al. 2004), liver lipids of cattle and pigs contain over 25 % stearic acid (St John et al. 1991; Brooks et al. 2011; Go et al. 2012).

The low rates of *SCD* gene expression and SCD activity in bovine, ovine, and porcine liver are consistent with very low rates of de novo fatty acid biosynthesis in the livers of livestock species, relative to adipose tissue depots (Hanson and Ballard 1967; Ingle et al. 1972; Mersmann et al. 1976; Go et al. 2012). Similarly, subcutaneous adipose tissue has approximately twice the SCD activity of marbling adipose tissue (Archibeque et al. 2005), which is consistent with the relative concentrations of MUFA in subcutaneous and marbling adipose tissues (Sturdivant et al. 1992; May et al. 1993; Brooks et al. 2011).

Within an animal, there is a good relationship between *SCD* gene expression and/or SCD activity and tissue fatty acid composition. However, these comparisons do not appear to hold when comparisons are made across breed types. Thus, although Wagyu cattle consistently exhibit higher MUFA concentrations in their muscle and adipose tissues than other breed types (Sturdivant et al. 1992; May et al. 1993; Zembayashi et al. 1995; Siebert et al. 2003), it has not been possible to establish differences in *SCD* gene expression or SCD activity between Wagyu and Angus steers (Cameron et al. 1994; Chung et al. 2007).

SCD Genetic Variation Across Livestock Species

SCD *Chromosomal Locations in Livestock Species*

Although several *SCD* isoforms have been reported for mice, the *SCD*-1 isoform is the most abundant in lipogenic tissues and is common to most species. The sole porcine *SCD* gene, which has been mapped to chromosome 14 (Uemoto et al. 2012), has >80 % homology with other mammalian *SCD* genes (Ren et al. 2004). Similarly, the ovine *SCD* gene, mapped to chromosome 22, is approximately 90 % homologous to bovine and porcine *SCD* genes (Lengli and Corl 2007). The bovine *SCD* gene was mapped to chromosome 26 (BTA26) (Chung et al. 2000; Campbell et al. 2001). Linkage analysis placed the *SCD* gene 3.6 cM distal to microsatellite HEL11 and 2.5 cM proximal to microsatellite BM1314 on chromosome 26 in a comparison of Angus and Brahman populations (Campbell et al. 2001). As in humans, the only other *SCD* isoform identified in livestock species is *SCD-5*, which has 90 % homology with human *SCD-5* (Lengli and Corl 2007; 2008). *SCD-5* gene expression is primarily limited to the brain, although low levels of *SCD-5* mRNA have been detected in liver and muscle.

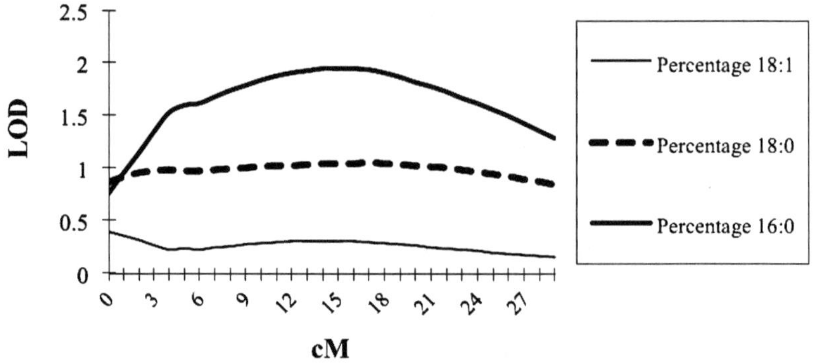

Fig. 12.2 Interval maps for fatty acid composition on BTA26. Palmitic acid approached an LOD score of 2.0, whereas stearic acid and oleic LOD scores were 1.0 and 0.3, indicating no significant quantitative trait loci for these fatty acids on BTA26. *SCD* maps to between 2.5 and 3.6 cm. S. B. Smith and E. M. Campbell, unpublished data

SCD *and Fatty Acid Quantitative Trait Loci*

Interval maps were generated for percentage palmitic acid (16:0), stearic acid, and oleic acid on BTA26 (Fig. 12.2). A quantitative trait locus (QTL) for palmitic acid on BTA26 approached an LOD score of 2.0, whereas LOD scores for stearic acid and oleic QTL were lower. An LOD score of 2.4 was equivalent to an overall *P*-value of 0.05 (Lander and Botstein 1989), so none of the QTL for fatty acids were significant in the region where we mapped the bovine *SCD* gene.

Earlier, we reported significant QTL for stearic acid and oleic acid 18 cM from the centromere on bovine chromosome BTA19 (Taylor et al. 1998). Individuals homozygous for Angus alleles possessed 2.4 % less stearic acid and 3.7 % more oleic acid in their subcutaneous adipose tissue than individuals homozygous for Brahman alleles. However, subcutaneous adipose tissue stearic acid and oleic acid composition did not differ between ¾ Angus:¼ Brahman progeny and ¼ Angus:¾ Brahman progeny from the same cattle population (Smith et al. 2009a, b). Thus, only small differences exist in fatty acid composition between *Bos indicus* and *Bos taurus* breed types.

SCD *Gene Polymorphisms in Asian Cattle*

Large variation in fatty acid composition has been reported between Japanese Black, Korean Hanwoo cattle and the US domestic cattle (Sturdivant et al. 1992; May et al. 1993; Smith et al. 2001, 2009a, b; Chung et al. 2006b). Japanese Black (Wagyu) cattle and Korean Hanwoo cattle share a common ancestry, and they both exhibit high MUFA:SFA ratios in their muscle and adipose tissues (Smith et al. 2009a, b). Although *SCD* gene expression and SCD catalytic activity do not differ between

Wagyu and Angus steers (Cameron et al. 1994; Chung et al. 2007), total MUFA proportions in Wagyu adipose tissue typically are 10 percentage points higher than in the US domestic cattle (reviewed in Smith et al. 2009a, b). Early research with Wagyu cattle reported the existence of *SCD* restriction site polymorphisms (Wilson et al. 1993), and we subsequently reported that significant microsatellite variation exists that is associated with adipose tissue fatty acid composition across geographically isolated populations of Japanese Black cattle (Smith et al. 2001).

Taniguchi et al. (2004) demonstrated single-nucleotide polymorphisms (SNPs) in the open-reading frame of an *SCD* cDNA generated from Japanese Black cattle, in which valine was replaced with alanine. The VA and AA genotypes contributed to higher MUFA and lower melting points in intramuscular adipose tissue than the more infrequent VV phenotype. Matsuhashi et al. (2011) and Yokota et al. (2012) subsequently demonstrated that *SCD* polymorphisms contributed to variations in fatty acid composition and varied across populations of Japanese Black cattle, confirming our earlier observations (Smith et al. 2001). Mannen (2011) confirmed that genetic polymorphisms of the *SCD* and fatty acid binding-protein genes are in part responsible for differences in fatty acid composition within the Japanese Black breed type. Similarly, *SCD* polymorphisms affect fatty acid composition in Japanese Holstein (dairy) cattle (Narukami et al. 2011) and Korean Hanwoo steers (Lee et al. 2008; Oh et al. 2011; Maharani et al. 2012).

SCD *Gene Polymorphisms in Pigs*

Similar associations of *SCD* polymorphisms with fatty acid composition have been reported for pigs (Bartz et al. 2013; Uemoto et al. 2012; Wang et al. 2013). Two groups (Bartz et al. 2013; Uemoto et al. 2012) have identified SNPs in the *SCD* promoter region in pigs that were associated with fatty acid composition, and Uemoto et al. (2012) identified QTL for 14:0, 18:0, and 18:1*n*-9 located near the porcine *SCD* gene.

Adipose Tissue Differentiation and *SCD* Gene Expression and Activity

Bovine Preadipocyte Culture Studies

Casimir and Ntambi (1996) first demonstrated that *SCD* gene expression increased immediately preceding lipid filling in murine 3T3-L1 preadipocytes, and we have demonstrated essentially identical results for bovine preadipocytes (Chung et al. 2006a; Fig. 12.3). Stromal-vascular cells were obtained by collagenase treatment of perirenal adipose tissue from mature Angus steers, were plated at a density of 10^4

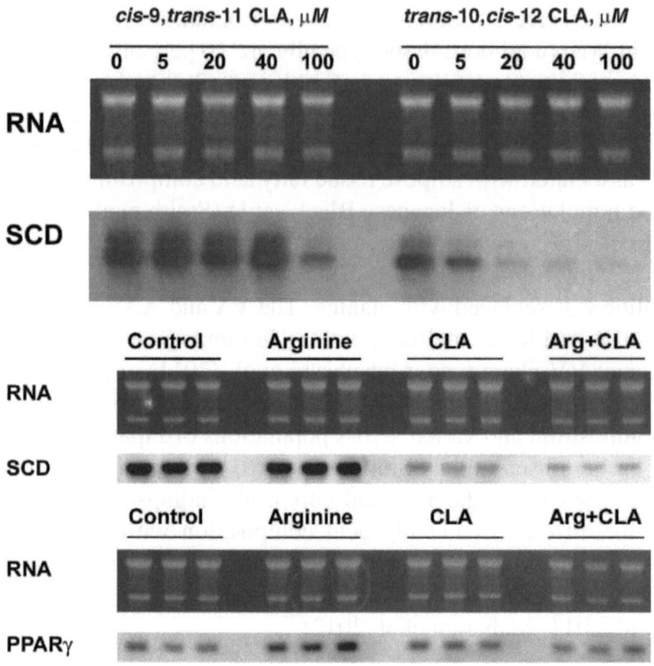

Fig. 12.3 Stearoyl-CoA desaturase gene expression in Angus preadipocytes differentiated in the presence of 5 μM pioglitazone, 10 μg/mL insulin, and DMEM. *Top panel*, preadipocytes were differentiated for 7 days with PI in the presence of *cis*-9, *trans*-11or *trans*-10, *cis*-12 CLA. *Bottom panel*, preadipocytes were differentiated in differentiation medium (Control) or with 5 mM arginine in the absence or presence of 40 μM *trans*-10, *cis*-12 CLA. From Chung et al. (2006a)

cells, and were grown to confluence. At confluence, the medium was supplemented either with insulin plus pioglitazone (a PPARγ agonist) or insulin alone. Prior to confluence, *SCD* mRNA was undetectable, but after 7 days of exposure to insulin plus pioglitazone, *SCD* mRNA and lipid-filled adipocytes were highly abundant.

Ntambi and coworkers demonstrated that the *trans*-10, *cis*-12 isomer of CLA strongly depresses *SCD* gene expression in 3T3-L1 preadipocytes (Choi et al. 2000). In our bovine preadipocyte cell line, *trans*-10, *cis*-12 CLA nearly abolished SCD gene expression, whereas *cis*-9, *trans*-11 CLA was without effect except at the highest concentrations (Fig. 12.3). Although *trans*-10, *cis*-12 CLA also strongly depressed lipid filling, it had no effect on *PPARγ* gene expression in bovine preadipocytes. Conversely, media arginine increased *PPARγ* gene expression, but not *SCD* gene expression, indicating that *PPARγ* and *SCD* gene expression are regulated independently (Chung et al. 2006a).

The depression of *SCD* gene expression by *trans*-10, *cis*-12 CLA is unusual in light of the fact that this CLA isomer is a product of rumen fermentation, and its accumulation would effectively block the conversion of *trans*-vaccenic acid (18:1*trans*-11, a primary product of ruminal fermentation) to *cis*-9, *trans*-11 CLA. Clearly, any production practice that profoundly increases the formation and

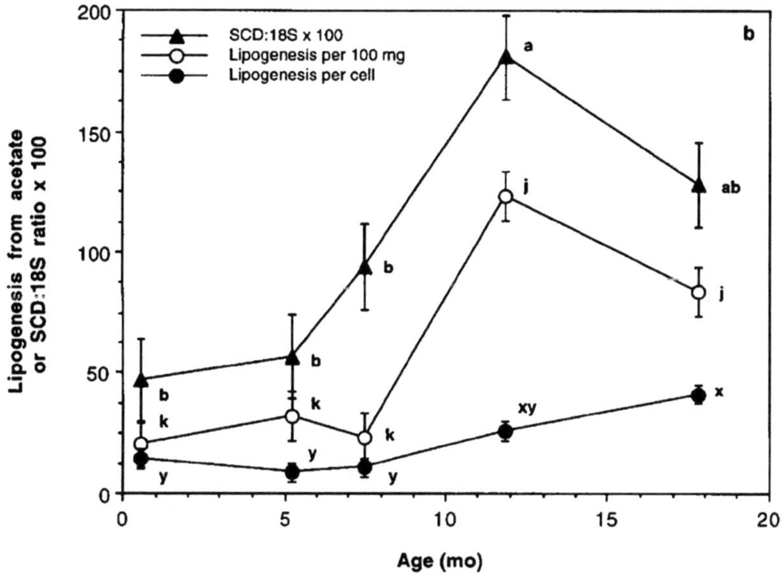

Fig. 12.4 Changes in *SCD* gene expression and lipogenesis during growth in preweaning (2.5 weeks to 7.5 months of age) and postweaning Angus steers. Overall standard errors of the means are affixed to the symbols. [ab,jk,xy]Values within a measurement with the same superscript are not different ($P > 0.05$). Reprinted with permission from Martin et al. (1999)

absorption of *trans*-10, *cis*-12 CLA from the digestive system will depress adipogenesis, and will cause adipose tissue lipids to become more saturated and contain less *cis*-9, *trans*-11 CLA.

Animal Studies

Previous research (Huerta-Leidenz et al. 1996; Chung et al. 2006b) demonstrated that adipose tissue MUFA increase over time. We first demonstrated that *SCD* gene expression attained a plateau at 12 months of age in subcutaneous adipose tissue of steers fed a corn-based finishing diet (Martin et al. 1999) (Fig. 12.4). The rate of de novo fatty acid biosynthesis increased gradually after weaning in adipose tissue of steers, but the rate lagged behind the elevation in *SCD* gene expression (Martin et al. 1999). As in bovine adipose tissue, *SCD* gene expression increased over time in contemporary and obese pigs (Smith et al. 1999) (Fig. 12.5). Adipose tissue *SCD* gene expression was less in contemporary pigs than in obese pigs, and it is less in pigs fed a low-fat, grain-based diet than in pigs fed a high-fat, milk-based diet. Changes in *SCD* gene expression over time were consistent with increases in subcutaneous adipocyte volume, such that there is a highly significant correlation between adipocyte volume and *SCD* gene expression ($R^2 = 0.79$, $P < 0.001$) (Fig. 12.5).

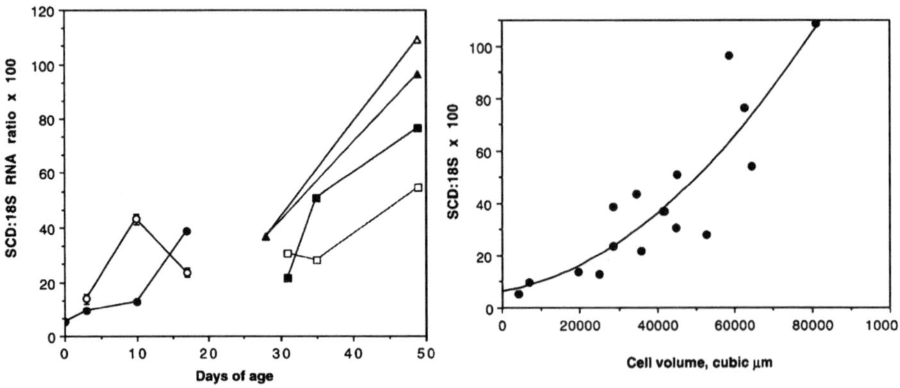

Fig. 12.5 Changes in *SCD* gene expression (*left panel*) and the relationship between *SCD* gene expression and adipocyte volume (*right panel*) in obese (*open symbols*) and contemporary crossbred pigs (*closed symbols*). *Circles*, suckling pigs; *triangles*, pigs fed a high-fat, milk-based diet; *squares*, pigs fed a low-fat, grain-based diet. There was a highly significant correlation between adipocyte volume and *SCD* gene expression ($R^2=0.79$, $P<0.001$). Reprinted with permission from Smith et al. (1999)

The results of these early studies indicated that *SCD* gene expression was coordinated with adipogenesis in cattle and pigs.

A subsequent study (Chung et al. 2007) demonstrated that SCD activity in bovine subcutaneous adipose tissue increased between 16 and 24 months of age in corn-fed steers but not in steers that had been fed a combination of hay and corn (Fig. 12.6). *SCD* gene expression did not change between 16 and 24 months of age in corn-fed steers, but de novo fatty acid biosynthesis from acetate declined sharply (Fig. 12.6). Thus, in this group of animals, fatty acid biosynthesis declined sharply while SCD activity increased. Although feeding the hay/corn depressed overall growth rate as compared to corn-fed steers (Lunt et al. 2005), de novo fatty acid synthesis actually increased between 20 and 28 months of age in the hay/corn-fed steers (Fig. 12.6). Conversely, SCD activity did not increase and *SCD* mRNA decreased. These cattle were raised to larger body weights (625 kg) and much greater adiposity than cattle used in the earlier study (490 kg; Martin et al. 1999), and the data indicates that the regulation of *SCD* gene expression becomes dissociated from de novo fatty acid biosynthesis in overly fattened cattle.

Lee et al. (2008) observed peak *SCD* mRNA at 12 months of age in muscle from Hanwoo steers, and Wang et al. (2013) demonstrated peak *SCD* gene expression at 20 and 25 months of age, respectively, in Piedmontese × Hereford and Wagyu × Hereford crossbred cattle. Smith et al. (2012) demonstrated that *SCD* gene expression attained a peak at 14 months of age and then declined by 16 months of age in brisket (cranial chest) adipose tissue but increased between 9 and 16 months of age in adipose tissue overlying the chuck (shoulder) (Fig. 12.7). *C/EBPβ*, *CPTI*, and *PPARγ* peak mRNA levels were observed at 14 months of age in both adipose tissue depots, and expression of these genes declined profoundly by 16 months of age. Collectively, these data and others (Lee et al. 2008; Brooks et al. 2011;

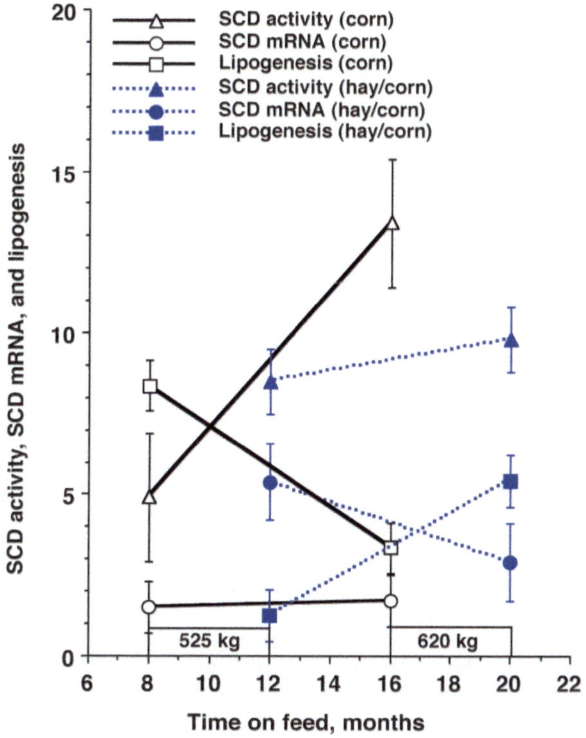

Fig. 12.6 Stearoyl-CoA desaturase catalytic activity (*triangles*), relative mRNA levels (*circles*), and lipogenesis from acetate (*squares*) in subcutaneous adipose tissue of Angus steers fed a corn-based diet (*open symbols*) or hay/corn diet (*filled symbols*). The corn-based diet was designed to yield an average daily gain of 1.36 kg/day and the hay-based diet was supplemented the same corn-based diet to yield 0.9 kg/day. By design, steers fed the corn-based diet for 8 months or the hay/corn-based diet for 12 months (16 and 20 months of age, respectively) weighed 525 kg at sampling. Steers fed the corn-based diet for 16 months or the hay/corn-based diet for 20 months (24 and 28 months of age, respectively) weighed 620 kg at sampling. SCD activity is expressed as nmol/(min·mg protein); *SCD* mRNA is the *SCD*:28S ratio; lipogenesis in nmol/(2 h·100 mg adipose tissue). Pooled standard errors are affixed to the means ($n=4$ for each data point). Adapted from Lunt et al. (2005) and Chung et al. (2007)

Wang et al. 2013) indicate that *SCD* gene expression declines in older, fatter cattle during that period in time in which adipose tissue becomes senescent.

Changes over time in the palmitoleic acid:stearic ratio, an index we have proposed to estimate SCD activity (Chung et al. 2006b; Smith et al. 2009a; Turk and Smith 2009; Smith et al. 2012), were essentially opposite to the pattern observed for *SCD* gene expression in adipose tissues overlying the brisket and chuck (Fig. 12.7). There is a highly significant, negative correlation between adipose tissue palmitoleic and stearic acid (Fig. 12.8), which indicates that the concentrations of palmitoleic acid and stearic acid in adipose tissue are coordinately controlled by SCD activity. Additionally, adipose tissue exhibiting high palmitoleic:stearic acid ratios

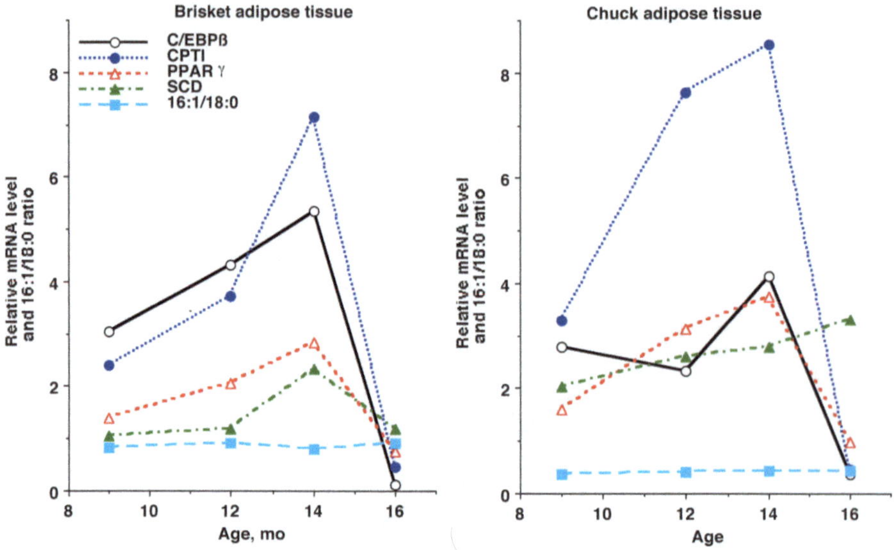

Fig. 12.7 Changes in relative gene expression and the palmitoleic:stearic acid ratio in brisket (*left panel*) and chuck (*right panel*) adipose tissue depots of Angus steers sampled at 9, 12, 14, and 16 months of age. For data pooled over time, *SCD* gene expression was twice as high ($P=0.01$) in chuck as in brisket, whereas the palmitoleic:stearic acid ratio was twice as high ($P=0.01$) in brisket as in chuck. There were no other adipose tissue main effects for relative mRNA amounts. Adapted from Smith et al. (2012)

also are enriched with total MUFA. However, *SCD* gene expression in brisket adipose tissue was only 50 % of the level observed in adipose tissue from the chuck, yet the palmitoleic:stearic acid ratio was twice as high in the brisket as in the chuck. Thus, in adipose tissue overlying the brisket, there is an apparent disconnect between this fatty acid index of SCD activity and *SCD* gene expression. We previously demonstrated that tailhead subcutaneous adipose tissue develops initially as brown adipose tissue (Landis et al. 2002), and brown adipose tissue is unusually concentrated in MUFA (Chen et al. 2007). Because it overlies vital organs, brisket subcutaneous adipose tissue may develop very early as brown adipose tissue, during which time it would have actively converted SFA to MUFA. Thus, the palmitoleic:stearic ratio, and probably other fatty acid indexes of SCD activity (Archibeque et al. 2005), provide at best an index of past *SCD* gene expression and SCD activity.

Regulation of SCD Activity by Fatty Acids

Polyunsaturated fatty acids (PUFAs) decrease *SCD* mRNA in mice and rats (Ntambi 1992; Landschulz et al. 1994). Similarly, feeding fish oil (high in *n*-3 PUFA) depresses *SCD* gene expression in beef cattle (Waters et al. 2009). In cattle and

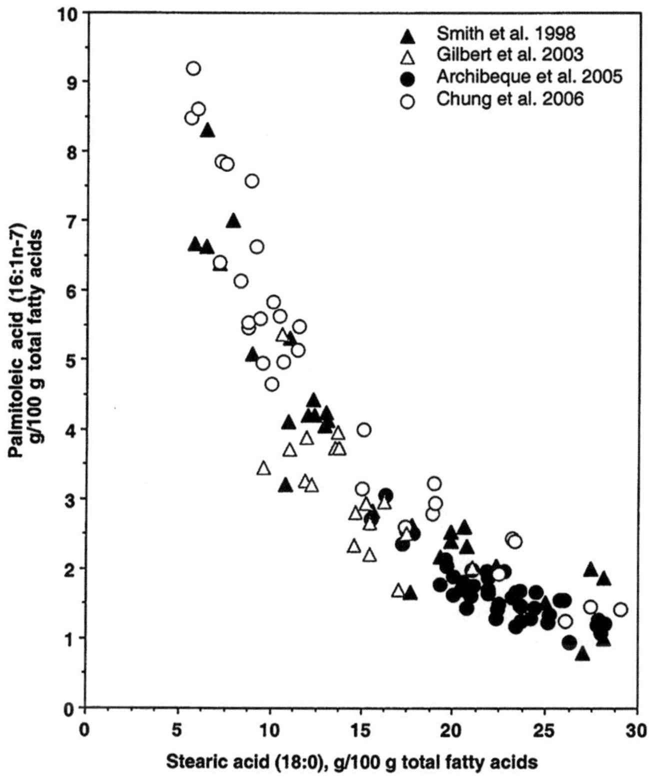

Fig. 12.8 Relationship between stearic acid (18:0) and palmitoleic acid (16:1*n*-7) in subcutaneous adipose tissue lipids of cattle raised in Australia and Japan (Smith et al. 1998) or in the USA (Gilbert et al. 2003; Archibeque et al. 2005; Chung et al. 2006b). Cattle raised in Australia were crossbred Murray Grey, Angus, and Grey Brahman steers fed a grain-based diet for a minimum of 400 days, whereas those raised in Japan were Murray Grey and Japanese Black fed for approximately 570 days. Cattle raised in the USA were Brangus (Gilbert et al. 2003), Angus (Archibeque et al. 2005; Chung et al. 2006b), or American Wagyu (Chung et al. 2006b)

sheep, changes in the ruminal environment in response to grain feeding also may contribute an increase in plasma and adipose tissue PUFA, thereby leading to a decrease in *SCD* gene expression in cattle. Kucuk et al. (2001) reported that increasing dietary forage increased duodenal flow of stearic acid and α-linolenic acid, but decreased the duodenal flow of oleic and linoleic acid. Conversely, feeding a high-corn diet could lower rumen pH sufficiently to decrease *trans*-vaccenic acid and palmitic acid by decreasing protozoal populations (Devillard et al. 2006; Or-Rashid et al. 2007). A prolonged reduction in rumen pH also could cause reductions in the population of bacteria responsible for rumen biohydrogenation (Fukuda et al. 2006; van de Vossenberg and Joblin 2003; Wallace et al. 2006), causing greater flow of linoleic acid (and less stearic acid) to the small intestine.

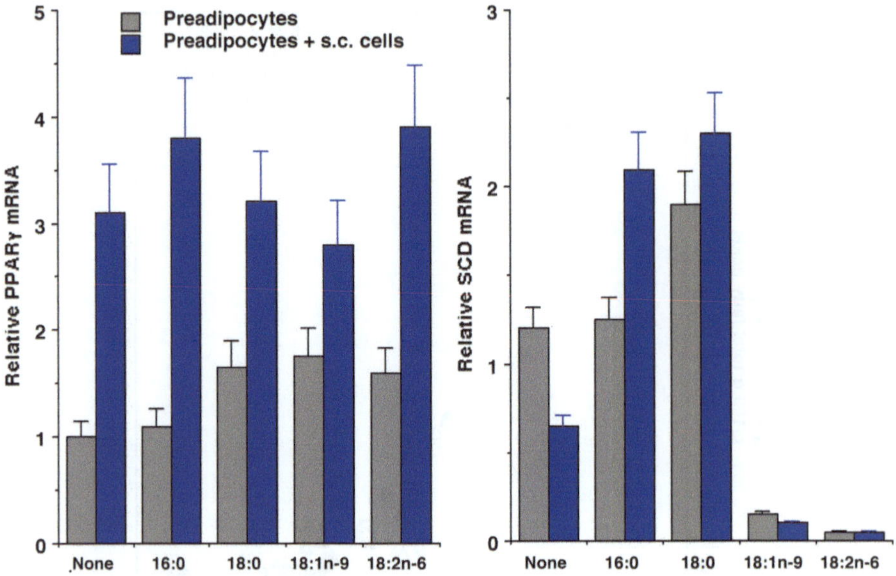

Fig. 12.9 Relative PPARγ (*left*) and SCD gene expression (*right*) in bovine subcutaneous preadipocytes cultured singly or co-cultured with satellite cells (myogenic precursor cells). Single or co-cultured cells were incubated for 7 days medium containing insulin plus ciglitazone or this medium containing 40 μM of the fatty acids indicated. Co-culture with subcutaneous (s.c.) cells increased PPARγ mRNA levels in subcutaneous preadipocytes. In co-cultured preadipocytes, palmitic acid (16:0) and stearic acid (18:0) significantly increased *SCD* mRNA levels, whereas oleic acid (18:1*n*-9) and linoleic acid (18:2*n*-6) strongly depressed *SCD* gene expression in single- and co-cultured preadipocytes (Smith, Johnson, and Choi, unpublished data)

Chung et al. (2006b) demonstrated that the duodenal concentrations of stearic acid and *trans*-vaccenic acid decreased, and linoleic acid and *cis*-9, *trans*-11 CLA increased, with increasing time on a corn-based, finishing diet. Thus, in cattle fed a corn-based, finishing diet, there is an apparent depression in the isomerization of linoleic acid to *cis*-9, *trans*-11 CLA as well as a decrease in the hydrogenation of *cis*-9, *trans*-11 CLA to *trans*-vaccenic acid and hence to stearic acid. These changes in duodenal fatty acid concentrations are reflected in tissue fatty acid compositions (Chung et al. 2006b).

It is possible that linoleic acid contributes to the depression in *SCD* gene expression observed in mature feedlot cattle (Brooks et al. 2011; Smith et al. 2012; Wang et al. 2013). Duodenal linoleic acid essentially doubles, and plasma linoleic acid increases by 50 % between 8 and 16 months of age in cattle fed grain-based diets (Brooks et al. 2011), leading to a significant increase in adipose tissue linoleic acid. Similar results were reported by Chung et al. (2006b). Linoleic acid nearly abolishes *SCD* gene expression in bovine subcutaneous preadipocytes (Fig. 12.9); therefore, linoleic acid could be the primary causative agent in depressing *SCD* gene expression in adipose tissue of mature feedlot cattle.

Regulation of *SCD* Gene Expression by Oleic Acid

It also is possible that circulating oleic acid and/or stearic acid may be responsible for the depression in *SCD* gene expression observed in mature cattle. Oleic acid was nearly as effective as linoleic acid (both 40 μM) in reducing *SCD* gene expression in bovine subcutaneous preadipocytes, although neither oleic nor linoleic acid affected *PPARγ* gene expression at this concentration (Fig. 12.9). In intramuscular preadipocytes, oleic acid as low as 1 μM increased *PPARγ*, whereas only 500 μM oleic acid increased *PPARγ* gene expression subcutaneous preadipocytes (Fig. 12.10). Oleic acid strongly depressed *SCD* gene expression in intramuscular and subcutaneous preadipocytes, as well as bovine muscle satellite cells (myoblasts) at concentrations above 10 μM (Fig. 12.10).

Ovine and porcine *SCD* genes contain completely conserved PUFA response elements (PPRE) (Zulkifli et al. 2010), and it is likely that an identical PPRE exists in the bovine *SCD* promoter region. Also, the PPRE appears to be nearly as sensitive to oleic acid as it is to linoleic acid. This indicates that increasing the adipose tissue concentration of oleic acid, either by endogenous synthesis or uptake from the circulation, would down-regulate *SCD* gene expression in livestock species.

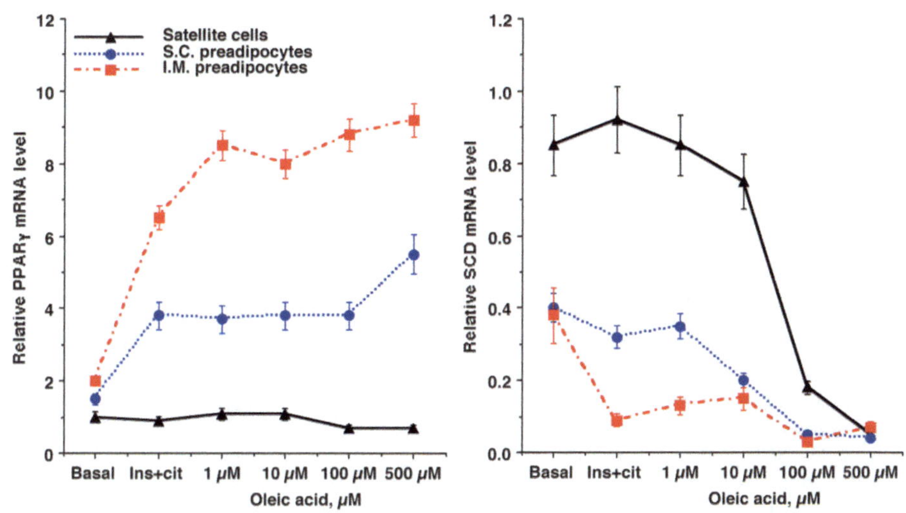

Fig. 12.10 Relative PPARγ (*left*) and SCD (*right*) gene expression in bovine muscle satellite cells (myogenic precursor cells), subcutaneous (SC) preadipocytes, and intramuscular (IM) preadipocytes. Cells were incubated for 7 days in basal medium, medium containing insulin plus ciglitazone (Ins+cit) or medium containing insulin plus ciglitazone and increasing concentrations of oleic. The scale for oleic acid is not linear (Smith, Johnson, and Chung, unpublished data)

Gene Expression in Co-cultured Preadipocytes and Myoblasts

Whereas oleic and linoleic acid depressed *SCD* gene expression, palmitic and stearic acid stimulated *SCD* gene expression in bovine subcutaneous preadipocytes (Fig. 12.9). However, this effect was apparent only in preadipocytes, which were co-cultured with skeletal muscle satellite cells. We demonstrated recently that co-culture of bovine subcutaneous preadipocytes and muscle satellite cells promoted increased *GPR43* gene expression in the preadipocytes and *C/EBPβ* and *PPARγ* gene in bovine myoblasts (Fig. 12.11) (Choi et al. 2013). The mechanism by which myoblasts increase *GPR43* gene expression in co-cultured preadipocytes is not known. We have proposed that oleic acid released from differentiating preadipocytes may promote *C/EBPβ* and *PPARγ* gene expression in the neighboring myoblasts, and this also may occur during growth in livestock species. This remains to be confirmed.

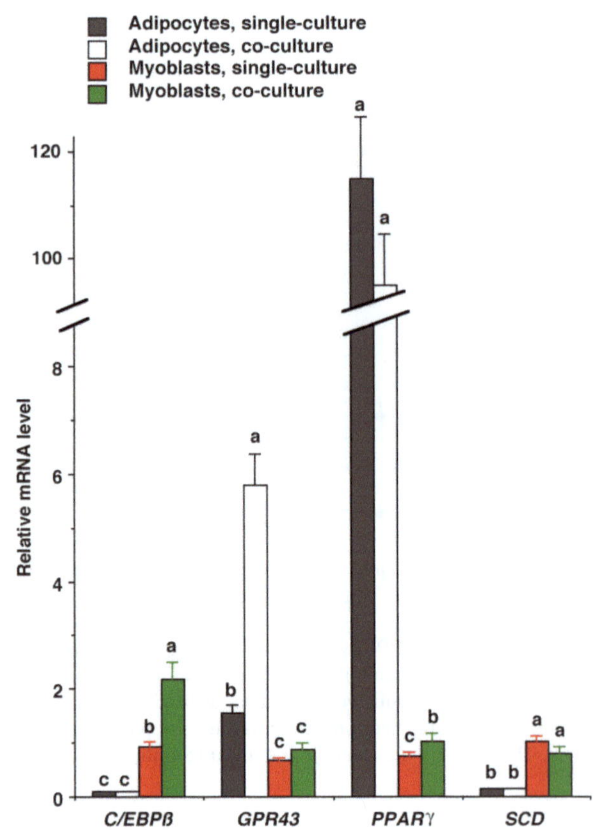

Fig. 12.11 Cell type × culture method interactions for relative mRNA levels in subcutaneous preadipocytes and myoblasts (muscle satellite cells) cultured singly or co-cultured. Co-culture of adipocytes with myoblasts increased *C/EBPβ* and *PPARγ* gene expression in myoblasts. Co-culture of adipocytes with myoblasts increased *GPR43* gene expression in adipocytes. *SCD* gene expression in myoblasts tended to be decreased by co-culture with adipoctyes ($P = 0.15$). Data are means ± SEM ($n = 12$). [abc]Means within a specific gene without common superscripts are different ($P < 0.05$). Adapted from Choi et al. (2013)

Summary

Hepatic *SCD* gene expression is strongly associated with elevated plasma triacylglycerols and obesity in laboratory species and may contribute to adipogenesis in livestock species. Because oleic acid has profound influences on the quality and healthfulness of meat products, strategies have been developed that promote rather than depress *SCD* gene expression and catalytic activity. Endogenously produced oleic acid may actually promote adipogenic gene expression in neighboring preadipocytes and myoblasts.

References

Adams TH, Walzem RL, Smith DR, Tseng S, Smith SB (2010) Hamburger high in total, saturated and *trans*-fatty acids decreases HDL cholesterol and LDL particle diameter, and increases plasma TAG, in mildly hypercholesterolaemic men. Br J Nutr 103:91–98

Archibeque SL, Lunt DK, Tume RK, Smith SB (2005) Fatty acid indices of stearoyl Co-A desaturase activity do not reflect actual stearoyl Co-A desaturase enzyme activity in adipose tissues of beef steers finished with corn-, flaxseed-, or sorghum-based diets. J Anim Sci 83:1153–1166

Bartz M, Szydlowski M, Kociucka B, Salamon S, Jelen HH, Switonski M (2013) Transcript level of the pig SCD gene has no effect on fatty acid composition in muscle and fat tissues, but its polymorphism with the putative miRNA target site is associated with daily gain and feed conversion ratio. J Anim Sci 91:10–19

Brooks MA, Choi CW, Lunt DK, Kawachi H, Smith SB (2011) Subcutaneous and intramuscular adipose tissue stearoyl-coenzyme A desaturase gene expression and fatty acid composition in calf- and yearling-fed Angus steers. J Anim Sci 89:2556–2570

Cameron PJ, Rogers M, Oman J, May SG, Lunt DK, Smith SB (1994) Stearoyl-CoA desaturase enzyme activity and mRNA levels are not different in subcutaneous adipose tissue from Angus and American Wagyu steers. J Anim Sci 72:2624–2628

Campbell EMG, Gallagher DS, Davis SK, Taylor JF, Smith SB (2001) Rapid communication: Mapping of the bovine stearoyl-coenzyme A desaturase (*SCD*) gene to BTA26. J Anim Sci 79:1954–1955

Casimir D, Ntambi JM (1996) cAMP activates the expression of stearoyl-CoA desaturase gene 1 during early preadipocyte differentiation. J Biol Chem 271:29847–29853

Chang JHP, Lunt DK, Smith SB (1992) Fatty acid composition and fatty acid elongase and stearoyl-CoA desaturase activities in tissues of steers fed high oleate sunflower seed. J Nutr 122:2074–2080

Chen C, Carstens GE, Gilbert CD, Theis CM, Archibeque SL, Kurz MW, Slay LJ, Smith SB (2007) Dietary supplementation high levels of saturated/monounsaturated fatty acids to ewes during late gestation reduces thermogenesis in newborn lambs by depressing fatty acid oxidation in perirenal brown adipose tissue. J Nutr 137:43–48

Choi SH, Chung KY, Johnson BJ, Go GW, Kim KH, Choi CW, Smith SB (2013) Co-culture of bovine muscle satellite cells with preadipocytes increases PPARγ and C/EBPβ gene expression in differentiated myoblasts and increases GPR43 gene expression in adipocytes. J Nutr Biochem 24:539–543

Choi Y, Kim YC, Han YB, Park Y, Pariza MW, Ntambi JM (2000) The trans-10, cis-12 isomer of conjugated linoleic acid downregulates stearoyl-COA desaturase1 gene expression in 3T3-L1 preadipocytes. J Nutr 130:1920–1924

Chung M, Ha S, Jeong S, Bok J, Cho K, Baik M, Choi Y (2000) Cloning and characterization of bovine stearoyl CoA desaturaseI cDNA from adipose tissues. Biosci Biotechnol Biochem 64:1526–1530

Chung KY, Choi CB, Kawachi H, Yano H, Smith SB (2006a) *Trans*-10, *cis*-12 conjugated linoleic acid antagonizes arginine-promoted differentiation of bovine preadipocytes. Adipocyte 2:93–100

Chung KY, Lunt DK, Choi CB, Chae SH, Rhoades RD, Adams TL, Booren B, Smith SB (2006b) Lipid characteristics of subcutaneous adipose tissue and *M. longissimus thoracis* of Angus and Wagyu steers fed to US and Japanese endpoints. Meat Sci 73:432–441

Chung KY, Lunt DK, Kawachi H, Yano H, Smith SB (2007) Lipogenesis and stearoyl-CoA desaturase gene expression and enzyme activity in adipose tissue of short- and long-fed Angus and Wagyu steers fed corn- or hay-based diets. J Anim Sci 85:380–387

Devillard E, McIntosh FM, Newbold CJ, Wallace RJ (2006) Rumen ciliate protozoa contain high concentrations of conjugated linoleic acids and vaccenic acid, yet do not hydrogenate linoleic acid or desaturate stearic acid. Br J Nutr 96:697–704

Duckett SK, Pratt SL, Pavan E (2009) Corn oil or corn grain supplementation to steers grazing endophyte-free tall fescue. II. Effects on subcutaneous fatty acid content and lipogenic gene expression. J Anim Sci 87:1120–1128

Ekeren PA, Smith DR, Lunt DK, Smith DR (1992) Ruminal biohydrogenation of fatty acids from high-oleate sunflower seeds. J Anim Sci 70:2574–2580

Enser M (1979) The role of insulin in the regulation of stearic acid desaturase activity in liver and adipose tissue from obese-hyperglycaemic (*ob*/*ob*) and lean mice. Biochem J 180:551–558

Fukuda S, Suzuki Y, Murai M, Asanuma N, Hino T (2006) Augmentation of vaccenate production and suppression of vaccenate biohydrogenation in cultures of mixed ruminal microbes. J Dairy Sci 89:1043–1051

Gilbert CD, Lunt DK, Miller RK, Smith SB (2003) Carcass, sensory, and adipose tissue traits of Brangus steers fed casein-formaldehyde-protected starch and/or canola lipid. J Anim Sci 81:2457–2468

Gilmore LA, Walzem RL, Crouse SF, Smith DR, Adams TH, Vaidyananthan V, Cao X, Smith SB (2011) Consumption of high-oleic acid ground beef increases HDL cholesterol concentration but both high- and low-oleic acid ground beef decrease HDL particle diameter in normocholesterolemic men. J Nutr 141:1188–1194

Go G, Wu G, Silvey DT, Choi SH, Li X, Smith SB (2012) Lipid metabolism in tissues of growing-finishing pigs fed supplemental conjugated linoleic acid and/or dietary arginine. Amino Acids 43:1713–1726

Hanson RW, Ballard FJ (1967) The relative significance of acetate and glucose as precursors for lipid synthesis in liver and adipose tissue from ruminants. Biochem J 105:529–536

Huerta-Leidenz NO, Cross HR, Savell JW, Belk KE, Lunt DK, Baker JF, Smith SB (1996) Fatty acid composition of adipose tissues of male calves at different stages of growth. J Anim Sci 74:1256–1264

Ingle DL, Bauman DE, Garrigus US (1972) Lipogenesis in the ruminant: in vivo site of fatty acid synthesis in sheep. J Nutr 102:617–624

Jiang Z, Michal JJ, Tobey DJ, Daniels TF, Rule DC, MacNeil MD (2008) Significant associations of stearoyl-CoA desaturase (scd1) gene with fat deposition and composition in skeletal muscle. Int J Biol Sci 4:345–351

Jones BH, Maher MA, Banz WJ, Zemel MB, Whelan JH, Smith PJ, Moustaid N (1996) Adipose tissue stearoyl-CoA desaturase mRNA is increased by obesity and decreased by polyunsaturated fatty acids. Am J Physiol 271:E44–E49

Kang K, Miyazaki M, Ntambi JM, Pariza MW (2004) Evidence that the anti-obesity effect of conjugated linoleic acid is independent of effects on stearoyl-CoA desaturase1 expression and enzyme activity. Biochem Biophys Res Commun 315:532–537

Klingenberg IL, Knabe DA, Smith SB (1995) Lipid metabolism in swine fed diets containing either high oleate sunflower oil or beef tallow. Comp Biochem Physiol 110:183–192

Kris-Etherton PM, Pearson TA, Wan Y, Hargrove RL, Moriarity K, Fishell V, Etherton TD (1999) High-monounsaturated fatty acid diets lower both plasma cholesterol and triacylglycerol concentrations. Am J Clin Nutr 70:1009–1015

Kucuk O, Hess BW, Luden PA, Rule DC (2001) Effect of forage:concentrate ratio on ruminal digestion and duodenal flow of fatty acids in ewes. J Anim Sci 79:2233–2240

Lander ES, Botstein D (1989) Mapping Mendelian factors underlying quantitative traits using RFLP linkage maps. Genetics 121:185–199

Landis MD, Carstens GE, McPhail EG, Randel RD, Green KK, Slay L, Smith SB (2002) Ontogenic development of brown adipose tissue in Angus and Brahman fetal calves. J Anim Sci 80:591–601

Landschulz KT, Jump DB, McDougald OA, Lane MD (1994) Transcriptional control of the stearoyl-CoA desaturase-1 gene by polyunsaturated fatty acids. Biochem Biophys Res Commun 200:763–768

Lee SH, Cho YM, Lee SH, Kim BS, Kim NK, Choy YH, Kim KH, Yoon D, Im SK, Oh SJ, Park EW (2008) Identification of marbling-related candidate genes in M. longissimus dorsi of high- and low marbled Hanwoo (Korean Native Cattle) steers. BMB Rep 41:846–851

Legrand P, Hermier D (1992) Hepatic Δ9 desaturation and plasma VLDL level in genetically lean and fat chickens. Int J Obes 16:289–294

Legrand P, Lemarchal P (1991) Stearyl-CoA desaturase activity and triglyceride secretion in isolated and cultured hepatocytes from genetically lean and fat chickens. Comp Biochem Physiol 102B:371–375

Lengli AJ, Corl BA (2007) Identification and characterization of a novel bovine stearoyl-CoA desaturase isoform with homology to human SCD5. Lipids 42:499–508

Lengli AJ, Corl BA (2008) Comparison of pig, sheep and chicken SCD5 homologs: evidence for an early gene duplication event. Comp Biochem Physiol B Biochem Mol Biol 150:440–446

Lunt DK, Chung KY, Choi CB, Smith SB (2005) Production characteristics and carcass quality of Angus and Wagyu steers fed to US and Japanese endpoints. J Anim Vet Adv 4:949–953

Maharani D, Jung Y, Jung WY, Jo C, Ryoo SH, Lee SH, Yeon SH, Lee JH (2012) Association of five candidate genes with fatty acid composition in Korean cattle. Mol Biol Rep 39:6113–6121

Mannen H (2011) Identification and utilization of genes associated with beef qualities. Anim Sci J 82:1–7

Martin GS, Lunt DK, Britain KG, Smith SB (1999) Postnatal development of stearoyl coenzyme A desaturase gene expression and adiposity in bovine subcutaneous adipose tissue. J Anim Sci 77:630–636

Matsuhashi T, Maruyama S, Uemoto Y, Kobayashi N, Mannen H, Abe T, Sakaguchi S, Kobayashi E (2011) Effects of bovine fatty acid synthase, stearoyl-coenzyme A desaturase, sterol regulatory element-binding protein 1, and growth hormone gene polymorphisms on fatty acid composition in Japanese Black cattle. J Anim Sci 89:12–22

May SG, Sturdivant CA, Lunt DK, Miller RK, Smith SB (1993) Comparison of sensory characteristics and fatty acid composition between Wagyu crossbred and Angus steers. Meat Sci 35:289–298

Mersmann HJ, Allen CD, Steffen DG, Brown LG, Danielson DM (1976) Effect of age, weaning and diet on swine adipose tissue and liver lipogenesis. J Anim Sci 43:140–150

Miyazaki M, Ntambi JM (2003) Role of stearoyl-coenzyme A desaturase in lipid metabolism. Prostaglandins Leukot Essent Fatty Acids 68:113–121

Narukami T, Sasazaki S, Oyama K, Nogi T, Taniguchi M, Mannen H (2011) Effect of DNA polymorphisms related to fatty acid composition in adipose tissue of Holstein cattle. Anim Sci J 82:406–411

Ntambi JM (1992) Dietary regulation of stearoyl-CoA desaturase 1 gene expression in mouse liver. J Biol Chem 267:10925–10930

Oh DY, Lee YS, Yeo JS (2011) Identification of the SNP (single nucleotide polymorphism) of the stearoyl-CoA desaturase (SCD) associated with unsaturated fatty acid in Hanwoo (Korean cattle). Asian-australas J Anim Sci 24:757–765

Or-Rashid MM, Odongo NE, McBride BW (2007) Fatty acid composition of ruminal bacteria and protozoa, with emphasis on conjugated linoleic acid, vaccenic acid, and odd-chain and branched-chain fatty acids. J Anim Sci 85:1228–1234

Page AM, Sturdivant CA, Lunt DK, Smith SB (1997) Dietary cottonseed depresses lipogenesis but has no effect on stearoyl-coenzyme A desaturase activity in bovine subcutaneous adipose tissue. Comp Biochem Physiol 118B:79–84

Payne E, Masters CJ (1971) The in vitro incorporation of ^{14}C-labelled fatty acids into ovine liver and omental adipose tissue. Int J Biochem 2:623–631

Ren J, Knorr C, Huang L, Brenig B (2004) Isolation and molecular characterization of the porcine desaturase gene. Gene 340:19–30

Siebert BD, Pitchford WS, Kurk ZA, Huchel H, Deland MP, Bettema CD (2003) Differences in delta9 desaturase activity between Jersey- and Limousin-sired cattle. Lipids 38:539–543

Smith SB, Yang A, Larsen TW, Tume RK (1998) Positional analysis of triacylglycerols from bovine adipose tissue lipids varying in degree of unsaturation. Lipids 33:197–207

Smith SB, Mersmann HJ, Smith EO, Britain KG (1999) Stearoyl coenzyme A desaturase gene expression during growth in adipose tissue from obese and crossbred pigs. J Anim Sci 77:1710–1716

Smith SB, Zembayashi M, Lunt DK, Sanders JO, Gilbert CD (2001) Carcass traits and microsatellite distributions in offspring of sires from three graphical regions of Japan. J Anim Sci 79:3041–3051

Smith SB, Gill CA, Lunt DK, Brooks MA (2009a) Regulation of fat and fatty acid composition in beef cattle. Asian-australas J Anim Sci 22:1225–1233

Smith SB, Kawachi H, Choi CB, Choi CW, Wu G, Sawyer JE (2009b) Cellular regulation of bovine intramuscular adipose tissue development and composition. J Anim Sci 87 (14 Suppl):E72–E82. doi:10.2527/jas.2008-1340

Smith SB, Go GW, Johnson BJ, Chung KY, Choi SH, Sawyer JE, Silvey DT, Gilmore LA, Ghahramany G, Kim KH (2012) Adipogenic gene expression and fatty acid composition in subcutaneous adipose tissue depots of Angus steers between 9 and 16 months of age. J Anim Sci 90:2505–2514

St John LC, Young CR, Knabe DA, Schelling GT, Grundy SM, Smith SB (1987) Fatty acid profiles and sensory and carcass traits of tissues from steers and swine fed an elevated monounsaturated fat diet. J Anim Sci 64:1441–1447

St John LC, Lunt DK, Smith SB (1991) Fatty acid elongase and desaturase activities in bovine liver and adipose tissue microsomes. J Anim Sci 69:1064–1073

Sturdivant CA, Lunt DK, Smith SB (1992) Fatty acid composition of longissimus muscle and subcutaneous and intramuscular adipose tissues of Japanese Wagyu cattle. Meat Sci 32:449–458

Taniguchi M, Utsugi T, Oyama K, Mannen H, Kobayashi M, Tanabe Y, Ogino A, Tsuji S (2004) Genotype of stearoyl-coA desaturase is associated with fatty acid composition in Japanese Black cattle. Mamm Genome 15:142–148

Taylor JF, Coutinho LL, Herring KL, Gallagher DS Jr, Brenneman RA, Burney N, Sanders JO, Turner JW, Smith SB, Miller RK, Savell JW, Davis SK (1998) Candidate gene analysis of *GH1* for effects on growth and carcass composition of cattle. Anim Genet 29:194–201

Turk SN, Smith SB (2009) Carcass fatty acid mapping. Meat Sci 81:658–663

Uemoto Y, Nakano H, Kikuchi T, Sato S, Ishida M, Shibata T, Kadowaki H, Kobayashi E, Suzuki K (2012) Fine mapping of porcine SSC14 QTL and SCD gene effects on fatty acid composition and melting point of fat in a Duroc purebred population. Anim Genet 43:225–228

van de Vossenberg JL, Joblin KN (2003) Biohydrogenation of C18 unsaturated fatty acids to stearic acid by a strain of *Butyrivibrio hungatei* from the bovine rumen. Lett Appl Microbiol 37:424–428

Wallace JR, Chaudhary LC, McKain N, McEwan NR, Richardson AJ, Vercoe PE, Walker ND, Paillard D (2006) Clostridium proteoclasticum: a ruminal bacterium that forms stearic acid from linoleic acid. FEMS Microbiol Lett 265:195–201

Wang W, Xue W, Jin B, Zhang X, Ma F, Xu X (2013) Candidate gene expression affects intramuscular fat content and fatty acid composition in pigs. J Appl Genet 54:113–118

Waters K, Ntambi JM (1994) Insulin and dietary fructose induce stearoyl-CoA desaturase 1 gene expression in liver of diabetic mice. J Biol Chem 267:10925–10930

Waters SM, Kelly JP, O'Boyle P, Moloney AP, Kenny DA (2009) Effect of level and duration of dietary n-3 polyunsaturated fatty acid supplementation on the transcriptional regulation of Δ^9-desaturase in muscle of beef cattle. J Anim Sci 87:244–252

Westerling DB, Hedrick HB (1979) Fatty acid composition of bovine lipids as influenced by diet, sex and anatomical location and relationship to sensory characteristics. J Anim Sci 48:1343–1348

Wilson JJ, Zembayashi M, Lunt DK, Smith SB (1993) Rapid communication: a *Taq* I restriction fragment length polymorphism in the *stearoyl-CoA desaturase* gene in DNA from purebred Japanese Black cattle. J Anim Sci 71:2575

Wood JD, Richardson RI, Nute GR, Fisher AV, Campo MM, Kasapidou E, Sheard PR, Enser M (2004) Effects of fatty acids on meat quality: a review. Meat Sci 66:21–32

Yokota S, Sugita H, Ardiyanti A, Shoji N, Nakajima H, Hosono M, Otomo Y, Suda Y, Katoh K, Suzuki K (2012) Contributions of FASN and SCD gene polymorphisms on fatty acid composition in muscle from Japanese Black cattle. Anim Genet 43:790–792

Zembayashi M, Nishimura K, Lunt DK, Smith SB (1995) Effect of breed type and sex on the fatty acid composition of subcutaneous and intramuscular lipids of finishing steers and heifers. J Anim Sci 73:3325–3332

Zulkifli RM, Parr T, Slater AM, Brameld JM (2010) Regulation of ovine and porcine stearoyl coenzyme A desaturase gene promoters by fatty acids and sterols. J Anim Sci 88:2565–2575

Chapter 13
Expression and Nutritional Regulation of Stearoyl-CoA Desaturase Genes in the Ruminant Mammary Gland: Relationship with Milk Fatty Acid Composition

L. Bernard, C. Leroux, and Y. Chilliard

Introduction

In lactating ruminants, stearoyl-CoA desaturase (*SCD*) gene expression in the mammary gland focused particular attention due to its implication in milk technological and nutritional qualities for human consumers. Indeed, the mammary enzyme SCD or Δ9-desaturase by introducing a cis double bond between carbons 9 and 10 of the fatty acids (FA) has a key role in the synthesis of milk monounsaturated FA and specific conjugated linoleic acid (CLA) isomers only found in ruminant products. Then, in milk, about 60 % of oleic (cis-9-18:1; a major milk FA), 50–56 % of palmitoleic (cis-9-16:1), 90 % of myristoleic (cis-9-14:1), and >60 % of the major isomer of CLA (cis-9, trans-11-18:2 or rumenic acid) originate from mammary gland synthesis by the action of SCD enzyme on circulating stearic (18:0), palmitic (16:0), myristic (14:0), and vaccenic acids (trans-11-18:1), respectively. This action is very important due to the impact of milk fat concentration and FA profile in determining milk nutritional quality. Indeed, certain saturated FA (mainly 12:0, 14:0 and 16:0) and trans-FA are considered to exert negative effects when consumed in excess, whereas others (4:0, anteiso-15:0, cis-9-18:1, 18:3*n*-3, and some CLA isomers) have potentially positive effects on human health (Parodi 2005; Shingfield et al. 2008b). For example, cis-9, trans-11-18:2, the major isomer of CLA in ruminant milk, exhibits anticarcinogenic and antiatherogenic (Wahle et al. 2004) properties in animal models. Thus, by contributing to the synthesis of FA that are beneficial from the point of view of human nutrition, e.g., cis-9, trans-11-CLA and to a lesser extent cis-9-18:1, SCD contribute to improve the nutritional quality of milk fat. In addition, by introducing a cis-9 double bond to FA,

L. Bernard (✉) • C. Leroux • Y. Chilliard
INRA, UMR1213 Herbivores, Saint-Genès-Champanelle 63122, France

Clermont Université, VetAgro Sup, UMR Herbivores, Clermont-Ferrand, BP 10448 63000, France
e-mail: Laurence.Bernard@clermont.inra.fr

J.M. Ntambi (ed.), *Stearoyl-CoA Desaturase Genes in Lipid Metabolism*, DOI 10.1007/978-1-4614-7969-7_13, © Springer Science+Business Media New York 2013

SCD contribute to lower the milk fat melting point (Parodi 1982). Differences in mammary gland level of SCD protein may help in explaining the substantial variation in milk fat content of these fatty acids and thus studies on the regulation of SCD activity and gene expression in the mammary gland have focused major interest in the last decades. Overall, different experiments have been conducted in dairy ruminants for studying the effect of nutrition on milk fat secretion and composition in relation with the regulation of mammary lipogenic gene expression and in particular of *SCD* gene. For this purpose measures were performed at the level of the mammary tissue and milk fat and different variables were determined as follows: (1) mammary *SCD* transcript abundance (*SCD1* and *SCD5* mRNA), (2) mammary SCD activity, (3) the in vivo Δ9-desaturation by a chemical tracer technique, and (4) milk FA Δ-9 desaturation ratios (4 milk ratios of the pairs of FA that represent a product/substrate relationship for SCD: cis-9-14:1/14:0, cis-9-16:1/16:0, cis-9-18:1/18:0, cis-9, trans-11-CLA/trans-11-18:1), which are less invasive indicators commonly used as a proxy for Δ9-desaturase activity due to the difficulty in measuring the activity.

This chapter reviews the present knowledge on *SCD* genes expression in the ruminant mammary gland, in particular, the known effects of nutritional factors in relation with milk FA composition and the effect of specific FA studied using in vitro models. In addition, the known or putative molecular mechanisms underlying these regulations are presented.

Functions of SCD in the Mammary Gland of Ruminant

Milk fat is composed of ca. 98 % of triglycerides (TG) whose FA have a dual origin: (1) either they are de novo synthesized in the mammary gland (Fig. 13.1) from circulating acetate and 3-hydroxybutyrate, produced by ruminal fermentation of carbohydrates and by rumen epithelium from absorbed butyrate, respectively, thus resulting in short and medium chain FA (C4:0 to C16:0) that represent 40–50 % of the FA secreted in milk or (2) they are imported from the plasma, where they are either released by the enzyme lipoprotein lipase (LPL) (Barber et al. 1997) from triglycerides circulating in chylomicra or very low density lipoprotein (VLDL), or derived from the plasma non-esterified fatty acids (NEFA) that circulate bound to albumin, for long-chain FA (≥C18), as well as ca. one half of the 16:0 and very low amount of 14:0 depending on the diet composition. Whatever their origin, these FA may be desaturated by SCD, but not elongated, in the secretory mammary epithelial cells (MECs) (Chilliard et al. 2000). In lactating ruminants, mammary glands have high levels of *SCD* mRNA (Ward et al. 1998; Bernard et al. 2005a) and activity (Bickerstaffe and Annison 1970; Kinsella 1970; McDonald and Kinsella 1973; Bernard et al. 2005b, 2009a, b). The protein SCD is located in the endoplasmic reticulum and catalyses the Δ-9 desaturation, introducing a cis double bond, of a spectrum of fatty acyl-CoA substrates, mainly from C14 to C19. This reaction requires NADH, oxygen, and an electron transport sequence comprising NADH-cytochrome *b*5 reductase, cytochrome *b*5, and SCD. The preferred substrate of SCD

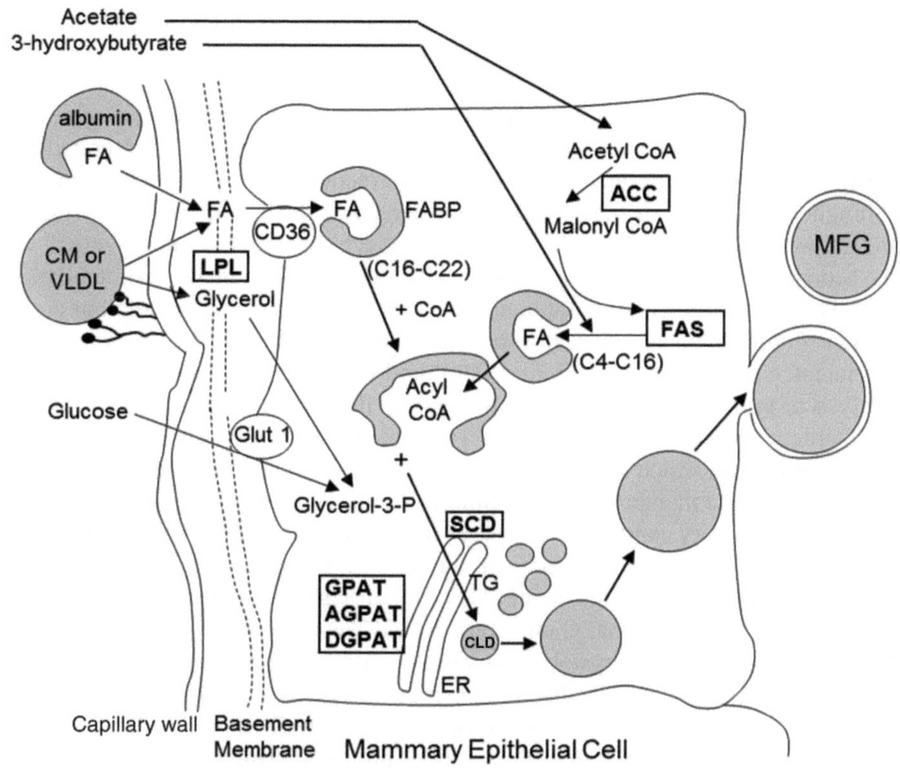

Fig. 13.1 Milk fat synthesis in the ruminant mammary epithelial cell (from Bernard et al. 2008). *ACC* acetyl CoA carboxylase, *AGPAT* acyl glycerol phosphate acyl transferase, *CD36* cluster of differentiation 36, *CLD* cytoplasmic lipid droplet, *CoA* coenzyme A, *CM* chylomicron, *DGAT* diacyl glycerol acyl transferase, *ER* endoplasmic reticulum, *FA* fatty acid, *FABP* fatty acid binding protein, *FAS* fatty acid synthase, *Glut 1* glucose transporter1, *GPAT* glycerol-3 phosphate acyl transferase, *LPL* lipoprotein lipase, *MFG* milk fat globule, *SCD* stearoyl-CoA desaturase, *TG* triglyceride, *VLDL* very low density lipoprotein

is stearic acid leading to the synthesis of oleic acid, the major unsaturated fatty acid found in milk triacylglycerol. In bovine, a number of other saturated acyl-CoA also serve as substrates for SCD including 10:0, 12:0, 14:0, 15:0, and 17:0 (Shingfield et al. 2008b) as well as trans-FA such as trans-7- and trans-12-18:1 (Palmquist et al. 2005; Shingfield et al. 2007). Thus, trans-7, cis-9-18:2 (CLA) in milk originates almost exclusively from the action of Δ9-desaturase on trans-7-18:1 in the mammary gland (Corl et al. 2002). In addition, recent results also suggest that conjugated linolenic acids (CLnA) such as cis-9, trans-11, trans-13-18:3, and cis-9, trans-11, trans-15-18:3 in milk are synthesized endogenously via the action of Δ9-desaturase on trans-11, trans-13-18:2, and trans-11, trans-15-18:2 in the mammary gland, respectively (Lerch et al. 2012). All these trans-18:1 or trans-18:2 FA that may serve as substrate for mammary SCD originate from the ruminal biohydrogenation of the major FA found in the diet delivered to ruminant.

The unsaturation of fatty acid chain is a major determinant of the melting temperature of triglycerides (TG), thus determining the fluidity of milk fat. The fundamental function of mammary SCD in maintaining the physical properties of TG is even increased in ruminants due to the large biohydrogenation processes of dietary PUFA that occur in the rumen which led to the synthesis of saturated FA (mainly 18:0) as well as of numerous intermediates, in particular, trans-FA (Shingfield et al. 2010) that could be further taken up by the mammary gland. Indeed, whereas it is well established that endogenous synthesis of cis-9-18:1 via the action of SCD on 18:0 in the mammary gland is an important point of regulation in milk TG synthesis and maintenance of milk fat fluidity, the presence of a cis- or trans-double bond as well as of short-chain FA and saturated FA also intervene in the maintenance of milk fat fluidity (Chilliard et al. 2000). In this way, a meta-analysis of milk FA yield in relation to duodenal flows (Glasser et al. 2008) showed an interdependence of C18 and C4 to C16 probably due to the esterification step of milk fat synthesis, which involves a putative balance between long-chain and de novo synthesized FA to maintain milk TG fluidity. In the same way it was proposed that, under certain dietary conditions, decreases in the availability of 18:0 for endogenous synthesis of cis-9-18:1 combined with an increase in trans-18:1 at the mammary gland could be a possible mechanism to explain part of the decrease in milk fat secretion (Loor et al. 2005a; Gama et al. 2008). Thus, the ability to maintain milk fat melting point within physiological values could play a role in the regulation of milk fat secretion when milk FA composition varies, for example, in response to feeding factors. In this context, the hypothesis of a contribution of milk fat fluidity to the regulation of milk fat secretion has been proposed (Timmen and Patton 1988; Chilliard et al. 2000) and was confirmed recently in a meta-analysis on cow, goat, and ewe milk FA composition and estimated melting point (Toral et al. 2013a). In addition, studies in goats (Chilliard et al. 2006) on the effect of the CSN1S1 genotype on milk FA profile have reported higher Δ9-desaturation ratios in defective vs. high genotypes suggesting higher expression of SCD in defective genotypes in which the synthesis of short-chain FA is lower. This result led to the hypothesis that higher SCD expression could compensate for the concomitant decrease in short-chain FA and thus to maintain the milk fat melting point in defective genotypes.

Characterization of SCD Genes

Multiple isoforms of *SCD* have been found in most species examined to date. Some rodent genomes (as mice) contain four *SCD* isoforms (*SCD1–4*) with different expression profiles according to tissues. For example, murine *SCD1* is expressed predominantly in lipogenic tissues (Ntambi et al. 1988), while *SCD2* is preferentially expressed in brain and neuronal tissues (Kaestner et al. 1989), *SCD3* is expressed in sebocytes, preputial gland, and Harderian gland (Zheng et al. 2001), and *SCD4* expression is restricted to the heart (Miyazaki et al. 2003). Up to now, the genomes of many other species, including chickens, pigs, cows, and humans, have been found to contain two *SCD* isoforms. Then, in humans: *SCD1* is highly expressed in adipose tissue and liver while a second *SCD* gene, termed *SCD5*, is highly expressed in brain and pancreas (Wang et al. 2005). In addition to tissue-specific expression profile, there is evidence to suggest that some SCD protein isoforms may differ in their preferred substrate specificity. For example, murine SCD1, SCD2, and SCD4 have been shown to desaturate both palmitoyl-CoA and stearoyl-CoA, while murine SCD3 desaturates palmitoyl-CoA but not stearoyl-CoA (Miyazaki et al. 2006). The tissue and substrate specificity of SCD isoforms suggests that each may have a distinct physiological role.

In ruminants earlier studies reported only one *SCD* gene, generating a transcript of 5 kb in size which was characterized in sheep (Ward et al. 1998), cows (Chung et al. 2000), and goat (Bernard et al. 2001). When compared, the coding region of the caprine *SCD* mRNA shares 98, 95, 90, and 86 % identity with the ovine, bovine, porcine, and human *SCD1* mRNAs, respectively. This *SCD* isoform was highly expressed in lipogenic tissues, adipose tissue, and the lactating mammary gland and was recently renamed *SCD1* since the discovery of the *SCD5* isoform characterized in ruminant species (Lengi and Corl 2007, 2008). In these species *SCD5* is mainly expressed in the brain (Lengi and Corl 2008). In addition to *SCD1* a significant expression of *SCD5* was reported in the lactating mammary gland in cows (Gervais et al. 2009; Jacobs et al. 2011) and goats (Bernard et al. 2010) with probably different regulation for these two isoforms (Gervais et al. 2009; Jacobs et al. 2011). However, both the regulation of the expression and the contribution of *SCD5* isoform to FA Δ-9 desaturation process are still unknown.

SCD1 transcript in goat presents an unusually long (3.8 kb) 3′-UTR sequence which derives from a single exon (Bernard et al. 2001), as observed for human (Zhang et al. 1999), rat (Mihara 1990), and mouse (Ntambi et al. 1988). In the caprine, the 3′-UTR is characterized by the presence of several AU-rich elements as observed in the 3′-UTR region of mouse (Kaestner et al. 1989; Sessler et al. 1996), rat (Thiede and Strittmatter 1985; Mihara 1990), and human (Zhang et al. 1999) *SCD1* transcripts, which could be mRNA destabilization sequences as suggested by Sessler et al. (1996). These AU-rich elements have been shown to play active roles in the mRNA degradation (Vakalopoulou et al. 1991) which is responsible for their turnover. In addition, these motifs could be possible targets of polyunsaturated fatty acids effects on *SCD* mRNA, as proposed by Ntambi (1999) for *SCD1* and *SCD2* mRNA in mouse adipocytes. Otherwise, SCD5 transcript in bovine and ovine was of 2.6 kb in length (Lengi and Corl 2007, 2008) with a 3′UTR corresponding to half the mRNA length.

SCD1 gene was localized on bovine and caprine chromosomes 26q21 and on ovine chromosome 22q21 (Bernard et al. 2001), and *SCD5* was localized on bovine chromosome 6 (Lengi and Corl 2007) and on ovine chromosome 6 (Lengi and Corl 2008). In caprine *SCD1* gene spans about 15 kb and has six exons varying in size from 131 (third exon) to 4,047 (sixth exon) bp, and five introns varying in size from 600 to 3,700 bp (Bernard et al. 2001; Fig. 13.2). This structural organization is similar to *SCD1* gene reported in human (Zhang et al. 1999) and rodents (Ntambi et al. 1988; Kaestner et al. 1989; Mihara 1990). The genomic structure of bovine

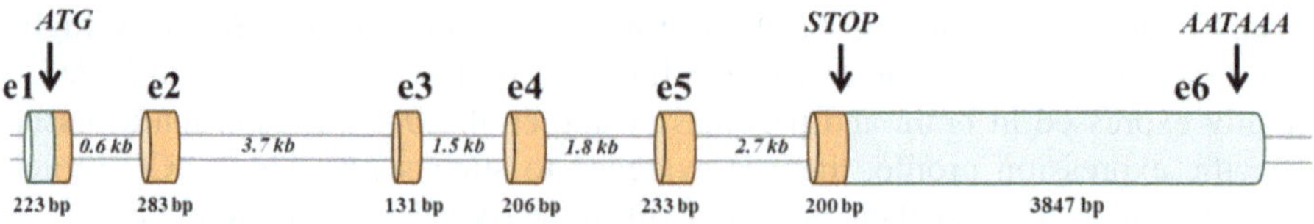

Fig. 13.2 Structural organization of the caprine SCD1 transcription unit. Overall structural organization of the caprine SCD1 gene deduced from PCR analysis. *Open bars* represent introns and exons are depicted by *large*, *green* (UTR) or *orange* (TR) *boxes* (from Bernard et al. 2001)

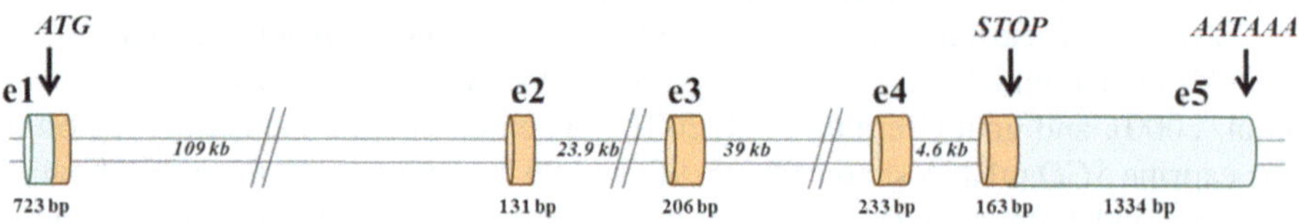

Fig. 13.3 Structural organization of the bovine SCD5 transcription unit. Overall structural organization of the bovine SCD5 gene adapted from Lengi and Corl (2007). *Open bars* represent introns and exons are depicted by *large*, *green* (UTR) or *orange* (TR) *boxes*

SCD5 deduced from known bovine genomic DNA sequence showed that *SCD5* is organized into five exons and four introns in a fragment tenfold longer than *SCD1* spanning more than 175 kb (Lengi and Corl 2007) and is similar to human *SCD5* (Fig. 13.3). Despite the length difference of their genes, the SCD1 and SCD5 encoded proteins are of similar size and are, respectively, composed of 359 and 335 amino acids (AA). Sequence alignment shows 65 % identity between bovine SCD1 and SCD5 at the amino acid level. Both proteins share highly conserved three histidine motifs that are catalytically essential for the desaturase activity. Conversely to SCD1 protein, the N-terminus of the SCD5 protein lacks the conserved PEST (peptide sequence rich in proline, glutamic acid, serine, and threonine). This sequence is present in proteins that have a short intracellular half-time with the hypothesis that PEST sequence acts as a signal peptide for protein degradation (mediated by proteasome or calpaïn). These data suggest that the regulation of SCD1 and SCD5 proteins may differ.

Little is known in ruminant on the molecular mechanisms controlling *SCD* genes expression (see section "Knowledge on the Signalling Pathways Mediating Nutritional Regulation of SCD Gene Expression in Ruminants"). The promoter region of the bovine *SCD1* gene have been studied (Keating et al. 2005) and an area of 36 bp in length was identified as having a critical role in the transcriptional activation and was designated the *SCD* transcriptional enhancer element (STE). Otherwise, a recent study regarding the possible common transcription factors between bovine *SCD5*, expressed in brain and nervous tissue, and rodent *SCD2*, expressed in nervous tissue, demonstrated that early growth response 2 (EGR2) and sterol regulatory element binding protein 1a (SREBP-1a) may bind to the same DNA site in the bovine *SCD5* promoter (Lengi and Corl 2012).

In other respects, in order to explain individual differences in milk fat yield and FA composition, in particular in cis-9 monounsaturated FA, studies in ruminants were focused on the genetic variability of the region containing *SCD* gene. A lot of programs were dedicated to quantitative trait loci (QTL) detection in dairy bovine breeds on milk parameters. In particular, genome-scan for bovine milk fat composition revealed a QTL for the monounsaturated FA and their unsaturation indices on Bos taurus autosome (BTA) 26 corresponding to the chromosome where *SCD1* gene is located (Stoop et al. 2009). These authors have strongly suggested that this BTA26 QTL was caused by a mutation in the *SCD1* gene. Among the single nucleotide polymorphisms (SNP) detected in *SCD1* gene, a nonsynonymous one is located in exon 5 causing a substitution of valine (V) with alanine (A) at position 293 (A293V; Kgwatalala et al. 2007; Schennink et al. 2008). Several studies in different bovine breeds reported significant association between A293V genotypes and milk FA composition with milk of cows carrying the SCD *AA* genotype having a greater content of cis-9-18:1 and a higher cis-9-14:1/14:0 ratio than did milk of cows carrying the SCD *VV* genotype (Kgwatalala et al. 2009b; Mele et al. 2007; Conte et al. 2010), whereas others (Moioli et al. 2007) could not confirm this association. These latter authors suggested a role of regulatory factors acting on the promoter region of *SCD* gene or through activation of the enzyme. In that way, whereas an earlier study on *SCD1* promoter has not allowed the detection of SNP (Keating et al. 2005), a more recent study highlighted a SNP (133A>C) creating a new consensus site for the transcription factor SP1 binding site, which could be responsible for a difference in the milk FA desaturation ratio (Pauciullo et al. 2012). Moreover, different SNPs were also detected in intronic or exonic regions without AA substitution in the protein. However, their linkage with milk yield was particularly studied in haplotypic configuration suggesting that they could be used as potential genetic markers to improve milk performance (Alim et al. 2012). For example, SNP haplotypes identified in the 3′UTR region were reported to be associated with different 10:0 and 12:0 Δ-9 desaturation ratios in bovine milk (Kgwatalala et al. 2009a).

In small ruminants, although less documented, *SCD1* gene polymorphism has also been studied. The first characterization of the caprine *SCD1* mRNA has revealed a polymorphism involving the deletion of a nucleotide triplet (TGT) in the 3′UTR region (Bernard et al. 2001). The analyses of this mutation in haplotype-based configuration with two other 3′UTR SNP were suggestively associated with lower milk stearic FA content and with higher percentages of PUFA (Zidi et al. 2010). An in silico analysis of the secondary structure of *SCD1* mRNA pointed out the 3nt deletion as being involved in a considerable change in the secondary structure of the *SCD1* mRNA. This change could alter the amount of *SCD1* transcripts in MECs (Zidi et al. 2010). In sheep, a QTL analysis for the FA profile showed a QTL for the ratio of rumenic/vaccenic acid in ovine chromosome 22 which coincided with the position of the *SCD* gene (Carta et al. 2008). However, no association of four SNP previously detected (Garcia-Fernandez et al. 2009) has been observed with the ratio of rumenic/vaccenic acid, whereas a possible effect of the *SCD* polymorphism on milk fat percentage has been suggested (Garcia-Fernandez et al. 2010).

Due to its recent discovery, the polymorphism of *SCD5* gene, located on BTA6, has not been studied to the same extend. A recent study revealed that polymorphisms in the *SCD5* gene were among the most representative markers associated with unsaturated/saturated FA ratio in milk (Rincon et al. 2012). In this study, two SNP were detected within exon 3 which were, respectively, associated with an increase in saturated FA concentration combined with a decrease in monounsaturated FA, and with a decrease in saturated FA combined with an increase in PUFA. Moreover, an intronic mutation named G allele was shown to be associated with a decrease in saturated FA/unsaturated FA ratio and an increase in monounsaturated FA and Δ-9 desaturation ratios for 14:0 and trans-11-18:1 content in Holstein milk (Rincon et al. 2012). However, the mechanism by which this mutation acts needs to be clarified.

Regulation of Mammary *SCD* Genes Expression by Dietary Factors

Effect of Dietary Factors

In ruminants, milk FA composition is greatly influenced by feeding factors (Jensen 2002; Dewhurst et al. 2006; Chilliard et al. 2007). Thus, modification of milk fat content and FA composition by dietary manipulation has been investigated and few studies were performed to relate the effect of diet on milk FA profile to mammary lipid metabolism considering the major lipogenic genes, in particular, *SCD*. The response of *SCD* gene expression to nutritional changes involves the control of events that could occur at transcriptional (e.g., through transcription factors), post-transcriptional (e.g., such as mRNA stability), translational (e.g., its initiation or through microRNA action), and post-translational (e.g., via turnover or regulation of the activity of the enzymatic protein) levels. However, it is often unclear whether the regulatory factors are the dietary components themselves or their metabolites or hormonal changes produced in response to the nutritional changes. Moreover, in most of the studies, it is difficult to conclude on the level of regulation involved since measurements of *SCD* mRNA of the enzyme protein content and activity are not studied simultaneously. Only few studies have investigated the role of diet on the regulation of *SCD* gene expression and activity in ruminants.

As previously mentioned, prior to the characterization of the bovine *SCD5* gene (Lengi and Corl 2007), as only one *SCD* isoform was thought to be present in this species, the gene was simply referred to as *SCD* for *SCD1*. The in vivo trials reporting the nutritional regulation of mammary *SCD1* and lipogenic genes expression have been carried out in mid-lactation cows and goats and mainly with lipid supplements (Table 13.1). In cows, studies have been primarily realized with diets that induce milk fat depression (MFD). These diets belongs to three types: (1) diets rich in starch without addition of lipid supplements (e.g., high grain/low forage diets), but containing a minimal amount of PUFA in dietary feedstuffs; (2) diets with low level of fiber associated with supplemental PUFA of plant origin; (3) diets associated with dietary supplements of marine oils (fish oils, fish meals, oils from marine

Table 13.1 Effect of dietary lipid supplements on mammary *SCD* and lipogenic genes expression and/or enzyme activity in ruminants

Biochemical pathway[a]	Species	Transcript/protein[b]	Response (%)[c]	Nature of lipid supplement	Inclusion rate (g/kg dry matter)	Response of milk fat yield (%)[c]	Reference
Transcription factors							
SREBF1/S14	Bovine	mRNA	–45/–60	Soyabean oil + fish oil[d]	30 + 15	–38	Harvatine and Bauman (2006)
SREBF1	Bovine	mRNA	–30	Sunflower oil + marine algae[e]	27 + 4	–30	Angulo et al. (2012)
	Bovine	mRNA	–31	Linseed oil + marine algae[e]	27 + 4	–31	Angulo et al. (2012)
SREBF1/PPARG	Bovine	mRNA	NS/NS	Soybean oil + fish oil	25 + 10	–50	Invernizzi et al. (2010)
Desaturation and de novo lipogenesis							
SCD1/ACC	Bovine	mRNA	NS	Rapeseed	33	NS	Delbecchi et al. (2001)
SCD1/ACC/FAS	Bovine	mRNA	(–29)/–46/–64	Fish oil	37	–44	Ahnadi et al. (2002)
SCD1	Caprine	mRNA	–43	Oleic sunflower oil	35	+16	Bernard et al. (2005b)
	Caprine	mRNA	–54	Formaldehyde-linseed	112	+8	Bernard et al. (2005b)
SCD	Caprine	Activity	–30	Sunflower oil	55	+17	Bernard et al. (2009b)
	Caprine	Activity	–22	Oleic sunflower oil	35	+16	Bernard et al. (2005b)
	Caprine	Activity	–27	Linseed oil	55	+15	Bernard et al. (2009b)
	Caprine	Activity	–35	Formaldehyde-linseed	112	+8	Bernard et al. (2005b)
SCD1/FAS	Bovine	mRNA	(–25)/–45	Soyabean oil + fish oil[d]	30 + 15	–38	Harvatine and Bauman (2006)
	Bovine	mRNA	–34.5/–31	Sunflower oil + marine algae[e]	27 + 4	–30	Angulo et al. (2012)
	Bovine	mRNA	–45.5/–33	Linseed oil + marine algae[e]	27 + 4	–31	Angulo et al. (2012)
SCD/ACC/FAS	Caprine	Activity	(–17/+12)/+81	Sunflower oil	61	(–3)	Bernard et al. (2009a)
SCD1/ACC/FAS	Caprine	mRNA	(+15/+19/+12)	Rapeseed	146	(+5)	Ollier et al. (2009)
	Bovine	mRNA	(+39/+38/+36)	Safflower seed	135	NR	Murrieta et al. (2006)
	Caprine	mRNA	+166/+72/+74	Safflower oil	50	+26	Li et al. (2012)
	Bovine	mRNA	(–35)/–28/–41	Sunflower oil[d]	10	–27	Peterson et al. (2003)
	Caprine	mRNA	(–7/–8/+5)	Sunflower oil	61	(+7)	Bernard et al. (2009a)
	Caprine	mRNA	(+10/+15/+6)	Sunflower oil	44	+20	Ollier et al. (2009)
	Caprine	mRNA	(+17/+10/+9)	Sunflower oil	55	+17	Bernard et al. (2009b)

(continued)

Table 13.1 (continued)

Biochemical pathway[a]	Species	Transcript/protein[b]	Response (%)[c]	Nature of lipid supplement	Inclusion rate (g/kg dry matter)	Response of milk fat yield (%)[c]	Reference
	Caprine	mRNA	(+22/+2/+10)	Linseed oil	55	+15	Bernard et al. (2009b)
	Caprine	mRNA	(−16/+4/−4)	Linseed oil	62	+14	Bernard et al. (2009a)
	Caprine	mRNA	+121/−26/+52	Linseed oil	50	+34	Li et al. (2012)
	Caprine	Activity	(+11/−3/+64)	Linseed oil	62	+14	Bernard et al. (2009a)
ACC	Bovine	mRNA	−68	Soyabean oil[d]	50	−43	Piperova et al. (2000)
FAS	Caprine	Activity	+68	Sunflower oil	61	(+7)	Bernard et al. (2009a)
	Bovine	Activity	−61/−44	Soyabean oil[d]	50	−43	Piperova et al. (2000)
	Caprine	mRNA	(−21)	Oleic sunflower oil	35	+16	Bernard et al. (2005a)
	Caprine	mRNA	(−1)	Formaldehyde-linseed	112	+8	Bernard et al. (2005a)
	Caprine	Activity	(+27)	Sunflower oil	55	+17	Bernard et al. (2009b)
	Caprine	Activity	(0)	Oleic sunflower oil	35	+16	Bernard et al. (2005b)
	Caprine	Activity	(+30)	Linseed oil	55	+15	Bernard et al. (2009b)
	Caprine	Activity	(−31)	Formaldehyde-linseed	112	+8	Bernard et al. (2005b)
Fatty acid uptake/processing							
LPL	Caprine	mRNA	(−0.03)	Rapeseed	146	(+5)	Ollier et al. (2009)
	Bovine	mRNA	+50	Safflower seed	135	NR	Murrieta et al. (2006)
	Caprine	mRNA	+43	Safflower oil	50	+26	Li et al. (2012)
	Bovine	mRNA	−61	Soyabean oil + fish oil[d]	30+15	−38	Harvatine and Bauman (2006)
	Bovine	mRNA	−30	Sunflower oil[d]	10	−27	Peterson et al. (2003)
	Caprine	mRNA	(+17)	Sunflower oil	61	(+7)	Bernard et al. (2009a)
	Caprine	mRNA	(+23)	Sunflower oil	44	+20	Ollier et al. (2009)
	Caprine	mRNA	(+1)	Sunflower oil	55	+17	Bernard et al. (2009b)
	Caprine	mRNA	+51	Oleic sunflower oil	35	+16	Bernard et al. (2005b)
	Caprine	mRNA	(−1)	Linseed oil	55	+15	Bernard et al. (2009b)
	Caprine	mRNA	(+26)	Formaldehyde-linseed	112	+8	Bernard et al. (2005b)
	Caprine	mRNA	(+13)	Linseed oil	62	+14	Bernard et al. (2009a)

	Caprine	mRNA	+54	Linseed oil	50	+34	Li et al. (2012)
	Bovine	mRNA	−34	Sunflower oil + marine algae[e]	27 + 4	−30	Angulo et al. (2012)
	Bovine	mRNA	−45	Linseed oil + marine algae[e]	27 + 4	−31	Angulo et al. (2012)
	Caprine	Activity	(+42)	Sunflower oil	61	(+7)	Bernard et al. (2009a)
	Caprine	Activity	(+35)	Sunflower oil	55	+17	Bernard et al. (2009b)
	Caprine	Activity	(+13)	Oleic sunflower oil	35	+16	Bernard et al. (2005b)
	Caprine	Activity	(+21)	Linseed oil	55	+15	Bernard et al. (2009b)
	Caprine	Activity	(−17)	Linseed oil	62	+14	Bernard et al. (2009a)
FACL	Bovine	mRNA	−34	Sunflower oil[d]	10	−27	Peterson et al. (2003)
Esterification							
AGPAT/GPAT	Bovine	mRNA	−27/−52	Sunflower oil[d]	10	−27	Peterson et al. (2003)
GPAM	Bovine	mRNA	−10	Sunflower oil + marine algae[e]	27 + 4	−30	Angulo et al. (2012)
	Bovine	mRNA	−26	Linseed oil + marine algae[e]	27 + 4	−31	Angulo et al. (2012)
FA transport							
FABP	Bovine	mRNA	(−25)	Sunflower oil[d]	10	−27	Peterson et al. (2003)
FABP3/FABP4	Caprine	mRNA	(+11)/+36	Sunflower oil	44	+20	Ollier et al. (2009)
	Caprine	mRNA	(+21)/(+11)	Rapeseed	146	NS	Ollier et al. (2009)

Adapted from Shingfield et al. (2013)

[a]*ACC* acetyl CoA carboxylase, *AGPAT* acylglycerol phosphate acyl transferase, *FAS* fatty acid synthase, *FABP* fatty acid binding protein, *FACL* fatty acyl-CoA ligase, *GPAM* glycerol-3-phosphate acyltransferase 1, *GPAT* glycerol phosphate acyl transferase, *LPL* lipoprotein lipase, *SCD* stearoyl-CoA desaturase, *S14* thyroid hormone responsive spot 14, *SREBF1* sterol response element binding protein

[b]Measurement of tissue transcript abundance (mRNA) or protein activity (Activity)

[c]Response reported when treatment effects were significant ($P < 0.10$) and calculated as [((Treatment − Control)/Control) × 100]. Values in parenthesis are not significant at $P < 0.05$. NS, values not reported because unsignificant

[d]Lipid supplementation also accompanied by decreases in dietary forage:concentrate ratio; NR, not reported

[e]Control treatment, supplement of rumen-stable fractionated palm fat rich in 16:0

mammals, and/or algae) which induce MFD whatever the level of starch or fiber in the diet. In these dietary conditions, variable responses of mammary *SCD1* mRNA abundance have been reported in cows: either no variation with high concentrate diets containing sunflower oil (Peterson et al. 2003) or fish oil and soya bean oil (Harvatine and Bauman 2006); or decreases with diets supplemented with plant oils rich in either *n*-3 or *n*-6 fatty acids (FA) plus docosahexaenoic acid (DHA)-rich algae (Angulo et al. 2012) or with rumen "protected" fish oil (Ahnadi et al. 2002); or surprisingly increases with diets supplemented with fish oil and soya bean oil (Invernizzi et al. 2010). From these studies it is difficult to conclude on the effect of the dietary treatments that induce MFD on *SCD1* mRNA abundance and probably that the forage:concentrate ratio and the nature and the level of the PUFA into the diet may intervene to explain such differences. Otherwise, studies in goats have shown that supplementing diets based on grass hay with sunflower-seed oil rich in oleic acid (Bernard et al. 2005b), sunflower-seed oil (18:2 rich), or linseed oil (18:3 rich) decreased SCD activity (Bernard et al. 2009b) but had no effect on mammary *SCD1* mRNA abundance, whereas supplementing grass hay-based diets with formaldehyde-treated linseed (FLS) decreased mammary *SCD1* mRNA without effect on the SCD activity (Bernard et al. 2005b). In contrast, inclusion of sunflower-seed oil and linseed oil in maize silage-based diets (Bernard et al. 2009a) and of soya beans in lucerne hay-based diets (Bernard et al. 2005a) had no effect on caprine mammary *SCD1* mRNA abundance and/or SCD activity. Lastly, in goat, addition of fish oil and starch to a diet containing sunflower-seed oil has no effect on mammary *SCD1* and *SCD5* mRNA abundances (Bernard et al. 2010).

Overall, these observations (Table 13.1) highlight the importance of interactions between the composition of the basal diet and lipid supplement in the goat with the implication that specific PUFA escaping metabolism in the rumen or specific biohydrogenation intermediates may inhibit SCD activity via transcriptional or post-transcriptional regulatory mechanisms.

In vivo Δ-9 desaturase activity has often been estimated by the milk ratios for the pairs of FA that represent a product/substrate relationship for SCD (Bauman et al. 2001) due to the fact that the in vitro activity assay needs fresh materials and is laborious (Legrand et al. 1997). Moreover, ex vivo activity assay is done in optimal conditions (pH, substrate, cofactors) and may differ from in vivo conditions. From the few studies in goats reporting mammary SCD activity together with milk fatty acid composition, the response to dietary lipids of the four FA pair ratios that represent a proxy for SCD activity was related to the response of the SCD activity itself (Bernard et al. 2008). In some experiments with goats, lipid supplements in the diet have been shown to reduce milk FA desaturation ratios but have no effect on SCD activity (Bernard et al. 2009a, b, c). However, it was shown that the milk ratio of myristoleic acid to myristic acid (cis-9-14:1/14:0) gave the best estimation for the response of mammary SCD activity (Fig. 13.4; from Bernard et al. 2008), probably due to the fact that almost all the myristoleic acid present in the milk is likely to be synthesized in the mammary gland from de novo synthesis of myristic acid followed by desaturation by SCD. Indeed, 14:0 and 14:1 even less are poorly represented in feedstuffs used for ruminants, including lipid supplements (except for 14:0

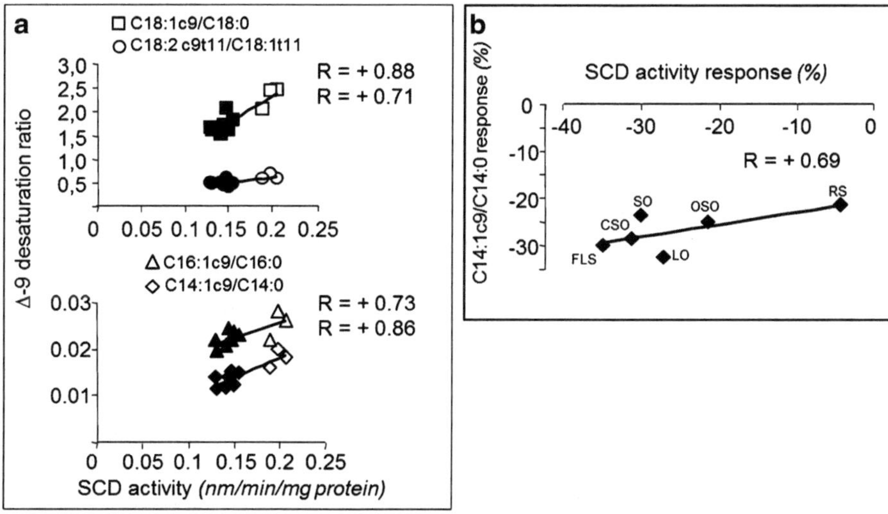

Fig. 13.4 Relationships between milk FA desaturation ratios and stearoyl-CoA desaturase (SCD) activity in 43 goats fed hay-based diet supplemented, or not, with lipids. The lipid supplements were either 3.6 % of lipids from oleic sunflower oil (OSO) or formaldehyde-treated linseed (FLS) (Bernard et al. (2005b)), or 5.8 % lipids from linseed oil (LO) or sunflower oil (SO) (Bernard et al. (2009a)), or 6.5 % whole rapeseed (RS) or 4.5 % of sunflower oil (CSO) (Ollier et al. (2009)) (from Bernard et al. 2008). (**a**) Results are means of the seven lipid-supplemented groups (*black symbols*) and the three control groups (*white symbols*). (**b**) Relationship between the responses (% of the control group) of SCD activity and milk cis-9-14:1/14:0 to the six lipid supplements

in specific lipid supplements such as coconut or krabo oils), and the circulating 14:1 in ruminant is very low (Christie 1981). Otherwise, variable proportions of palmitoleic, oleic, and rumenic acids come from absorption from the digestive tract and/or mobilization of body fat reserves. Consequently, the ratios that involve these latter FA are less indicative of SCD activity even though stearoyl-CoA and palmitoyl-CoA are the preferred substrates for SCD. Indeed, activity of SCD was estimated to contribute to approximately 90 % of cis-9-14:1 compared to 50–56 % of cis-9-16:1, 60 % of cis-9-18:1, and >60 % of cis-9, trans-11-18:2 in bovine milk fat (Enjalbert et al. 1998; Mosley and McGuire 2007; Glasser et al. 2008). Moreover, differential uptake, turnover, and use of different FA of the four pair ratios by the mammary tissue itself may occur.

In other respects, several experiments in goats have provided evidence to support the involvement of post-translational mechanisms in the regulation of SCD activity. Inclusion of cis-9-18:1 enriched sunflower oil (Bernard et al. 2005b), sunflower oil, or linseed oil (Bernard et al. 2009b) in the diet of goats fed grass hay-based diets has been shown to lower mammary SCD activity measured ex vivo without changing mammary *SCD1* mRNA abundance. Otherwise, in goats fed maize silage-based diet, addition of plant oils had no effect on mammary *SCD1* mRNA or SCD activity (Bernard et al. 2009a). However, in four in vivo studies in goats (Bernard et al.

2005b, 2009a, b; Ollier et al. 2009), the relationship between the responses of *SCD1* mRNA and enzyme activity (means per treatment) to the eight lipid supplements (expressed relative to the control) was relatively weak ($r=0.55$), which may indicate the existence of putative post-traductional regulatory events. In the same way, a recent study in goats using a proteomic analysis suggested an increased protein expression of PSMA5 (proteasome subunit α type 5) and a higher mRNA abundance of *SCD* in the mammary tissue following trans-11-18:1 treatment (by intravenous bolus injection) (Jin et al. 2012). SCD is a short-lived integral membrane protein of the endoplasmic reticulum (Oshino and Sato 1972), and PSMA5 as a part of the ubiquitin proteasome system may be involved in constitutive SCD degradation, and thus could play a role in SCD regulation as further suggested in MAC-T cells (Jin et al. 2012).

Otherwise, in order to determine the in vivo SCD metabolic activity, different methods have been developed and used in ruminants, in particular, to estimate the in vivo conversion of stearic to oleic acid or vaccenic to rumenic acid in the mammary gland. These methods include direct methods, by means of a tracer (for example, [1-^{13}C]trans-11 18:1; Mosley et al. 2006; Bernard et al. 2010), and indirect methods, through quantification of duodenal and milk FA flows (Shingfield et al. 2007; Glasser et al. 2008) or inhibition of the desaturase system using sterculic acid (Griinari et al. 2000; Corl et al. 2001; Kay et al. 2004; Bichi et al. 2012). Studies using these methods provided clear evidence that the endogenous desaturation of stearic and vaccenic acids are the main sources of milk oleic and rumenic acids, respectively. Indeed, about 80 % (Glasser et al. 2008), 80–84 % (Annison et al. 1974), and 63 % (Bichi et al. 2012) of oleic acid comes from stearic acid, respectively, in cows, goats, and ewes. In the same way, 64–97 % (Mosley et al. 2006; Glasser et al. 2008), 63–73 % (Bernard et al. 2010), and 74 % (Bichi et al. 2012) of rumenic acid comes from vaccenic acid, respectively, in cows, goats, and ewes. In addition, few of them provided an estimation of stearic or vaccenic acid desaturation in the mammary gland: for stearic acid it was of 59 % (Mosley and McGuire 2007) and 54 % (Glasser et al. 2008) in cows; for vaccenic acid it was of 25.7 % (Mosley et al. 2006), 28.9 % (Shingfield et al. 2007), 19.8 % (Qiu et al. 2004), 21 % (Glasser et al. 2008) in cows and of 31.6 % in goats (Bernard et al. 2010).

Ruminant Species Specificities

Indirect comparisons from experiments examining in cows and goats the effect of nutrition on milk fatty acid composition outlined several differences in milk cis-9-FA and Δ-9 desaturation ratios responses to lipid supplements among these two species (Chilliard et al. 2007). As a large proportion of the cis-9-FA secreted in milk is synthesized endogenously via the action of mammary SCD, differences in their concentration may suggest specificities in mammary SCD regulation among species. Indeed, in experiments with goats fed maize silage- or grassland hay-based supplemented with plant oils (Chilliard et al. 2007; Bernard et al. 2009c), the

Table 13.2 Effect of inclusion of sunflower oil (rich in cis-9, cis-12-18:2) into a maize silage-based diet on milk fatty acid secretion and composition (including trans-10, cis-12-, trans-10, trans-12-, trans-9, trans-11-CLA) and milk fatty acid desaturation ratio responses in cows and goats

Species	Bovine		Caprine	
Reference	Roy et al. (2006)		Bernard et al. (2009c)	
Major forage of the diet	Maize silage		Maize silage	
Treatment	Control	Sunflower	Control	Sunflower
Oil inclusion rate (g/kg dry matter)	0	52	0	63
Diet composition				
Concentrate (g/kg dry matter)	520	537	612	545
Neutral detergent fiber (g/kg dry matter)	285	266	310	295
Milk fat content and secretion				
Milk fat output (g/d or response)[a]	1,380	(−44.2 %)	107	(+6.5 %)
14:0 (g/d or response)	156.0	(−67.5 %)	12.0	(−28.5 %)
18:0 (g/d or response)	111.1	(−11.9 %)	4.9	(+96.7 %)
trans-10-18:1 (g/d or response)	5.54	(+837 %)	0.44	(+682 %)
Fatty acid (g/100 g fatty acids)				
10:0	3.46	1.2	10.58	6.91
12:0	3.94	1.65	5.72	3.12
14:0	12.12	7.06	12.07	8.10
16:0	32.32	18.93	29.85	18.78
18:0	8.63	13.62	4.88	9.01
trans-10-18:1	0.43	7.22	0.44	3.23
trans-10, cis-12-CLA	0.003	0.024	0.004	0.064
trans-9, trans-11-CLA	0.006	0.008	0.015	0.059
trans-10, trans-12-CLA	0.004	0.015	0.003	0.024
Milk fatty acid desaturation ratios				
cis-9-14:1/14:0	0.094	0.132	0.019	0.012
cis-9-18:1/18:0	1.92	2.08	2.81	1.74
cis-9, trans-11-CLA/trans-11-18:1	0.40	0.53	0.70	0.50

Adapted from Shingfield et al. (2010)
[a]Response calculated as [((Treatment − Control)/Control) × 100]

relative increases in milk cis-9-18:1 concentrations were lower than observed when comparable diets were fed to lactating cows (Roy et al. 2006; Chilliard et al. 2007). These interspecies differences could be related to a greater sensitivity of the SCD enzyme to the inhibitory effects of PUFA in the goat than the cow (Bernard et al. 2008) and/or to larger increases in the enrichment of trans-FA that may act as inhibitor of SCD in the goat compared to the cows receiving similar diets (such as trans-9, trans-11-CLA in goat milk in Bernard et al. (2009c) compared to bovine milk in Roy et al. (2006); Table 13.2; see section "In Vivo Studies with Specific Fatty Acid Infusion or Injection"). Moreover, the balance between (1) the level and nature of unsaturated FA taken up by the mammary gland and (2) the mammary synthesis of short-chain FA and desaturase activity, which play a role in the regulation of milk fat fluidity (Toral et al. 2013a), may intervene in explaining these interspecies differences. Indeed, it may be hypothesized that the goat which has higher concentrations of milk short-chain FA (Table 13.2) would require lower level of Δ-9 desaturation for maintaining the milk fat melting point. However, in the absence of direct comparisons between bovine and small ruminant species with similar dietary conditions and stage of lactation, it is difficult to draw clear conclusions on species specificities of SCD regulation.

Mechanisms of the Nutritional Regulation of Mammary SCD

In Vivo Studies with Specific Fatty Acid Infusion or Injection

A key role for specific fatty acids on mammary lipogenesis and desaturation was showed from in vivo studies (Shingfield et al. 2010). In particular, studies with specific dietary conditions inducing a dramatic MFD (Bauman and Griinari 2003) underlined the role of specific trans-FA that act as potent inhibitors of mammary lipid synthesis and desaturation process. Indeed, as diet-induced MFD was associated with an increase in milk trans-10-18:1 content (Griinari et al. 1998; Piperova et al. 2000; Loor et al. 2005b) and trans-10, cis-12-CLA (Baumgard et al. 2000), these FA were proposed as candidates for inhibition of mammary lipogenesis (Bauman and Griinari 2001). Studies with post-ruminal infusion of trans-10, cis-12-CLA clearly demonstrated the anti-lipogenic effect of this CLA isomer which was however not always accompanied by a reduction of milk desaturation ratios. Then, post-ruminal infusion studies in lactating cows have established that in amounts above 5 g/day, trans-10, cis-12-CLA typically decreased milk FA desaturation ratios consistent with a decrease in *SCD1* mRNA abundance (Table 13.3). In the bovine, in vivo studies indicated that injection or infusion of trans-10, cis-12-CLA decreased mammary *SCD1* gene transcription (Baumgard et al. 2002; Gervais et al. 2009), whereas no effect was observed on *SCD5* (Gervais et al. 2009). In cows, the responses of mRNA abundances of the lipogenic genes involved in other pathways than desaturation (de novo FA synthesis, FA uptake, transport and esterification) to either diets that induce MFD (Table 13.1) or trans-10, cis-12-CLA infusion (Tables 13.3 and 13.4) showed a large decrease that occurred prior to a decrease in *SCD1* mRNA. It is likely that the responses of lipogenic genes and *SCD1* to these factors reflect a two-step phenomenon which could be related to differences in the response to the transcription factors involved in these regulations (see section "Knowledge on the Signalling Pathways Mediating Nutritional Regulation of SCD Gene Expression in Ruminants"). Otherwise, in the caprine, administration of trans-10, cis-12-CLA directly at the duodenum (Andrade and Schmidely 2006) or fed as calcium salts (Shingfield et al. 2009) was reported to lower milk FA desaturation ratios in the absence (Andrade and Schmidely 2006) or with a small decrease (Shingfield et al. 2009; Table 13.3) in milk fat secretion. Similarly, in the ovine, distribution of trans-10, cis-12-CLA as encapsulated lipids was reported to lower milk fat secretion and FA desaturation ratios (Lock et al. 2006).

Table 13.3 Summary of CLA isomers acting as inhibitors of milk fat synthesis and/or milk fatty acid desaturase ratios in ruminants

Treatment[a]	Species	Dose (g/d)	Duration (h)	Change in milk fat secretion (%)[b]	Milk desaturation ratio response[c] C14/C16/C18/C18:1 t11	Transfer efficiency into milk (%)	Reference
t10c12-CLA							
Post-rumen	Bovine	13.6	5	−48	−32/NS/−17/−24	12	Baumgard et al. (2002)
	Bovine	5	4×24	−28	−10/NS/NS/−10	28	Perfield et al. (2006)
	Bovine	4.2	4×24	−20	−14/NS/−32/−27	19	Sæbø et al. (2005)
	Bovine	5	5×24	−27	−6/−4/−4/−8	22	Perfield et al. (2007)
Intravenous	Bovine	10	5×24	−32	NS		Gervais et al. (2009)
Post-rumen	Caprine	2	3×24	NS	NR/−21/−50/−45	18	Andrade and Schmidely (2006)
Calcium salt	Caprine	7.5	10×24	−20	−70/−53/−57/−17	2.37	Shingfield et al. (2009)
	Caprine	15	10×24	−28	−74/−50/−63/−2	2.38	
	Caprine	22	10×24	−32	−66/−34/−66/−14	2.19	
Lipid encapsulated	Ovine	2.4	10×24	−16	NS/−24/−6/−8	3.8	Lock et al. (2006)
t10t12-CLA							
Post-rumen	Bovine	5	4×24	NS	−19/−18/−10/−15	17	Perfield et al. (2006)
	Bovine	4.7	4×24	NS	−14/NS/−20/−15	23	Sæbø et al. (2005)
t9t11-CLA							
Post-rumen	Bovine	5	5×24	NS	−29/−11/−8/−13	21	Perfield et al. (2007)

[a]Treatment consists of either infusion, abomasal, duodenal, or intravenous of CLA or addition of CLA to the diet as lipid encapsulated or calcium salt

[b]Response reported when treatment effects were significant ($P<0.10$) and calculated as [((Treatment − Control)/Control) × 100]

[c]Response of the milk desaturation ratio (14:1-cis9/14:0) reported when treatment effects were significant ($P<0.10$); NS, values not reported because unsignificant; NR, not reported

Table 13.4 Effect of infusion or injection of trans-10, cis-12-CLA and trans-11-18:1 on mammary mRNA abundances of *SCD1* and lipogenic genes in ruminants

Treatment[a]	Species	Transcript[b]	Dose (g/d)	Duration (h)	Response (%)[c]	Milk desaturation ratio response[d] C9-14:1/14:0	Change in milk fat secretion (%)[c]	Reference
t10c12-CLA								
Post-rumen	Bovine	SCD1	13.6	5	−50	−32	−48	Baumgard et al. (2002)
		FAS/ACC			−40/−39			
		LPL/FABP			−48/−54			
		GPAT/AGPAT			−42/−41			
Intravenous	Bovine	SCD1	10	3×24	NS	NR	−24	Harvatine and Bauman (2006)
		SREBF1/Insig1/S14			−30/−45/−40			
		FAS/LPL			−25/−25			
Intravenous[e]	Bovine	SCD1/SCD5	10	5×24	Tendancy/NS	NS	−32	Gervais et al. (2009)
		ACC/FAS/LPL			−46/−57/tendancy			
		SREBF1			−59			
t11-18:1	Caprine	SCD1	0.3	4×24	+27	+66 (c9t11/t11)	ND	Jin et al. (2012)
Intravenous		ACC/FAS/LPL			−40/−64/NS			

[a]Treatment consists of either post-ruminal infusion and/or intravenous injection of trans-10, cis-12-CLA and trans-11-18:1

[b]Measurement of tissue transcript abundance (mRNA) of *SCD* stearoyl-CoA desaturase 1 and/or 5, *ACC* acetyl CoA carboxylase, *AGPAT* acylglycerol phosphate acyl transferase, *FAS* fatty acid synthase, *FABP* fatty acid binding protein, *GPAT* glycerol phosphate acyl transferase, *Insig1* insulin induced gene 1, *LPL* lipoprotein lipase, *S14* thyroid hormone responsive spot 14, *SREBF1* sterol response element binding protein

[c]Response reported when treatment effects were significant ($P<0.10$) and calculated as [((Treatment − Control)/Control) × 100]

[d]Response of the milk desaturation ratio (14:1-cis9/14:0) reported when treatment effects were significant ($P<0.10$); NS, values not reported because unsignificant; NR, not reported

[e]Control treatment, infusion of cis-9, cis-12-C18:2

From data in ruminant species (Tables 13.1, 13.3, and 13.4), it may be hypothesized that whereas mammary lipogenic genes expression (at the level of the mRNA and protein) in caprine is less sensitive to the anti-lipogenic effect of the trans-10, cis-12-CLA than in cows, for SCD protein the figure could be different with a higher sensitivity in goats. Indeed, indirect comparison from studies in goats (Bernard et al. 2009c) and cows (Roy et al. 2006) with similar dietary conditions (diets based on maize silage and supplemented with 5 % sunflower oil) showed that the four milk FA desaturation ratios decreased in goats, whereas these ratios increased in cows (Table 13.2).

Moreover, post-ruminal infusion studies in cows have shown that in addition to trans-10, cis-12-CLA, trans-10, trans-12- (Sæbø et al. 2005) and trans-9, trans-11-CLA (Perfield et al. 2007) reduced milk FA desaturation ratios (Table 13.3), and therefore, these two latter biohydrogenation intermediates could be specific inhibitors of SCD activity since they were not associated with MFD. In particular, the high response of milk trans-9, trans-11-CLA in goats receiving sunflower oil could play a direct role in the typical decrease of milk FA desaturation ratios in this animal species (Table 13.2). Otherwise, abomasal infusions in cows and ewes (Griinari et al. 2000; Corl et al. 2001; Kay et al. 2004; Bichi et al. 2012) of sterculic oil, known as a strong inhibitor of SCD, and administration of Co-EDTA in the rumen or per os (Shingfield et al. 2006, 2008a; Taugbøl et al. 2008, 2010; Karlengen et al. 2012) decreased the milk FA desaturation ratios without alteration of mammary *SCD1* mRNA abundance (Karlengen et al. 2012) and without decreasing milk fat secretion. It was hypothesized (Shingfield et al. 2008a) that Cobalt interferes with the transfer of electrons from cytochrome b5 to the di-iron protein center of the terminal desaturase protein, thus altering the oxidation–reduction reactions of the conversion of the acyl-CoA substrates. In bovine the concomitant effects of trans-10, cis-12-CLA on mammary *SCD1* mRNA abundance and milk FA desaturation ratios (Table 13.4) suggest an effect at the level of *SCD1* transcription for this CLA isomer. Otherwise, the mechanisms by which trans-10, trans-12 CLA and trans-9, trans-11 CLA decreased milk FA desaturation ratios are not clear and could involve an inhibition of *SCD1* gene transcription or activity.

In Vitro Studies on the Effect of Fatty Acids on Mammary SCD

Few in vitro systems have been developed to investigate the role of specific FA on the regulation of mammary lipogenesis (Table 13.5), including: (1) dispersed bovine mammary epithelial cells (bMECs) (Hansen et al. 1986; Hansen and Knudsen 1987) or primary bMECs (Matitashvili and Bauman 2000); (2) bMEC lines (MAC-T (Jayan and Herbein 2000; Peterson et al. 2004; Kadegowda et al. 2009; Ma and Corl 2012), BME-UV (McFadden et al. 2008)) or the cloned bMECs (Liu et al. 2006; Yonezawa et al. 2008); and (3) mammary slices (Matitashvili et al. 2001; Bernard et al. 2012). Although they are a simplification of physiological conditions, these

Table 13.5 Effect of fatty acid and of agonist or silencer of transcription factor on the regulation of mammary SCD1 and lipogenic gene mRNA abundance and/or enzyme activity in vitro

Biochemical pathway[a]	Transcript/protein[b]	In vitro model	FA supplement/no FA or treatment	Concentration (μM)	Duration treatment (h)	Change in expression or activity (%)[c]	Reference
Nuclear receptors							
SREBF1	mRNA	MAC-T	trans-10, cis-12-CLA	75	48	NS	Peterson et al. (2004)
			cis-9, trans-11-CLA	75	48	NS	
	Precursor protein (125 kDa)		trans-10, cis-12-CLA	75	48	NS	
			cis-9, trans-11-CLA	75	48	−11	
	Mature protein (68 kDa)		trans-10, cis-12-CLA	75	48	−26	
			cis-9, trans-11-CLA	75	48	NS	
SREBF1/ PPARG/S14	mRNA	MAC-T	16:0	100	12	NS/NS/NS	Kadegowda et al. (2009)
			18:0	100	12	NS/NS/NS	
			cis-9-18:1	100	12	−139/NS/NS	
			trans-10-18:1	100	12	−48/NS/NS	
			trans-10, cis-12-CLA	100	12	−189/NS/-73	
			20:5	100	12	−125/NS/NS	
			Rosiglitazone	50	12	+100/NS/NS	
SREBF1	mRNA	BME-UV	T0901317 (T09)	2	8	+416	McFadden and Corl (2010)
	Precursor protein		LXR agonist	2	8	+240	
	Mature protein			2	8	+210	
SREBF1	mRNA	Goat MEC	T0901317 (T09)	0.01/0.1/1	24	+132/+269/+381	Wang et al. (2012)
FAS	mRNA		LXR agonist	0.01/0.1/1	24	+127/+157/+223	
LXRA	mRNA			0.01/0.1/1	24	NS/NS/NS	

SREBF1/S14	mRNA	MAC-T	SREBF1-SiRNA[d]	0.1	48	−90/−95	Ma and Corl (2012)
SREBF1	Precursor protein			0.1	48	Undetectable	
	Mature protein			0.1	48	Undetectable	
<u>*Desaturation and de novo lipogenesis*</u>							
SCD1	mRNA	MAC-T	trans-11-18:1	50	(90)+6	NS	Jin et al. (2012)
PSMA5	mRNA	MAC-T	trans-11-18:1	50	(90)+6	+340	
SCD1	mRNA	bovMEC	trans-11-18:1	200	6 or 72	+200	Matitashvili and Bauman (2000)
			cis-11-18:1			NS	
SCD1	mRNA	MAC-T	trans-10, cis-12-CLA	75	48	−40	Peterson et al. (2004)
			cis-9, trans-11-CLA	75	48	NS	
SCD1	mRNA	MAC-T	16:0	100	12	+95	Kadegowda et al. (2009)
			18:0	100	12	NS	
			cis-9-18:1	100	12	−428	
			trans-10-18:1	100	12	−100	
			trans-10, cis-12-CLA	100	12	−357	
			20:5	100	12	−205	
			Rosiglitazone	50	12	NS	
SCD	Activity	MAC-T	cis-9-18:1/18:0	100	72	−21	Jayan and Herbein (2000)
			trans-11-18:1/18:0	100	72	+72	
ACC/FAS	Activity	MAC-T	cis-9-18:1/18:0	100	72	−27/−35	Jayan and Herbein (2000)
			trans-11-18:1/18:0	100	72	−58/−43	
SCD1	mRNA	MAC-T	SREBF1-SiRNA[d]	0.1	48	−60	Ma and Corl (2012)
ACC/FAS/IDH1	mRNA					−40/−65/−50	
ACSS1/ACSS2	mRNA					NS/−60	

(continued)

Table 13.5 (continued)

Biochemical pathway[a]	Transcript/protein[b]	In vitro model	FA supplement/no FA or treatment	Concentration (μM)	Duration treatment (h)	Change in expression or activity (%)[c]	Reference
SCD1/ACC	Protein					−90/−47	
FAS	mRNA	BME-UV	T0901317 (T09)	2	8	+185	McFadden and Corl (2010)
FAS	mRNA	Goat MEC	T0901317 (T09)	0.01/0.1/1	24	+127/+157/+223	Wang et al. (2012)
SCD	Activity	Goat cell fraction[e]	Oleic (>lin >lnn)[f]	0.8	1	−53	Bickerstaffe and Annison (1970)
Long-chain activation and transport							
ACSL1/FABP3	mRNA	MAC-T	SREBF1-SiRNA[d]	0.1	48	NS/−76	Ma and Corl (2012)
Esterification and protein of milk fat globule							
GPAM/AGPAT6	mRNA	MAC-T	SREBF1-SiRNA[d]	0.1	48	−20/+22	Ma and Corl (2012)
DGAT1						±70	
Protein of the milk fat globule							
PLIN2/BTN1A1/LPIN1	mRNA	MAC-T	SREBF1-SiRNA[d]	0.1	48	−250/NS/−90	Ma and Corl (2012)

[a]*SCD* stearoyl-CoA desaturase, *ACC* acetyl CoA carboxylase, *FAS* fatty acid synthase, ACSS*1* acyl-CoA synthetase short chain family member 1, *ACSS2* acyl-CoA synthetase short chain family member 2, *ACSL1* acyl-CoA synthetase long chain family member 1, *ACSL2* acyl-CoA synthetase long chain family member 2, *IDH1* isocitrate dehydrogenase 1, *S14* thyroid hormone responsive spot 14, *SREBF1* sterol response element binding protein, *PPARG* peroxisome proliferator receptor gamma, *LPIN1* lipin1, *PLIN2* perilipin 2, *BTN1A1* butyrophilin subfamily 1 member A1, *AGPAT6* acylglycerol phosphate acyl transferase 6, *FABP3* fatty acid binding protein 3, *GPAM* glycerol-3-phosphate acyltransferase 1, *DGAT1* diacylglycerol acyl transferase 1

[b]Measurement of tissue transcript abundance (mRNA) or protein activity (Activity)

[c]Response reported when treatment effects were significant ($P < 0.10$) and calculated as [((Treatment − Control)/Control) × 100]; NS, values not reported because unsignificant

[d]MACT-T transfected with sterol regulatory element binding protein SREBP-1 small interfering (si) RNA

[e]Goat subcellular fraction of mammary tissue (post-mitochondrial fraction)

[f]lin linoleic acid, lnn linolenic acid

in vitro systems offer an advantage over in vivo nutritional studies because the effects of specific FA in small amount can be evaluated under controlled conditions. Then, several years ago looking at the effects of specific FA on bovine mammary cell line (MAC-T cells), Jayan and Herbein (2000) showed that, compared to stearic acid, oleic acid decreased SCD activity. Similarly, still using MAC-T cell, Kadegowda et al. (2009) demonstrated that oleic acid, among other long-chain FA including 20:5*n*-3, was the most potent inhibitor of *SCD* gene expression, which is suggestive of a feedback-inhibition mechanism. Furthermore, earlier studies (Bickerstaffe and Annison 1970) observed negative effects of oleic, linolcic, and linolenic acids on goat mammary SCD activity measured on the subcellular fraction of goat mammary gland, which have been partly confirmed by in vivo studies on goat (Bernard et al. 2005b, 2009b). Regarding the effect of trans-FA, only two trans-18:1 isomers have been tested in vitro (Table 13.5). Incubation of trans-11-18:1 with bovine MEC increased *SCD1* mRNA abundance (Matitashvili and Bauman 2000) and SCD activity in bovine MAC-T cell line (Jayan and Herbein 2000), whereas trans-10-18:1 reduced *SCD1* mRNA abundance in bovine MAC-T cell line (Kadegowda et al. 2009). Similarly, in goats, intravenous bolus injection of trans-11-18:1 increased both the expression of *SCD* in mammary tissue (Jin et al. 2012) and milk cis-9, trans-11-CLA, suggesting an upregulation of the expression at the protein level. Otherwise, trans-10, cis-12-CLA (and to a lesser extent cis-9, trans-11-CLA) caused a significant reduction in *SCD* transcriptional activity (Keating et al. 2006) in bovine MAC-T cells, due to the *SCD* transcriptional enhancer element region (STE; see section "Characterization of SCD Genes"). The same study also showed that bovine *SCD* promoter was upregulated by insulin and downregulated by oleic acid, whereas linoleic, linolenic, stearic, and vaccenic acids had no effect. Elsewhere, using bovine and caprine mammary slices, Bernard et al. (2012) did not evidenced significant effect of trans-10, cis-12-CLA on *SCD1* mRNA abundance after 20 h incubation time. Conversely, in bovine MAC-T cell, Kadegowda et al. (2009) demonstrated that trans-10, cis-12-CLA reduced expression of *SCD1* together with *SREBF1*. These two latter studies outline differences of response of different in vitro models for studying lipogenic genes expression.

So far, the developed in vitro systems present several limitations: MECs and cultured mammary slices have a limited lifetime, modified cell lines often have abnormal characteristics and low secretory activity. Nevertheless, use of these approaches has allowed to decipher, at least in part, the underlying mechanisms responsible for the inhibitory effects of trans-10, cis-12-CLA on lipogenesis and lipogenic gene expression in primary culture of bMEC (Matitashvili and Bauman 2000), MEC lines (Peterson et al. 2004; McFadden et al. 2008; Kadegowda et al. 2009), and mammary explants (Matitashvili et al. 2001) (see section "Knowledge on the Signalling Pathways Mediating Nutritional Regulation of SCD Gene Expression in Ruminants"). Further research is necessary in ruminants to identify the major regulators of FA desaturation which is hampered by the lack of pure trans-18:1 and 18:2 isomers and by the difficulty to obtain an in vitro functional model for lipid synthesis and secretion (Barber et al. 1997; Bernard et al. 2012).

Knowledge on the Signalling Pathways Mediating Nutritional Regulation of SCD *Gene Expression in Ruminants*

Whereas the signalling pathways involved in the regulation of *SCD* gene expression by dietary factors (e.g., the inhibitory effect of PUFA) have been comprehensively described in rodent tissues (Nakamura and Nara 2004; Ntambi et al. 2004), still little is known about these mechanisms in ruminants. Among the class of transcription factors known as major regulators of lipid synthesis are the sterol response element binding protein (SREBP) family (Eberlé et al. 2004) and peroxisome proliferators-activated receptors (PPARs) (Clarke 2001).

Because different SREBP isoforms have different roles in regulating lipid synthesis, these isoforms have been studied extensively in different species (Brown and Goldstein 1997).

SREBP family is composed of three members: two isoforms of SREBP-1 (a and c) and SREBP-2. Both SREBP-1a and SREBP-1c, which can be differentially expressed in various tissues, are produced from a single gene (*SREBF-1*, located on bovine chromosome 17) presenting alternative splicing. *SREBF1c* and *SREBF1a* use different first exons and both regulate FA synthesis. SREBP-2 is transcribed from a separate gene (*SREBF2*, located on bovine chromosome 5) and regulates the cholesterol biosynthetic pathway (Horton 2002).

Multiple proteins are involved in the SREBP pathway because (1) SREBP are translated as precursor proteins which need maturation steps to become mature that involve several proteins (such as SCAP, INSIG1, 2, S1P, S2P) and (2) the SREBP mature proteins recruit other proteins acting as co-activators to activate the transcription of lipogenic genes to whom belong *LPL*, *ACACA*, *FASN*, and *SCD1* genes.

Regarding PPARs, three different isoforms of PPAR (PPARα, PPARβ/δ, and PPARγ), have been identified and are encoded by three different genes; they exhibit significant differences in tissue distribution and ligands but all function by dimerizing with RXR and binding to the PPAR response element (PPRE) DNA sequence (Sampath and Ntambi 2005).

In lactating cows, in vivo studies have shown the involvement of SREBP-1 and/or PPAR during diet-induced changes in mammary lipogenic gene expression (Angulo et al. 2012; Bionaz and Loor 2008) or in response to 3–5 d intravenous infusions of trans-10, cis-12-CLA (Harvatine and Bauman 2006; Gervais et al. 2009) with the implication that these transcription factors are central elements in the overall regulation of milk fat synthesis.

However, little is known in the ruminant mammary gland on the signalling pathways involved in *SCD* gene expression. Among six in vivo studies (either nutritional or infusion studies) in cows with MFD that measured the mammary mRNA abundances of both lipogenic genes and *SCD1*, only two reported a simultaneous decrease in mRNA abundance of lipogenic genes and *SCD1* (Baumgard et al. 2002; Angulo et al. 2012). Conversely, in the others, either nutritional studies (Ahnadi et al. 2002; Harvatine and Bauman 2006; Peterson et al. 2003) or with intravenous

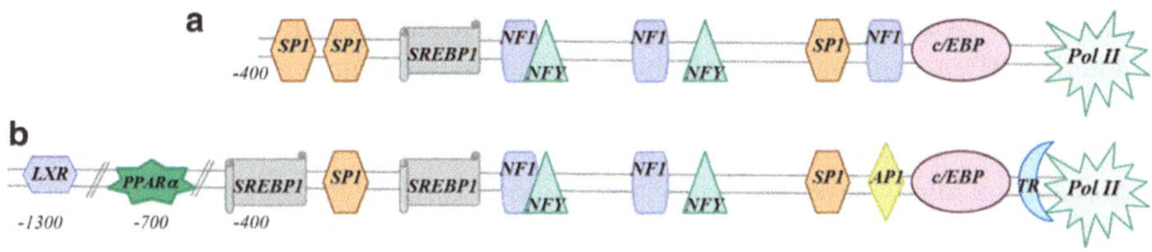

Fig. 13.5 Diagram of the SCD1 gene promoter. The diagram shows the different transcription factors involved in the regulation of SCD1 gene transcription. (**a**) Represents the binding sites of transcription factors deduced by sequence homology between bovine sequence (AY241932) and pig (AY487830) and human (AF 320307). (**b**) Represents elements characterized in human, mouse, and chicken promoters from Mauvoisin and Mounier (2011)

infusion of trans-10, cis-12-CLA (Gervais et al. 2009), a coordinate decrease in the abundance of lipogenic genes was observed with only a tendency to decrease for *SCD1* mRNA suggesting that the transcription of this gene varies less or less rapidly in response to the diet. Probably the other lipogenesis-related transcription factors (e.g., PPARs, SP1, NFY, CEBPs; Fig. 13.5) expressed in the ruminant mammary gland (Bionaz and Loor 2008; Toral et al. 2013b) are involved in the mediation of diet-induced effects on mammary *SCD1* gene expression that may counteract for the effect of *SREBF1*.

Otherwise, in an in vitro study using bovine MAC-T cell line, addition of trans-10, cis-12-CLA and 20:5*n*-3 decreased both *SREBF1* and *SCD1* mRNA abundances with variable effects on the other lipogenic genes and without changing *PPARG* mRNA (Kadegowda et al. 2009). Recently, it was reported that the bovine *SREBF1* gene contained in its promoter responsive elements for both *SREBF1* (SRE) that could allow for a positive feedback regulation by SREBP-1 itself and for liver X receptor (LXR-RE; Lengi and Corl 2010). Therefore, the SREBP-1 responsiveness to LXR activation was demonstrated in BME-UV cell line (McFadden and Corl 2010) using T0901317 (T09), an LXR agonist that resulted in the upregulation of transcription, translation, and proteolytic cleavage of SREBP-1. Because both SRE and LXR-RE have been characterized in *SCD1* gene promoter in human, mouse, or chicken (Mauvoisin and Mounier 2011) and are conserved between species, probably they may also be present in bovine *SCD1* gene promoter. Using bovine MAC-T cell line, Peterson et al. (2004) have demonstrated that the inhibitory effect of trans-10, cis-12-CLA acts through inhibition of the proteolytic activation of SREBP-1, resulting in a reduction in the transcriptional activation of lipogenic genes (*ACACA* and *FASN*) and *SCD1*, without increasing the *SREBF1* expression. The role of SREBP-1 in regulating milk fat synthesis had not been directly evaluated in ruminants until a recent study in MAC-T cells (Ma and Corl 2012) using RNA interference to knock down *SREBF1* (the *SREBF1*-specific siRNA targeted a region of the transcript identical between *SREBF1a* and *SREBF1c* and did not distinguish the two isoforms 1a and 1c) to investigate the specific role of SREBP-1 in affecting the expression of genes encoding key enzymes of milk fat synthesis including *SCD1*. In this study,

FABP3 and *SCD1 mRNAs* were both decreased when SREBP-1 mRNA decreased (and SREBP protein was undetectable), indicating that transport and desaturation of long-chain FA could be regulated transcriptionally by SREBP-1.

Altogether these data underlined the necessity to better characterize the molecular mechanisms involved in the dietary regulation of *SCD* gene expression in ruminant species, and in particular to precise the role of SREBP-1 and PPARs. The observed differences in the response of *SCD* and other lipogenic genes expression to decreased *SREBF1* mRNA abundance (Table 13.1) may suggest a lower threshold response for *SCD* to SREBP-1 level compared to other lipogenic genes which would be in line with the hypothesis of a two-step mechanisms of regulation for these genes, with the implication of other transcription factors in these regulations.

Conclusions

A balance between the endogenously synthesized short-chain FA, the exogenous unsaturated long-chain FA, and the SCD desaturation products must be maintained within the mammary gland in order to preserve the fluidity of milk fat, which emphasizes the importance of knowledge on SCD regulation. Over the last years, the development of studies in ruminants on the nutritional regulation of a few "candidate" genes involved in mammary lipogenesis including Δ9-desaturation (*SCD*) has shown that *SCD* gene regulation differ from other lipogenic genes. Indeed, in cows with dietary conditions that induce a dramatic MFD, a decrease in the mRNA abundance of mammary lipogenic genes was observed which was not always observed for *SCD*. It was hypothesized that the responses of lipogenic genes and *SCD1* to dietary factors that induce MFD in cows occur as a two-step phenomenon, with a first downregulation of lipogenic genes followed by *SCD1*, which could be related to the observed differences in the response to *SREBF1* or to other transcription factors involved in these regulations. In addition, these studies showed that the response of mammary *SCD* mRNA or activity to nutritional factors do not always match with the responses of milk FA Δ-9 desaturation ratios. In goats, *SCD* is regulated at a transcriptional and/or post-transcriptional level, depending on the lipid supplements. Conversely, in cows, data converged to demonstrate that *SCD* mRNA varied little in response to diet except for a decrease when "protected" fish oil or combination of PUFA and DHA-rich algae was fed. Indirect comparison of data from bovine and caprine suggested that SCD in caprine could be more sensitive to the effect of dietary PUFA or their ruminal biohydrogenation intermediates than in cows and that trans-9, trans-11-CLA could play a specific role in the regulation of mammary Δ-9 desaturation in goats receiving PUFA-rich diets. Future experiments examining changes in mammary *SCD* and other lipogenic genes expression in the different ruminant species in response to dietary treatments as well as ruminant *SCD1* and *SCD5* promoter sequencing would provide a more comprehensive insight into the regulation of these genes and their impact in the synthesis of milk fat.

References

Ahnadi CE, Beswick N, Delbecchi L, Kennelly JJ, Lacasse P (2002) Addition of fish oil to diets for dairy cows. II. Effects on milk fat and gene expression of mammary lipogenic enzymes. J Dairy Res 69(4):521–531

Alim MA, Fan YP, Wu XP, Xie Y, Zhang Y, Zhang SL, Sun DX, Zhang Y, Zhang Q, Liu L et al (2012) Genetic effects of stearoyl-coenzyme A desaturase (SCD) polymorphism on milk production traits in the Chinese dairy population. Mol Biol Rep 39(9):8733–8740

Andrade PVD, Schmidely P (2006) Effect of duodenal infusion of trans10, cis12-CLA on milk performance and milk fatty acid profile in dairy goats fed high or low concentrate diet in combination with rolled canola seed. Reprod Nutr Dev 46(1):31–48

Angulo J, Mahecha L, Nuernberg K, Nuernberg G, Dannenberger D, Olivera M, Boutinaud M, Leroux C, Albrecht E, Bernard L (2012) Effects of polyunsaturated fatty acids from plant oils and algae on milk fat yield and composition are associated with mammary lipogenic and SREBF1 gene expression. Animal 6:1–12

Annison EF, Bickerstaffe R, Linzell JL (1974) Glucose and fatty acid metabolism in cows producing milk of low fat content. J Agric Sci 82:87–95

Barber MC, Clegg RA, Travers MT, Vernon RG (1997) Lipid metabolism in the lactating mammary gland. Biochim Biophys Acta 1347(2–3):101–126

Bauman DE, Corl BA, Baumgard LH, Griinari JM (2001) Conjugated linoleic acid (CLA) and the dairy cow. In: Recent advances in animal nutrition, Garnsworthy PC, Wiseman J (eds), Nottingham university press, Nottingham, p 221–250

Bauman DE, Griinari JM (2001) Regulation and nutritional manipulation of milk fat: low-fat milk syndrome. Livest Prod Sci 70(1–2):15–29

Bauman DE, Griinari JM (2003) Nutritional regulation of milk fat synthesis. Annu Rev Nutr 23:203–227

Baumgard LH, Corl BA, Dwyer DA, Sæbø A, Bauman DE (2000) Identification of the conjugated linoleic acid isomer that inhibits milk fat synthesis. Am J Physiol Regul Integr Comp Physiol 278(1):R179–R184

Baumgard LH, Matitashvili E, Corl BA, Dwyer DA, Bauman DE (2002) Trans-10, cis-12 conjugated linoleic acid decreases lipogenic rates and expression of genes involved in milk lipid synthesis in dairy cows. J Dairy Sci 85(9):2155–2163

Bernard L, Bonnet M, Leroux C, Shingfield KJ, Chilliard Y (2009a) Effect of sunflower-seed oil and linseed oil on tissue lipid metabolism, gene expression, and milk fatty acid secretion in Alpine goats fed maize silage-based diets. J Dairy Sci 92(12):6083–6094

Bernard L, Leroux C, Bonnet M, Rouel J, Martin P, Chilliard Y (2005a) Expression and nutritional regulation of lipogenic genes in mammary gland and adipose tissues of lactating goats. J Dairy Res 72(2):250–255

Bernard L, Leroux C, Chilliard Y (2008) Expression and nutritional regulation of lipogenic genes in the ruminant lactating mammary gland. Adv Exp Med Biol 606:67–108

Bernard L, Leroux C, Faulconnier Y, Durand D, Shingfield KJ, Chilliard Y (2009b) Effect of sunflower-seed oil or linseed oil on milk fatty acid secretion and lipogenic gene expression in goats fed hay-based diets. J Dairy Res 76(241–248)

Bernard L, Leroux C, Hayes H, Gautier M, Chilliard Y, Martin P (2001) Characterization of the caprine stearoyl-CoA desaturase gene and its mRNA showing an unusually long 3′-UTR sequence arising from a single exon. Gene 281(1–2):53–61

Bernard L, Montazer Torbati MB, Graulet B, Leroux C, Chilliard Y (2012) Long-chain fatty acids differentially alter lipogenesis in bovine and caprine mammary slices. J Dairy Res 80(89–95)

Bernard L, Mouriot J, Rouel J, Glasser F, Capitan P, Pujos-Guillot E, Chardigny JM, Chilliard Y (2010) Effects of fish oil and starch added to a diet containing sunflower-seed oil on dairy goat performance, milk fatty acid composition and in vivo delta 9-desaturation of [C-13]vaccenic acid. Br J Nutr 104(3):346–354

Bernard L, Rouel J, Leroux C, Ferlay A, Faulconnier Y, Legrand P, Chilliard Y (2005b) Mammary lipid metabolism and milk fatty acid secretion in alpine goats fed vegetable lipids. J Dairy Sci 88(4):1478–1489

Bernard L, Shingfield KJ, Rouel J, Ferlay A, Chilliard Y (2009c) Effect of plant oils in the diet on performance and milk fatty acid composition in goats fed diets based on grass hay or maize silage. Br J Nutr 101(2):213–224

Bichi E, Toral PG, Hervás G, Frutos P, Gómez-Cortés P, Juárez M, de la Fuente MA (2012) Inhibition of Δ9-desaturase activity with sterculic acid: effect on the endogenous synthesis of cis-9 18:1 and cis-9, trans-11 18:2 in dairy sheep. J Dairy Sci 95(9):5242–5252

Bickerstaffe R, Annison EF (1970) The desaturase activity of goat and sow mammary tissue. Comp Biochem Physiol 35:653–665

Bionaz M, Loor JJ (2008) Gene networks driving bovine milk fat synthesis during the lactation cycle. BMC Genomics 9:366

Brown MS, Goldstein JL (1997) The SREBP pathway: regulation of cholesterol metabolism by proteolysis of a membrane-bound transcription factor. Cell 89(3):331–340

Carta A, Casu S, Usai MG, Addis M, Fiori M, Fraghi A, Miari S, Mura L, Piredda G, Schibler L et al (2008) Investigating the genetic component of fatty acid content in sheep milk. Small Ruminant Res 79(1):22–28

Chilliard Y, Ferlay A, Mansbridge RM, Doreau M (2000) Ruminant milk fat plasticity: nutritional control of saturated, polyunsaturated, trans and conjugated fatty acids. Ann Zootechnol 49(3):181–205

Chilliard Y, Glasser F, Ferlay A, Bernard L, Rouel J, Doreau M (2007) Diet, rumen biohydrogenation and nutritional quality of cow and goat milk fat. Eur J Lipid Sci Technol 109(8):828–855

Chilliard Y, Rouel J, Leroux C (2006) Goat's alpha-s1 casein genotype influences its milk fatty acid composition and delta-9 desaturation ratios. Anim Feed Sci Technol 131(3–4):474–487

Christie WW (1981) The composition, structure and function of lipids in the tissues of ruminant animals. In: Christie WW (ed) Lipid metabolism in ruminant animals. Pergamon press, Oxford, pp 95–192

Chung M, Ha S, Jeong S, Bok J, Cho K, Baik M, Choi Y (2000) Cloning and characterization of bovine stearoyl CoA desaturasel cDNA from adipose tissues. Biosci Biotechnol Biochem 64(7):1526–1530

Clarke SD (2001) Polyunsaturated fatty acid regulation of gene transcription: a molecular mechanism to improve the metabolic syndrome. J Nutr 131(4):1129–1132

Conte G, Mele M, Chessa S, Castiglioni B, Serra A, Pagnacco G, Secchiari P (2010) Diacylglycerol acyltransferase 1, stearoyl-CoA desaturase 1, and sterol regulatory element binding protein 1 gene polymorphisms and milk fatty acid composition in Italian Brown cattle. J Dairy Sci 93(2):753–763

Corl BA, Baumgard LH, Dwyer DA, Griinari JM, Phillips BS, Bauman DE (2001) The role of delta(9)-desaturase in the production of cis-9, trans-11 CLA. J Nutr Biochem 12(11):622–630

Corl BA, Baumgard LH, Griinari JM, Delmonte P, Morehouse KM, Yuraweczc MP, Bauman DE (2002) Trans-7, cis-9 CLA is synthesized endogenously by delta(9)-desaturase in dairy cows. Lipids 37(7):681–688

Delbecchi L, Ahnadi CE, Kennelly JJ, Lacasse P (2001) Milk fatty acid composition and mammary lipid metabolism in Holstein cows fed protected or unprotected canola seeds. J Dairy Sci 84(6):1375–1381

Dewhurst RJ, Shingfield KJ, Lee MRF, Scollan ND (2006) Increasing the concentrations of beneficial polyunsaturated fatty acids in milk produced by dairy cows in high-forage systems. Anim Feed Sci Technol 131(3–4):168–206

Eberlé D, Hegarty B, Bossard P, Ferre P, Foufelle F (2004) SREBP transcription factors: master regulators of lipid homeostasis. Biochimie 86(11):839–848

Enjalbert F, Nicot MC, Bayourthe C, Moncoulon R (1998) Duodenal infusions of palmitic, stearic or oleic acids differently affect mammary gland metabolism of fatty acids in lactating dairy cows. J Nutr 128(9):1525–1532

Gama MAS, Garnsworthy PC, Griinari JM, Leme PR, Rodrigues PHM, Souza LWO, Lanna DPD (2008) Diet-induced milk fat depression: association with changes in milk fatty acid composition and fluidity of milk fat. Livest Sci 115(2–3):319–331

Garcia-Fernandez M, Gutierrez-Gill B, Garcia-Gamez E, Arranz JJ (2009) Genetic variability of the stearoyl-CoA desaturase gene in sheep. Mol Cell Probes 23(2):107–111

Garcia-Fernandez M, Gutierrez-Gil B, Garcia-Gamez E, Sanchez JP, Arranz JJ (2010) Detection of quantitative trait loci affecting the milk fatty acid profile on sheep chromosome 22: role of the stearoyl-CoA desaturase gene in Spanish Churra sheep. J Dairy Sci 93(1):348–357

Gervais R, McFadden JW, Lengi AJ, Corl BA, Chouinard PY (2009) Effects of intravenous infusion of trans-10, cis-12 18: 2 on mammary lipid metabolism in lactating dairy cows. J Dairy Sci 92(10):5167–5177

Glasser F, Ferlay A, Doreau M, Schmidely P, Sauvant D, Chilliard Y (2008) Long-chain fatty acid metabolism in dairy cows: a meta-analysis of milk fatty acid yield in relation to duodenal flows and de novo synthesis. J Dairy Sci 91(7):2771–2785

Griinari JM, Corl BA, Lacy SH, Chouinard PY, Nurmela KVV, Bauman DE (2000) Conjugated linoleic acid is synthesized endogenously in lactating dairy cows by delta(9)-desaturase. J Nutr 130(9):2285–2291

Griinari JM, Dwyer DA, McGuire MA, Bauman DE, Palmquist DL, Nurmela KV (1998) Trans-octadecenoic acids and milk fat depression in lactating dairy cows. J Dairy Sci 81(5):1251–1261

Hansen HO, Knudsen J (1987) Effect of exogenous long-chain fatty acids on lipid biosynthesis in dispersed ruminant mammary gland epithelial cells: esterification of long-chain exogenous fatty acids. J Dairy Sci 70(7):1344–1349

Hansen HO, Tornehave D, Knudsen J (1986) Synthesis of milk specific fatty acids and proteins by dispersed goat mammary-gland epithelial cells. Biochem J 238(1):167–172

Harvatine KJ, Bauman DE (2006) SREBP1 and thyroid hormone responsive spot 14 (S14) are involved in the regulation of bovine mammary lipid synthesis during diet-induced milk fat depression and treatment with CLA. J Nutr 136(10):2468–2474

Horton JD (2002) Sterol regulatory element-binding proteins: transcriptional activators of lipid synthesis. Biochem Soc Trans 30:1091–1095

Invernizzi G, Thering BJ, McGuire MA, Savoini G, Loor JJ (2010) Sustained upregulation of stearoyl-CoA desaturase in bovine mammary tissue with contrasting changes in milk fat synthesis and lipogenic gene networks caused by lipid supplements. Funct Integr Genomics 10(4):561–575

Jacobs AAA, van Baal J, Smits MA, Taweel HZH, Hendriks WH, van Vuuren AM, Dijkstra J (2011) Effects of feeding rapeseed oil, soybean oil, or linseed oil on stearoyl-CoA desaturase expression in the mammary gland of dairy cows. J Dairy Sci 94(2):874–887

Jayan GC, Herbein JH (2000) “Healthier” dairy fat using trans-vaccenic acid. Nutr Food Sci 30(6):304–309

Jensen RG (2002) The composition of bovine milk lipids: January 1995 to December 2000. J Dairy Sci 85(2):295–350

Jin YC, Li ZH, Hong ZS, Xu CX, Han JA, Choi SH, Yin JL, Zhang QK, Lee KB, Kang SK et al (2012) Conjugated linoleic acid synthesis-related protein proteasome subunit α 5 (PSMA5) is increased by vaccenic acid treatment in goat mammary tissue. J Dairy Sci 95(8):4286–4297

Kadegowda AKG, Bionaz M, Piperova LS, Erdman RA, Loor JJ (2009) Peroxisome proliferator-activated receptor-gamma activation and long-chain fatty acids alter lipogenic gene networks in bovine mammary epithelial cells to various extents. J Dairy Sci 92(9):4276–4289

Kaestner KH, Ntambi JM, Kelly TJJ, Lane MD (1989) Differentiation-induced gene expression in 3T3-L1 preadipocytes. A second differentially expressed gene encoding stearoyl-CoA desaturase. J Biol Chem 264(25):14755–14761

Karlengen IJ, Harstad OM, Taugbol O, Berget I, Aastveit AH, Vage DI (2012) The effect of excess cobalt on milk fatty acid profiles and transcriptional regulation of SCD, FASN, DGAT1 and DGAT2 in the mammary gland of lactating dairy cows. J Anim Physiol Anim Nutr 96(6):1065–1073

Kay JK, Mackle TR, Auldist MJ, Thomson NA, Bauman DE (2004) Endogenous synthesis of cis-9, trans-11 conjugated linoleic acid in dairy cows fed fresh pasture. J Dairy Sci 87(2):369–378
Keating AF, Kennelly JJ, Zhao FQ (2006) Characterization and regulation of the bovine stearoyl-CoA desaturase gene promoter. Biochem Biophys Res Commun 344(1):233–240
Keating AF, Stanton C, Murphy JJ, Smith TJ, Ross RP, Cairns MT (2005) Isolation and characterization of the bovine stearoyl-CoA desaturase promoter and analysis of polymorphisms in the promoter region in dairy cows. Mamm Genome 16(3):184–193
Kgwatalala PM, Ibeagha-Awemu EM, Hayes JF, Zhao X (2007) Single nucleotide polymorphisms in the open reading frame of the stearoyl-CoA desaturase gene and resulting genetic variants in Canadian Holstein and Jersey cows. DNA Seq 18(5):357–362
Kgwatalala PM, Ibeagha-Awemu EM, Hayes JF, Zhao X (2009a) Stearoyl-CoA desaturase 1 3′UTR SNPs and their influence on milk fatty acid composition of Canadian Holstein cows. J Anim Breed Genet 126(5):394–403
Kgwatalala PM, Ibeagha-Awemu EM, Mustafa AF, Zhao X (2009b) Stearoyl-CoA desaturase 1 genotype and stage of lactation influences milk fatty acid composition of Canadian Holstein cows. Anim Genet 40(5):609–615
Kinsella JE (1970) Stearic acid metabolism by mammary cells. J Dairy Sci 53(12):1757–1765
Legrand P, Catheline D, Fichot MC, Lemarchal P (1997) Inhibiting delta9-desaturase activity impairs triacylglycerol secretion in cultured chicken hepatocytes. J Nutr 127(2):249–256
Lengi AJ, Corl BA (2007) Identification and characterization of a novel bovine stearoyl-CoA desaturase isoform with homology to human SCD5. Lipids 42(6):499–508
Lengi AJ, Corl BA (2008) Comparison of pig, sheep and chicken SCD5 homologs: evidence for an early gene duplication event. Comp Biochem Physiol B Biochem Mol Biol 150(4):440–446
Lengi AJ, Corl BA (2010) Short communication: identification of the bovine sterol regulatory element binding protein-1c promoter and its activation by liver X receptor1. J Dairy Sci 93(12):5831–5836
Lengi AJ, Corl BA (2012) Regulation of the bovine SCD5 promoter by EGR2 and SREBP1. Biochem Biophys Res Commun 421(2):375–379
Lerch S, Shingfield KJ, Ferlay A, Vanhatalo A, Chilliard Y (2012) Rapeseed or linseed in grass-based diets: effects on conjugated linoleic and conjugated linolenic acid isomers in milk fat from Holstein cows over 2 consecutive lactations. J Dairy Sci 95(12):7269–7287
Li XZ, Yan CG, Lee HG, Choi CW, Song MK (2012) Influence of dietary plant oils on mammary lipogenic enzymes and the conjugated linoleic acid content of plasma and milk fat of lactating goats. Anim Feed Sci Technol 174(1–2):26–35
Liu W, Degner SC, Romagnolo DF (2006) Trans-10, cis-12 conjugated linoleic acid inhibits prolactin-induced cytosolic NADP + –dependent isocitrate dehydrogenase expression in bovine mammary epithelial cells. J Nutr 136(11):2743–2747
Lock AL, Teles BM, Perfield JW, Bauman DE, Sinclair LA (2006) A conjugated linoleic acid supplement containing trans-10, cis-12 reduces milk fat synthesis in lactating sheep. J Dairy Sci 89(5):1525–1532
Loor JJ, Doreau M, Chardigny JM, Ollier A, Sébédio JL, Chilliard Y (2005a) Effects of ruminal or duodenal supply of fish oil on milk fat secretion and profiles of trans-fatty acids and conjugated linoleic acid isomers in dairy cows fed maize silage. Anim Feed Sci Technol 119(3–4):227–246
Loor JJ, Ferlay A, Ollier A, Chilliard Y (2005b) Relationship among trans and conjugated fatty acids and bovine milk fat yield due to dietary concentrate and linseed oil. J Dairy Sci 88(2):726–740
Ma L, Corl BA (2012) Transcriptional regulation of lipid synthesis in bovine mammary epithelial cells by sterol regulatory element binding protein-1. J Dairy Sci 95(7):3743–3755
Matitashvili E, Bauman DE (2000) Effect of different isomers of C18:1 and C18:2 fatty acids on lipogenesis in bovine mammary epithelial cells. J Anim Sci 78(Supplement 1):165
Matitashvili E, Baumgard LH, Bauman DE (2001) The effect of trans-10, cis-12 conjugated linoleic acid (CLA) infusion on milk fat synthesis and expression of lipogenic enzymes in the mammary gland of lactating cows. J Anim Sci 79(Suppl 1):310

Mauvoisin D, Mounier C (2011) Hormonal and nutritional regulation of SCD1 gene expression. Biochimie 93(1):78–86

McDonald TM, Kinsella JE (1973) Stearyl-CoA desaturase of bovine mammary microsomes. Arch Biochem Biophys 156(1):223–231

McFadden JW, Corl BA (2010) Activation of liver X receptor (LXR) enhances de novo fatty acid synthesis in bovine mammary epithelial cells. J Dairy Sci 93(10):4651–4658

McFadden JW, Mullarky IK, Corl BA (2008) Inhibitory effect of unsaturated fatty acids on *de novo* fatty acid synthesis in bovine mammary epithelial cells. J Anim Sci 86(E-Suppl 2):566

Mele M, Conte G, Castiglioni B, Chessa S, Macciotta NPP, Serra A, Buccioni A, Pagnacco G, Secchiari P (2007) Stearoyl-Coenzyme A desaturase gene polymorphism and milk fatty acid composition in Italian Holsteins. J Dairy Sci 90(9):4458–4465

Mihara K (1990) Structure and regulation of rat liver microsomal stearoyl-CoA desaturase gene. J Biochem 108(6):1022–1029

Miyazaki M, Bruggink SM, Ntambi JM (2006) Identification of mouse palmitoyl-coenzyme A Delta 9-desaturase. J Lipid Res 47(4):700–704

Miyazaki M, Jacobson MJ, Man WC, Cohen P, Asilmaz E, Friedman JM, Ntambi JM (2003) Identification and characterization of murine SCD4, a novel heart-specific stearoyl-CoA desaturase isoform regulated by leptin and dietary factors. J Biol Chem 278(36):33904–33911

Moioli B, Contarini G, Avalli A, Catillo G, Orru L, De Matteis G, Masoero G, Napolitano F (2007) Short communication: effect of stearoyl-coenzyme A desaturase polymorphism on fatty acid composition of milk. J Dairy Sci 90(7):3553–3558

Mosley EE, McGuire MA (2007) Methodology for the in vivo measurement of the delta(9)-desaturation of myristic, palmitic, and stearic acids in lactating dairy cattle. Lipids 42:939–945

Mosley EE, Shafii B, Moate PJ, McGuire MA (2006) Cis-9, trans-11 conjugated linoleic acid is synthesized directly from vaccenic acid in lactating dairy cattle. J Nutr 136(3):570–575

Murrieta CM, Hess BW, Scholljegerdes EJ, Engle TE, Hossner KL, Moss GE, Rule DC (2006) Evaluation of milk somatic cells as a source of mRNA for study of lipogenesis in the mammary gland of lactating beef cows supplemented with dietary high-linoleate safflower seeds. J Anim Sci 84(9):2399–2405

Nakamura MT, Nara TY (2004) Structure, function, and dietary regulation of delta 6, delta 5, and delta 9 desaturases. Annu Rev Nutr 24:345–376

Ntambi JM (1999) Regulation of stearoyl-CoA desaturase by polyunsaturated fatty acids and cholesterol. J Lipid Res 40(9):1549–1558

Ntambi JM, Buhrow SA, Kaestner KH, Christy RJ, Sibley E, Kelly TJJ, Lane MD (1988) Differentiation-induced gene expression in 3T3-L1 preadipocytes. Characterization of a differentially expressed gene encoding stearoyl- CoA desaturase. J Biol Chem 263(33):17291–17300

Ntambi JM, Miyazaki M, Dobrzyn A (2004) Regulation of stearoyl-CoA desaturase expression. Lipids 39(11):1061–1065

Ollier S, Leroux C, de la Foye A, Bernard L, Rouel J, Chilliard Y (2009) Whole intact rapeseeds or sunflower oil in high-forage or high-concentrate diets affects milk yield, milk composition, and mammary gene expression profile in goats. J Dairy Sci 92(11):5544–5560

Oshino N, Sato R (1972) The dietary control of the microsomal stearyl CoA desaturation enzyme system in rat liver. Arch Biochem Biophys 149(2):369–377

Palmquist DL, Lock AL, Shingfield KJ, Bauman DE (2005) Biosynthesis of conjugated linoleic acid in ruminants and humans. In: Taylor SL (ed) Advances in food and nutrition research. Elsevier, San Diego, CA, pp 179–217

Parodi PW (1982) Positional distribution of fatty acids in the triglyceride classes of milk fat. J Dairy Res 49(1):73–80

Parodi PW (2005) Dairy product consumption and the risk of breast cancer. J Am Coll Nutr 24(6 Suppl):556S–568S

Pauciullo A, Cosenza G, Steri R, Coletta A, La Battaglia A, Di Berardino D, Macciotta NPP, Ramunno L (2012) A single nucleotide polymorphism in the promoter region of river buffalo stearoyl CoA desaturase gene (SCD) is associated with milk yield. J Dairy Res 79(04):429–435

Perfield JW, Delmonte P, Lock AL, Yurawecz MP, Bauman DE (2006) Trans-10, trans-12 conjugated linoleic acid does not affect milk fat yield but reduces delta9-desaturase index in dairy cows. J Dairy Sci 89(7):2559–2566

Perfield JW, Lock AL, Griinari JM, Sæbø A, Delmonte P, Dwyer DA, Bauman DE (2007) Trans-9, cis-11 conjugated linoleic acid reduces milk fat synthesis in lactating dairy cows. J Dairy Sci 90(5):2211–2218

Peterson DG, Matitashvili EA, Bauman DE (2003) Diet-induced milk fat depression in dairy cows results in increased trans-10, cis-12 CLA in milk fat and coordinate suppression of mRNA abundance for mammary enzymes involved in milk fat synthesis. J Nutr 133(10):3098–3102

Peterson DG, Matitashvili EA, Bauman DE (2004) The inhibitory effect of trans-10, cis-12 CLA on lipid synthesis in bovine mammary epithelial cells involves reduced proteolytic activation of the transcription factor SREBP-1. J Nutr 134(10):2523–2527

Piperova LS, Teter BB, Bruckental I, Sampugna J, Mills SE, Yurawecz MP, Fritsche J, Ku K, Erdman RA (2000) Mammary lipogenic enzyme activity, trans fatty acids and conjugated linoleic acids are altered in lactating dairy cows fed a milk fat-depressing diet. J Nutr 130(10):2568–2574

Qiu X, Eastridge ML, Firkins JL (2004) Effects of dry matter intake, addition of buffer, and source of fat on duodenal flow and concentration of conjugated linoleic acid and trans-11 C18:1 in milk. J Dairy Sci 87(12):4278–4286

Rincon G, Islas-Trejo A, Castillo AR, Bauman DE, German BJ, Medrano JF (2012) Polymorphisms in genes in the SREBP1 signalling pathway and SCD are associated with milk fatty acid composition in Holstein cattle. J Dairy Res 79(01):66–75

Roy A, Ferlay A, Shingfield KJ, Chilliard Y (2006) Examination of the persistency of milk fatty acid composition responses to plant oils in cows given different basal diets, with particular emphasis on *trans*-C18:1 fatty acids and isomers of conjugated linoleic acid. Anim Sci 82:479–492

Sæbø A, Sæbø PC, Griinari JM, Shingfield KJ (2005) Effect of abomasal infusions of geometric isomers of 10, 12 conjugated synthesis linoleic acid on milk fat in dairy cows. Lipids 40(8): 823–832

Sampath H, Ntambi JM (2005) Polyunsaturated fatty acid regulation of genes of lipid metabolism. Annu Rev Nutr 25:317–340

Sessler AM, Kaur N, Palta JP, Ntambi JM (1996) Regulation of stearoyl-CoA desaturase 1 mRNA stability by polyunsaturated fatty acids in 3T3-L1 adipocytes. J Biol Chem 271(47): 29854–29858

Schennink A, Heck JML, Bovenhuis H, Visker MHPW, van Valenberg HJF, van Arendonk JAM (2008) Milk fatty acid unsaturation: genetic parameters and effects of stearoyl-CoA desaturase (SCD1) and acyl CoA: diacylglycerol acyltransferase 1 (DGAT1). J Dairy Sci 91(5): 2135–2143

Shingfield KJ, Ahvenjarvi S, Toivonen V, Vanhatalo A, Huhtanen P (2007) Transfer of absorbed cis-9, trans-11 conjugated linoleic acid into milk is biologically more efficient than endogenous synthesis from absorbed vaccenic acid in lactating cows. J Nutr 137(5):1154–1160

Shingfield KJ, Arola A, Ahvenjarvi S, Vanhatalo A, Toivonen V, Griinari JM, Huhtanen P (2008a) Ruminal infusions of cobalt-EDTA reduce mammary delta 9-desaturase index and alter milk fatty acid composition in lactating cows. J Nutr 138(4):710–717

Shingfield KJ, Chilliard Y, Toivonen V, Kairenius P, Givens DI (2008b) Trans fatty acids and bioactive lipids in ruminant milk. Adv Exp Med Biol 606:3–65

Shingfield KJ, Bernard L, Leroux C, Chilliard Y (2010) Role of trans fatty acids in the nutritional regulation of mammary lipogenesis in ruminants. Animal 4(7):1140–1166

Shingfield KJ, Bonnet M, Scollan ND (2013) Recent developments in altering the fatty acid composition of ruminant-derived foods. Animal 7(Supplements1):132–162

Shingfield KJ, Rouel J, Chilliard Y (2009) Effect of calcium salts of a mixture of conjugated linoleic acids containing trans-10, cis-12 in the diet on milk fat synthesis in goats. Br J Nutr 101(7):1006–1019

Shingfield KJ, Toivonen V, Vanhatalo A, Huhtanen P, Griinari JM (2006) Short communication: indigestible markers reduce the mammary delta9-desaturase index and alter the milk fatty acid composition in cows. J Dairy Sci 89(8):3006–3010

Stoop WM, Bovenhuis H, Heck JML, van Arendonk JAM (2009) Effect of lactation stage and energy status on milk fat composition of Holstein-Friesian cows. J Dairy Sci 92(4):1469–1478

Taugbøl O, Karlengen IJ, Bolstad T, Aastveit AH, Harstad OM (2008) Cobalt supplied per os reduces the mammary {delta}9-desaturase index of bovine milk. J Anim Sci 86(11):3062–3068

Taugbøl O, Karlengen IJ, Salbu B, Aastveit AH, Harstad OM (2010) Intravenous injections of cobalt reduce fatty acid desaturation products in milk and blood of lactating cows. J Anim Physiol Anim Nutr 94(5):635–640

Thiede MA, Strittmatter P (1985) The induction and characterization of rat liver stearyl-CoA desaturase mRNA. J Biol Chem 260(27):14459–14463

Timmen H, Patton S (1988) Milk fat globules: fatty acid composition, size and in vivo regulation of fat liquidity. Lipids 23(7):685–689

Toral PG, Bernard L, Chilliard Y, Glasser F (2013a) Diet-induced variations in milk fatty acid composition have minor effects on the estimated melting point of milk fat in cows, goats, and ewes: insights from a meta-analysis. J Dairy Sci 96:1232–1236

Toral PG, Bernard L, Delavaud C, Gruffat D, Leroux C, Chilliard Y (2013b) Effects of fish oil and additional starch on tissue fatty acid profile and lipogenic gene mRNA abundance in lactating goats fed a diet containing sunflower-seed oil. Animal 7(6):948–956

Vakalopoulou E, Schaack J, Shenk T (1991) A 32-kilodalton protein binds to AU-rich domains in the 3′ untranslated regions of rapidly degraded mRNAs. Mol Cell Biol 11(6):3355–3364

Wahle KWJ, Heys SD, Rotondo D (2004) Conjugated linoleic acids: are they beneficial or detrimental to health? Prog Lipid Res 43(6):553–587

Wang W, Luo J, Zhong Y, Lin XZ, Shi HB, Zhu JJ, Li J, Sun YT, Zhao WS (2012) Goat liver X receptor alpha, molecular cloning, functional characterization and regulating fatty acid synthesis in epithelial cells of goat mammary glands. Gene 505(1):114–120

Wang J, Yu L, Schmidt RE, Su C, Huang X, Gould K, Cao G (2005) Characterization of HSCD5, a novel human stearoyl-CoA desaturase unique to primates. Biochem Biophys Res Commun 332(3):735–742

Ward RJ, Travers MT, Richards SE, Vernon RG, Salter AM, Buttery PJ, Barber MC (1998) Stearoyl-CoA desaturase mRNA is transcribed from a single gene in the ovine genome. Biochim Biophys Acta 1391(2):145–156

Yonezawa T, Haga S, Kobayashi Y, Katoh K, Obara Y (2008) Regulation of hormone-sensitive lipase expression by saturated fatty acids and hormones in bovine mammary epithelial cells. Biochem Biophys Res Commun 376(1):36–39

Zhang L, Ge L, Parimoo S, Stenn K, Prouty SM (1999) Human stearoyl-CoA desaturase: alternative transcripts generated from a single gene by usage of tandem polyadenylation sites. Biochem J 340(Pt 1):255–264

Zheng Y, Prouty SM, Harmon A, Sundberg JP, Stenn KS, Parimoo S (2001) Scd3–a novel gene of the stearoyl-CoA desaturase family with restricted expression in skin. Genomics 71(2): 182–191

Zidi A, Fernández-Cabanás VM, Urrutia B, Carrizosa J, Polvillo O, González-Redondo P, Jordana J, Gallardo D, Amills M, Serradilla JM (2010) Association between the polymorphism of the goat stearoyl-CoA desaturase 1 (SCD1) gene and milk fatty acid composition in Murciano-Granadina goats. J Dairy Sci 93(9):4332–4339

[illegible]

Chapter 14
Physiological Functions and Regulation of *C. elegans* Stearoyl-CoA Desaturases

Jennifer L. Watts

The *Caenorhabditis elegans* Model for Studies of Lipid Synthesis and Function

The nematode *C. elegans* is an excellent model for genetic studies of complex biological phenomena, and research using this organism has contributed significantly to elucidating the mechanisms of diverse biological processes (Barr 2003). Recent work in *C. elegans* has identified many regulatory proteins and downstream effector genes responsible for lipid homeostasis (Watts 2009; Vrablik and Watts 2012). While *C. elegans* stores lipids in intestinal and hypodermal cells rather than dedicated adipose tissue, other aspects of worm biochemistry and regulation of fat metabolism closely parallel humans. Because of its small size (1.5 mm), rapid life cycle, and ease of laboratory cultivation, *C. elegans* offers great potential for genetic analysis (Riddle et al. 1997). In addition to the traditional forward genetic approaches, reverse genetic studies are popular and powerful because the activity of a target gene can be inactivated in vivo by feeding *C. elegans* bacteria producing double stranded RNA for that gene, a technique referred to as RNA interference, or RNAi (Fire et al. 1998). During feeding the double stranded RNA is absorbed by the intestine and distributed throughout the animal (Timmons and Fire 1998). The construction of RNAi libraries that consist of *E. coli* strains expressing double stranded RNA that correspond to nearly every *C. elegans* gene has allowed for whole-genome screening approaches to identify genes involved in diverse biological processes. These processes include embryogenesis (Sonnichsen et al. 2005), neuronal specification (Poole et al. 2011), and the regulation of longevity (Hamilton et al. 2005).

J.L. Watts (✉)
School of Molecular Biosciences, Washington State University, Pullman, WA 99164, USA
e-mail: jwatts@vetmed.wsu.edu

J.M. Ntambi (ed.), *Stearoyl-CoA Desaturase Genes in Lipid Metabolism*,
DOI 10.1007/978-1-4614-7969-7_14,

C. elegans contains genes encoding enzymes required for fatty acid synthesis, elongation, and desaturation as well as those encoding peroxisomal and mitochondrial β-oxidation enzymes. Fatty acids are esterified to glycerol to form triacylglycerols (TAGs), which are stored in lipid droplets and yolk. In *C. elegans*, TAGs are a vital energy source during embryogenesis, periods of low food availability, and for the specialized non-feeding dauer larval stage. Like other eukaryotes, *C. elegans* obtains fatty acids from its diet and also synthesizes them de novo from acetyl-CoA (Perez and Van Gilst 2008). For de novo synthesis, all of the enzyme activities necessary for the synthesis of palmitic acid (16:0) from acetyl-CoA are encoded by two multifunctional enzymes: acetyl-CoA carboxylase and fatty acid synthase (Wakil 1989; Rappleye et al. 2003; Chirala and Wakil 2004; Wakil and Abu-Elheiga 2008). Palmitic acid can be integrated into TAGs or phospholipids, or it can be modified by fatty acid elongases and desaturases to form a variety of long chain polyunsaturated fatty acids (PUFAs; Wallis et al. 2002; Watts and Browse 2002). Long chain PUFAs are preferentially incorporated into membrane phospholipids, where they affect membrane properties including fluidity, temperature sensitivity, and signaling. In addition, *C. elegans* produces monomethyl branched-chain fatty acids (mmBCFAs), which are essential for growth and embryonic development (Kniazeva et al. 2004, 2012).

C. elegans Encodes Three SCD Orthologs

Due to the abundant and vital roles of lipids, the synthesis and breakdown of fatty acids are subject to many levels of regulation. A critical control point regulating lipid synthesis is the production of monounsaturated fatty acids. *C. elegans* encodes three stearoyl-CoA desaturases (SCDs), also known as Δ9 desaturases, that synthesize monounsaturated fatty acids from saturated fatty acids (Watts and Browse 2000). These enzymes, named FAT-5, FAT-6, and FAT-7, share significant amino acid similarity to rat and mouse SCDs, as well as to the yeast Δ9 desaturase OLE1. By expressing the SCD genes in the yeast *ole1* mutant, the substrate specificities of the three enzymes were determined. The FAT-5 desaturase is specific for palmitic acid (16:0), whereas the FAT-6 and FAT-7 desaturases preferentially introduce a double bond into stearic acid (18:0) (Watts and Browse 2000).

In contrast to mammals, oleic acid produced by the FAT-6 and FAT-7 SCDs is used as a substrate to produce a range of PUFAs. This reaction is mediated by a Δ12 desaturase, an enzyme normally found only in algae and plants (Wallis et al. 2002). Δ12 desaturation of oleic acid produces linoleic acid (18:2), which is further acted on by an omega-3 desaturase FAT-1 and a Δ6 desaturase FAT-3, which, together with elongation and Δ5 desaturation, produces a range of C18 and C20 omega-6 and omega-3 PUFAs (Fig. 14.1) (Watts and Browse 2002).

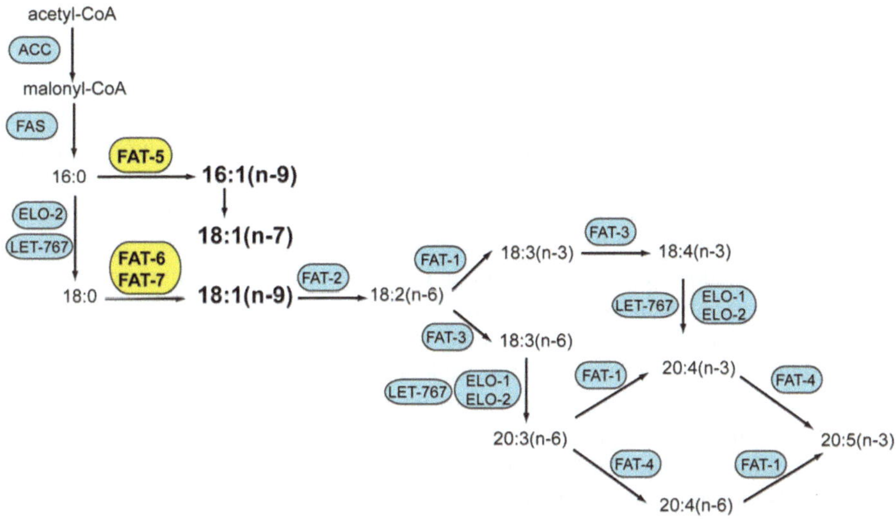

Fig. 14.1 Pathway of synthesis of unsaturated fatty acids in *C. elegans*. A variety of long chain polyunsaturated fatty acids can be synthesized from acetyl-CoA. The SCDs FAT-5, FAT-6, and FAT-7 are highlighted in *yellow*. *ACC* acetyl-CoA carboxylase, *FAS* fatty acid synthase, *ELO* fatty acid elongase, *LET* lethal, *FAT* fatty acid desaturase

Loss of One SCD Isoform Is Compensated for by Up-regulation of the Remaining SCDs

To determine the consequences of depleted SCD activity in *C. elegans*, loss of function mutants in each gene were isolated and analyzed. Gene knockdowns of individual SCD genes using RNAi have also been reported, but due to high sequence similarity between the three SCD genes, which lead to off-target effects, the phenotypes of strains carrying single and double mutations are more easily interpreted. The *fat-5;fat-6;fat-7* triple mutant strain was constructed and was found to be inviable, revealing that endogenous production of monounsaturated fatty acids is essential for survival (Brock et al. 2006). In contrast, the fatty acid composition changes and physiological consequences of the single desaturase mutants are subtle (Brock et al. 2006). The most highly expressed of the three Δ9 desaturases is the FAT-6 SCD, which is expressed in the intestine and hypodermal (skin) tissues (Brock et al. 2006). The *fat-6* mutants display increased stearic acid (18:0) and decreased unsaturated fatty acids, however, the change in fatty acid composition in the *fat-6* mutant worms is modulated by genetic compensation by the other SCDs. Real-time quantitative RT-PCR experiments showed that in the *fat-6* mutants, the *fat-7* and *fat-5* genes are up-regulated three to fivefold. Presumably, this compensation normalizes the fatty acid composition and allows the strain to grow, reproduce, and behave like wild type (Brock et al. 2006).

Similarly, the *fat-7* mutants have wild type fatty acid composition, fat stores, and reproductive success, even when combined with *klf-3* mutants, which lack a Kruppel-like transcription factor and regulate fat storage in *C. elegans* and in humans (Brock et al. 2006; Zhang et al. 2011). In *fat-7* mutants, both *fat-6* and *fat-5* are up-regulated 1.5–2-fold (Brock et al. 2006). Knockdown of *fat-6* or *fat-7* by RNAi was reported to show decreased resistance to oxidative stress and increased heat tolerance (Horikawa and Sakamoto 2009). However, given that high degree of DNA sequence similarity (86 % identity) between the *fat-6* and *fat-7* cDNA sequences, off-target reduction of the other SCD homologues confounds the interpretation of these results.

In addition to genetic compensation, defects in fatty acid desaturase genes can be compensated for by dietary input. For example, the *fat-5* mutant has only subtle fatty acid composition changes when grown on plate with dietary *E. coli*. This is because palmitoleic acid (16:1) and its elongation product cis-vaccenic acid (18:1*n*-7) are provided by the *E. coli* diet. When *fat-5* mutants are grown axenically in media lacking *E. coli*, the *fat-5* mutants have a more profound fatty acid composition defect, showing greatly increased palmitic acid (16:0) and severely reduced cis-vaccenic acid (18:1*n*-7) (Brock et al. 2006). A recent study examining the relationship between fat stores and dietary restriction in *C. elegans* showed that *fat-5* expression increased two to fourfold in nematodes exposed to dietary restriction during development (Palgunow et al. 2012). Dietary restriction during development resulted in reduced body size, but an increase in the size of lipid droplets and in overall TAG stores. In addition to increases in *fat-5* expression, other lipogenic genes such as those encoding DGAT and acetyl-CoA synthetase were also up-regulated, suggesting that *C. elegans* adapts metabolically to conserve fat stores during periods of suboptimal nutriention (Palgunow et al. 2012).

In spite of compensation to normalize overall fatty acid composition, LC–MS analysis revealed differences in various TAG and PC species among the *fat-5*, *fat-6*, and *fat-7* mutant strains, as well as evidence of increased absorption of dietary fatty acids in the *fat-6* mutant (Castro et al. 2012). Interestingly, metabolomic analysis revealed significant differences in non-lipid metabolites such as alanine and succinate, indicating an increase in catabolism of amino acids and an increase in the activity of the TCA cycle in the *C. elegans* SCD mutants (Castro et al. 2012). Metabolomic studies such as these promise to provide insights as to the complex compensatory mechanisms that occur in animals when lipid homeostasis is altered.

Double Mutant Strains Reveal Additional Requirements for SCDs

Examination of double mutant strains reveals additional roles for monounsaturated fatty acid synthesis in temperature adaptation and in the regulation of fat storage. The *fat-5;fat-7*, *fat-5;fat-6*, and *fat-6;fat-7* double mutant strains display significant changes in their fatty acid composition. The two strains lacking *fat-5* have increased palmitic acid (16:0) and severely reduced palmitoleic acid (16:1*n*-7) and cis-vaccenic acid (18:1*n*-7). Under standard conditions, these strains grow normally, but they have decreased survival in cold temperature (10 °C), indicating that the higher saturated and lower unsaturated fatty acid composition is detrimental to nematodes at low temperatures (Brock et al. 2007). In addition, the *fat-5;fat-6* double mutant strain was unable to survive starvation stress. After 4 days of starvation of L1 larvae in buffer without nutrients, wild type larvae are viable and maintain a normal fatty acid composition. However, the *fat-5;fat-6* double mutants die and have extremely high saturated fatty acid content in their membranes (Brock et al. 2007). The lack of dietary fatty acids, together with the observation that *fat-7* expression is highly repressed in the absence of food (Van Gilst et al. 2005b), reveals the roles of FAT-5 and FAT-6 in maintaining proper fatty acid composition during starvation.

The *fat-6;fat-7* Double Mutants Have Reduced Fat Stores and Increased Fat Oxidation, Similar to Mouse SCD1 Mutants

The *fat-6;fat-7* double mutant is unable to carry out normal PUFA synthesis, because it cannot synthesize the oleic acid precursor which is the substrate for the remaining PUFAs. However, unusual 18-carbon PUFAs are formed by elongation and desaturation of the FAT-5 product palmitoleic acid (16:1), which apparently partially substitute for vital functions performed by normal PUFAs (Brock et al. 2007). The *fat-6;fat-7* double mutants display slow growth and greatly reduced fertility. Dietary oleic acid partially, but not completely, rescues the growth and fertility deficiencies (Brock et al. 2007).

A striking phenotype of the *fat-6;fat-7* double mutant is greatly reduced fat stores compared to wild type (Brock et al. 2007). Real-time PCR analysis of fatty acid oxidation genes reveals that genes encoding components of mitochondrial fatty acid oxidation are expressed at higher levels in the *fat-6;fat-7* double mutant compared to wild type (Brock et al. 2007). Together with the SCD1 mouse observations, these studies demonstrate that Δ9 desaturase is an important determinant of TAG accumulation and that metabolic changes resulting from decreased SCD signaling elicit protection from obesity. Pharmacological manipulation of SCD activity might therefore benefit treatments of obesity, diabetes, and other diseases of metabolic syndrome. However, a better understanding of the downstream pathways affected by SCD1 signaling will be necessary in order to understand and anticipate potential side effects from such drugs (Dobrzyn and Ntambi 2005).

Nuclear Hormone Receptors Regulate Transcription of SCDs

Nuclear hormone receptors (NHRs) are important regulators of development and metabolism. Upon binding specific ligands, the NHRs dimerize and enter the nucleus where they activate or repress distinct target genes. Such regulation allows for precise, reversible responses to environmental, developmental, and nutritional signals. *C. elegans* contains a hugely expanded repertoire of NHRs, and sequence analysis reveals that 269 out of 284 *C. elegans* NHRs evolved from an ancient form of the hepatocyte nuclear factor 4 (HNF4) family (Robinson-Rechavi et al. 2005; Taubert et al. 2011). Several NHRs have been implicated in the regulation of fat metabolism in *C. elegans*, and the SCDs are key gene downstream effectors. NHR-80 and NHR-49 are required for expression of the *fat-5, fat-6,* and *fat-7* genes (Van Gilst et al. 2005a; Brock et al. 2006). Both mutant strains have a high stearic acid (18:0) content and reduced oleic acid (18:1*n*-9). The expression of *fat-6* and *fat-7* was also shown to be decreased in *nhr-13* mutants, although the overall fatty acid composition was not affected (Pathare et al. 2012). In contrast, a mutation in the aryl-hydrocarbon receptor, *ahr-1*, results in increased expression of the *fat-7* gene, resulting in small, but statistically significant changes in fatty acid composition (Aarnio et al. 2010).

Both NHR-49 and NHR-80 are required for the up-regulation of *fat-7* needed to compensate for *fat-6* deletion mutants, and the compensation is essential because the *nhr-49;fat-6* and *nhr-80;fat-6* strains are not viable (Brock et al. 2006). Despite the similarities in their effects on Δ9 desaturation, *nhr-49* deletion mutants display more severe growth and reproductive defects than *nhr-80* mutants, probably because NHR-49 also regulates other lipid metabolism genes, including genes involved in the biosynthesis of sphingolipids, genes involved in β-oxidation, and other genes required for the response to nutrient deprivation (Van Gilst et al. 2005b; Pathare et al. 2012).

While mutations in *nhr-49* and *nhr-80* have additive defects in combination with mutations in SCDs, depletion of the *nhr-64* gene results in improved physiology of SCD mutants. The low fat stores, slow growth, and embryonic lethality phenotypes in *fat-6;fat-7* double mutants were improved in combination with *nhr-64(RNAi)* (Liang et al. 2010). Likely gene targets of NHR-64 that facilitated these improvements include β-oxidation genes, such as *ech-5* and a thiolase. However, other gene changes in *nhr-64(RNAi)* strains were also noted, including reduced expression of several acylCoA synthetase genes and increased expression of *pod-2* (acetyl-CoA carboxylase). These studies illustrate the complexity of compensatory mechanisms regulated by NHRs that balance fat storage, growth, and reproductive efficiency (Liang et al. 2010).

A network of regulators of metabolic genes enriched in NHRs has been experimentally mapped in *C. elegans,* implicating numerous NHRs in control of metabolic processes (Arda et al. 2010). It has been proposed that the expansion of NHRs in *C. elegans* evolved to enable rapid and adaptive responses to environmental clues, including dietary nutrients. It is likely that changes in SCDs are important for the adaptation to diets encountered in nature, which may vary significantly in their lipid content. Thus, the regulation of SCDs by numerous NHRs signifies their important physiological roles.

Regulation of *C. elegans* SCDs by SREBP

The sterol regulatory element binding protein (SREBP) family of basic helix–loop–helix zipper transcription factors are critical regulators of cholesterol and fatty acid homeostasis in mammals (Osborne and Espenshade 2009; Jeon and Osborne 2012). Newly synthesized SREBPs reside in the endoplasmic reticulum (ER) membrane and are inactive. When specific cellular lipid levels are low, the proteins are transported to the Golgi, where they are acted on by site 1 and site 2 proteases released by the N-terminal fragment of SREBP, which then enter the nucleus and activate transcription of lipogenic genes. Mammals express a family of SREBP isoforms (SREBP-1a, -1c, and SREBP-2), in which different isoforms are responsible for the expression of a unique set of lipid metabolism genes. For example, SREBP-2 mainly activates genes responsible for cholesterol uptake and synthesis, while SREBP-1 isoforms regulate genes involved in the biosynthesis of fatty acids, phospholipids, and TAGs. *C. elegans* possesses a single SREBP gene, *sbp-1*. Knockdown of *sbp-1* by RNAi leads to low fat stores, high saturated fatty acid content, impaired growth, and reduced expression of lipogenic genes, including the *fat-5, fat-6,* and *fat-7* SCDs (Kniazeva et al. 2004; Yang et al. 2006). A deletion mutant in the *C. elegans* SREBP homolog, *sbp-1(ep79),* displays a similar slow growth and low fat stores as the RNAi treatment (Liang et al. 2010).

The translocation of SREBP from the ER to the nucleus is well understood for the cholesterol-regulated SREBP-2 isoform (Brown and Goldstein 1997). When cholesterol is present, it binds to a cholesterol-sensing protein called SCAP, which promotes association with INSIG (Yang et al. 2002). This complex maintains SREBP in the ER membrane. When cholesterol is absent, INSIG no longer is complexed with SREBP, facilitating trafficking to the Golgi where proteases cleave the transmembrane domains, allowing for translocation to the nucleus (Sun et al. 2007). *C. elegans* does not synthesize cholesterol, and it does not encode an INSIG homolog; however, recent studies have shed light on a conserved mechanism for the nuclear translocation of the SREBP-1 isoform.

The mechanism of nuclear localization of the SREBP-1 isoform depends on membrane phosphatidylcholine (PC) levels. RNAi knockdown of genes encoding s-adenyslyl methionine transferase (*sams-1*) and other genes contributing to the PC synthesis pathway led to increased nuclear localization of SREBP and induction of SREBP target genes (such as *fat-5* and *fat-7*) in *C. elegans* and in mammalian liver cell lines (Walker et al. 2011). When PC is low, the Golgi protease enzymes undergo retrograde transport to the ER, where the proteases cleave the SREBP protein, releasing the transmembrane domains. This liberates the DNA binding portion of SREBP, allowing it to translocate into the nucleus, where it promotes transcription of lipid synthesis genes, including SCDs (Walker et al. 2011). Thus, when PC synthesis is disrupted, SREBP-induced gene expression changes led to increased fat stores in nematodes and hepatic steatosis (fatty liver) in the mouse. An independent study confirmed that *C. elegans sams-1* and *pmt-1* mutants have high fat stores compared to wild type nematodes, and they express high levels of *fat-7* (Li et al. 2011).

During periods of food deprivation, lipid metabolism is shifted away from synthesis toward breakdown, and oxidation of fatty acids provides energy for the maintenance of the organism. Consistent with this, nuclear SREBP is rapidly depleted during fasting in the mouse liver (Horton et al. 1998) and in *C. elegans* (Walker et al. 2010). This is regulated by the NAD+-dependent deacetylase SIRT1, which directly deacetylates SREBP, leading to ubiquitination, protein degradation, and reduction in target gene expression. The importance of SREBP regulation of the FAT-7 SCDs is apparent, as the expression of the *fat-7* SCD is especially sensitive to fasting, with undetectable transcript levels in the absence of food (Walker et al. 2010).

MDT-15 Interacts with SBP-1 and NHRs to Regulate *C. elegans* SCDs

After binding to target sequences in specific promoters, SREBP and NHRs recruit multi-protein coregulator complexes containing transcriptional cofactors that link the transcriptional activator to the transcription initiation machinery (Blazek et al. 2005). One component, the ARC/mediator subunit MDT-15/MED15, interacts directly and specifically with the activation domain of SREBP (Yang et al. 2006). The *C. elegans* MDT-15 also binds specifically to NHR-49 and NHR-64 (Taubert et al. 2006), which both regulate the expression of genes involved in lipid metabolism. Accordingly, RNAi reduction of *mdt-15* expression results in slow growth, low fat stores, and reduced expression of *fat-7*. Dietary supplementation with oleic acid rescues many of the defects of *sbp-1* and *mdt-15* RNAi animals, indicating that the SCDs are key targets of these regulators (Yang et al. 2006). MDT-15 is also important for broader functions, including induction of detoxification genes in response to certain ingested xenobiotics and heavy metals, suggesting that it interacts with additional transcription factors (Taubert et al. 2008).

FOXO Regulates SCDs to Confer Long Lifespan and Cold Tolerance

Insulin/IGF-1 signaling modulates development, metabolism, stress resistance, and longevity in *C. elegans* (Lin et al. 1997, 2001; Henderson and Johnson 2001). Insulin/IGF-1 signaling during favorable growth conditions negatively regulates the DAF-16/FOXO transcription factor by a phosphorylation cascade that prevents nuclear localization. During unfavorable conditions, such as low nutrients, UV irradiation, certain toxins, and exposure to high temperatures, the phosphorylation cascade is inhibited such that the DAF-16/FOXO transcription factor enters the nucleus, leading to the expression of many protective genes, such as superoxide dismutase, drug metabolizing enzymes, and molecular chaperones (McElwee et al. 2003; Murphy et al. 2003). This altered gene expression confers long lifespan and stress resistance to the nematodes. SCDs have been identified as targets of DAF-16/FOXO (Murphy et al. 2003; Schuster et al. 2010), and *fat-7(RNAi)* leads to a shortened lifespan in the long-lived insulin/IGF-1 receptor mutant *daf-2* (Murphy et al. 2003). A recent study shows that the *age-1* mutant, a long-lived mutant disrupted in the insulin/IGF-1 signaling pathway, is cold-tolerant, and this tolerance depends on DAF-16/FOXO (Savory et al. 2011). While cold stress does not induce nuclear localization of DAF-16/FOXO, cold tolerance in the *age-1* mutants depends on functional SCDs. SCDs have previously been shown to be important for adaptation to cold temperatures in many organisms including plants (Browse and Xin 2001), fish (Tiku et al. 1996), and *C. elegans* (Brock et al. 2007; Murray et al. 2007). Increased SCD expression leads to higher unsaturated fatty acid content, resulting in increased membrane fluidity that is necessary for optimal membrane function in cold environments.

Fatty Acid Desaturation and Other Lipid Modifications in Germ Cells Regulate Lifespan

In *C. elegans*, ablation of germ cells results in a long lifespan. This can be achieved through laser ablation of germ cell precursors or by mutations that block the proliferation of germ cells (Hsin and Kenyon 1999; Arantes-Oliveira et al. 2002). Recent studies reveal that fat metabolism is altered in worms lacking germ cells and that the changes in fat metabolism are required for the increased longevity in these strains (Wang et al. 2008; O'Rourke et al. 2009; Goudeau et al. 2011). Specifically, the lipase *lipl-4* is induced in worms lacking a germline, suggesting that lipase activity might underlie the lifespan extension observed in germline-less animals. In agreement with this, depletion of *lipl-4* shortens the lifespan of worms lacking germ cells (Wang et al. 2008), while overexpression of *lipl-4* is sufficient to extend lifespan in wild type worms (Wang et al. 2008; Lapierre et al. 2011). Autophagy, the process of sequestering and degrading cytosolic components is also increased in germline-less *C. elegans*. The FOXA transcription factor PHA-4 is required for the induction of autophagy-related genes (Lapierre et al. 2011). The process of autophagy mediated through PHA-4 is required for extended lifespan of the germline-less worms, as well as for sustained induction of the *lipl-4* gene (Lapierre et al. 2011).

A recent study demonstrated that SCD activity under the control of NHR-80 is required for the extended lifespan of germline-less worms (Goudeau et al. 2011). The study found that a mutation in *nhr-80* blocks the extended lifespan of a germline-less *glp-1* mutant, while overexpression of *nhr-80* further increases the extended lifespan of *glp-1*. Neither the *nhr-80* mutation nor overexpression of *nhr-80* affected the lifespan of wild type worms (Goudeau et al. 2011). The SCD activity is required for the extension of lifespan because there was no lifespan extension in the *glp-1;fat-6;fat-7* triple mutants, even in strains where the *nhr-80* gene was ectopically overexpressed. This discovery delineates a link between signals from the reproductive system, SCD activity, and longevity. Further studies will be necessary to determine the key physiological roles of SCDs in longevity, where they may be necessary for optimal membrane fluidity, efficient energy storage, or as precursors of lipid signaling molecules.

Conclusions

SCD activity is essential for proper lipid homeostasis in *C. elegans*. SCDs play important physiological roles in ensuring proper membrane fluidity, synthesizing precursors of PUFAs and other lipid signaling molecules, and enabling efficient fat storage. Whole organism approaches have facilitated studies of the consequences and compensatory mechanisms that occur as a result of reduced SCD activity, including the importance of conserved SCD regulators such as nuclear receptors and SREBP. Future studies promise to clarify lipid-mediated mechanisms contributing to the complex processes of development, reproduction, stress responses, and aging.

Acknowledgements The author thanks the National Institutes of Health (R01-DK074114) for research support and Dr. Tracy Vrablik and Dr. Amy Walker for comments on the manuscript.

References

Aarnio V, Storvik M, Lehtonen M, Asikainen S, Reisner K, Callaway J et al (2010) Fatty acid composition and gene expression profiles are altered in aryl hydrocarbon receptor-1 mutant Caenorhabditis elegans. Comp Biochem Physiol C Toxicol Pharmacol 151:318–324

Arantes-Oliveira N, Apfeld J, Dillin A, Kenyon C (2002) Regulation of life-span by germ-line stem cells in Caenorhabditis elegans. Science 295:502–505

Arda HE, Taubert S, MacNeil LT, Conine CC, Tsuda B, Van Gilst M et al (2010) Functional modularity of nuclear hormone receptors in a Caenorhabditis elegans metabolic gene regulatory network. Mol Syst Biol 6:367

Barr MM (2003) Super models. Physiol Genomics 13:15–24

Blazek E, Mittler G, Meisterernst M (2005) The mediator of RNA polymerase II. Chromosoma 113:399–408

Brock TJ, Browse J, Watts JL (2006) Genetic regulation of unsaturated fatty acid composition in C. elegans. PLoS Genet 2:e108

Brock TJ, Browse J, Watts JL (2007) Fatty acid desaturation and the regulation of adiposity in Caenorhabditis elegans. Genetics 176:865–875

Brown MS, Goldstein JL (1997) The SREBP pathway: regulation of cholesterol metabolism by proteolysis of a membrane-bound transcription factor. Cell 89:331–340

Browse J, Xin Z (2001) Temperature sensing and cold acclimation. Curr Opin Plant Biol 4:241–246

Castro C, Sar F, Shaw WR, Mishima M, Miska EA, Griffin JL (2012) A metabolomic strategy defines the regulation of lipid content and global metabolism by Delta9 desaturases in Caenorhabditis elegans. BMC Genomics 13:36

Chirala SS, Wakil SJ (2004) Structure and function of animal fatty acid synthase. Lipids 39:1045–1053

Dobrzyn A, Ntambi JM (2005) Stearoyl-CoA desaturase as a new drug target for obesity treatment. Obes Rev 6:169–174

Fire A, Xu S, Montgomery MK, Kostas SA, Driver SE, Mello CC (1998) Potent and specific genetic interference by double-stranded RNA in *Caenorhabditis elegans*. Nature 391: 806–811

Goudeau J, Bellemin S, Toselli-Mollereau E, Shamalnasab M, Chen Y, Aguilaniu H (2011) Fatty acid desaturation links germ cell loss to longevity through NHR-80/HNF4 in C. elegans. PLoS Biol 9:e1000599

Hamilton B, Dong Y, Shindo M, Liu W, Odell I, Ruvkun G et al (2005) A systematic RNAi screen for longevity genes in C. elegans. Genes Dev 19:1544–1555

Henderson ST, Johnson TE (2001) daf-16 integrates developmental and environmental inputs to mediate aging in the nematode Caenorhabditis elegans. Curr Biol 11:1975–1980

Horikawa M, Sakamoto K (2009) Fatty-acid metabolism is involved in stress-resistance mechanisms of Caenorhabditis elegans. Biochem Biophys Res Commun 390:1402–1407

Horton JD, Bashmakov Y, Shimomura I, Shimano H (1998) Regulation of sterol regulatory element binding proteins in livers of fasted and refed mice. Proc Natl Acad Sci USA 95:5987–5992

Hsin H, Kenyon C (1999) Signals from the reproductive system regulate the lifespan of C. elegans. Nature 399:362–366

Jeon TI, Osborne TF (2012) SREBPs: metabolic integrators in physiology and metabolism. Trends Endocrinol Metab 23:65–72

Kniazeva M, Crawford QT, Seiber M, Wang CY, Han M (2004) Monomethyl branched-chain fatty acids play an essential role in Caenorhabditis elegans development. PLoS Biol 2:E257

Kniazeva M, Shen H, Euler T, Wang C, Han M (2012) Regulation of maternal phospholipid composition and IP(3)-dependent embryonic membrane dynamics by a specific fatty acid metabolic event in C. elegans. Genes Dev 26:554–566

Lapierre LR, Gelino S, Melendez A, Hansen M (2011) Autophagy and lipid metabolism coordinately modulate life span in germline-less C. elegans. Curr Biol 21:1507–1514

Li Y, Na K, Lee HJ, Lee EY, Paik YK (2011) Contribution of sams-1 and pmt-1 to lipid homoeostasis in adult Caenorhabditis elegans. J Biochem 149:529–538

Liang B, Ferguson K, Kadyk L, Watts JL (2010) The role of nuclear receptor NHR-64 in fat storage regulation in Caenorhabditis elegans. PLoS One 5:e9869

Lin K, Dorman JB, Rodan A, Kenyon C (1997) daf-16: an HNF-3/forkhead family member that can function to double the life-span of Caenorhabditis elegans. Science 278:1319–1322

Lin K, Hsin H, Libina N, Kenyon C (2001) Regulation of the Caenorhabditis elegans longevity protein DAF-16 by insulin/IGF-1 and germline signaling. Nat Genet 28:139–145

McElwee J, Bubb K, Thomas JH (2003) Transcriptional outputs of the Caenorhabditis elegans forkhead protein DAF-16. Aging Cell 2:111–121

Murphy CT, McCarroll SA, Bargmann CI, Fraser A, Kamath RS, Ahringer J et al (2003) Genes that act downstream of DAF-16 to influence the lifespan of Caenorhabditis elegans. Nature 424:277–283

Murray P, Hayward SA, Govan GG, Gracey AY, Cossins AR (2007) An explicit test of the phospholipid saturation hypothesis of acquired cold tolerance in Caenorhabditis elegans. Proc Natl Acad Sci USA 104:5489–5494

O'Rourke EJ, Soukas AA, Carr CE, Ruvkun G (2009) C. elegans major fats are stored in vesicles distinct from lysosome-related organelles. Cell Metab 10:430–435

Osborne TF, Espenshade PJ (2009) Evolutionary conservation and adaptation in the mechanism that regulates SREBP action: what a long, strange tRIP it's been. Genes Dev 23:2578–2591

Palgunow D, Klapper M, Doring F (2012) Dietary restriction during development enlarges intestinal and hypodermal lipid droplets in Caenorhabditis elegans. PLoS One 7:e46198

Pathare PP, Lin A, Bornfeldt KE, Taubert S, Van Gilst MR (2012) Coordinate regulation of lipid metabolism by novel nuclear receptor partnerships. PLoS Genet 8:e1002645

Perez CL, Van Gilst MR (2008) A 13C isotope labeling strategy reveals the influence of insulin signaling on lipogenesis in C. elegans. Cell Metab 8:266–274
Poole RJ, Bashllari E, Cochella L, Flowers EB, Hobert O (2011) A genome-wide RNAi screen for factors involved in neuronal specification in Caenorhabditis elegans. PLoS Genet 7:e1002109
Rappleye CA, Tagawa A, Le Bot N, Ahringer J, Aroian RV (2003) Involvement of fatty acid pathways and cortical interaction of the pronuclear complex in Caenorhabditis elegans embryonic polarity. BMC Dev Biol 3:8
Riddle DL, Blumenthal T, Meyer BJ, Priess JR (1997) Introduction to C. elegans. In: Riddle DL, Blumenthal T, Meyer BJ, Priess JR (eds) *C. elegans*, vol II. Cold Spring Harbor Laboratory Press, Plainview, pp 1–22
Robinson-Rechavi M, Maina CV, Gissendanner CR, Laudet V, Sluder A (2005) Explosive lineage-specific expansion of the orphan nuclear receptor HNF4 in nematodes. J Mol Evol 60:577–586
Savory FR, Sait SM, Hope IA (2011) DAF-16 and Delta9 desaturase genes promote cold tolerance in long-lived Caenorhabditis elegans age-1 mutants. PLoS One 6:e24550
Schuster E, McElwee JJ, Tullet JM, Doonan R, Matthijssens F, Reece-Hoyes JS et al (2010) DamID in C. elegans reveals longevity-associated targets of DAF-16/FoxO. Mol Syst Biol 6:399
Sonnichsen B, Koski LB, Walsh A, Marschall P, Neumann B, Brehm M et al (2005) Full-genome RNAi profiling of early embryogenesis in Caenorhabditis elegans. Nature 434:462–469
Sun LP, Seemann J, Goldstein JL, Brown MS (2007) Sterol-regulated transport of SREBPs from endoplasmic reticulum to Golgi: Insig renders sorting signal in Scap inaccessible to COPII proteins. Proc Natl Acad Sci USA 104:6519–6526
Taubert S, Van Gilst MR, Hansen M, Yamamoto KR (2006) A mediator subunit, MDT-15, integrates regulation of fatty acid metabolism by NHR-49-dependent and -independent pathways in C. elegans. Genes Dev 20:1137–1149
Taubert S, Hansen M, Van Gilst MR, Cooper SB, Yamamoto KR (2008) The Mediator subunit MDT-15 confers metabolic adaptation to ingested material. PLoS Genet 4:e1000021
Taubert S, Ward JD, Yamamoto KR (2011) Nuclear hormone receptors in nematodes: evolution and function. Mol Cell Endocrinol 334:49–55
Tiku PE, Gracey AY, Macartney AI, Beynon RJ, Cossins AR (1996) Cold-induced expression of delta 9-desaturase in carp by transcriptional and posttranslational mechanisms. Science 271:815–818
Timmons L, Fire A (1998) Specific Interference by ingested dsRNA. Nature 395:854
Van Gilst MR, Hadjivassiliou H, Jolly A, Yamamoto KR (2005a) Nuclear hormone receptor NHR-49 controls fat consumption and fatty acid composition in C. elegans. PLoS Biol 3:e53
Van Gilst MR, Hadjivassiliou H, Yamamoto KR (2005b) A Caenorhabditis elegans nutrient response system partially dependent on nuclear receptor NHR-49. Proc Natl Acad Sci USA 102:13496–13501
Vrablik TL, Watts JL (2012) Emerging roles for specific fatty acids in developmental processes. Genes Dev 26:631–637
Wakil SJ (1989) Fatty acid synthase, a proficient multifunctional enzyme. Biochemistry 28:4523–4530
Wakil SJ, Abu-Elheiga LA (2008) Fatty acid metabolism: target for metabolic syndrome. J Lipid Res 50(Suppl):S138–S143
Walker AK, Yang F, Jiang K, Ji JY, Watts JL, Purushotham A et al (2010) Conserved role of SIRT1 orthologs in fasting-dependent inhibition of the lipid/cholesterol regulator SREBP. Genes Dev 24:1403–1417
Walker AK, Jacobs RL, Watts JL, Rottiers V, Jiang K, Finnegan DM et al (2011) A conserved SREBP-1/phosphatidylcholine feedback circuit regulates lipogenesis in metazoans. Cell 147:840–852
Wallis JG, Watts JL, Browse J (2002) Polyunsaturated fatty acid synthesis: what will they think of next? Trends Biochem Sci 27:467
Wang MC, O'Rourke EJ, Ruvkun G (2008) Fat metabolism links germline stem cells and longevity in C. elegans. Science 322:957–960

Watts JL (2009) Fat synthesis and adiposity regulation in Caenorhabditis elegans. Trends Endocrinol Metab 20:58–65
Watts JL, Browse J (2000) A palmitoyl-CoA-specific delta9 fatty acid desaturase from Caenorhabditis elegans. Biochem Biophys Res Commun 272:263–269
Watts JL, Browse J (2002) Genetic dissection of polyunsaturated fatty acid synthesis in Caenorhabditis elegans. Proc Natl Acad Sci USA 99:5854–5859
Yang T, Espenshade PJ, Wright ME, Yabe D, Gong Y, Aebersold R et al (2002) Crucial step in cholesterol homeostasis: sterols promote binding of SCAP to INSIG-1, a membrane protein that facilitates retention of SREBPs in ER. Cell 110:489–500
Yang F, Vought BW, Satterlee JS, Walker AK, Jim Sun ZY, Watts JL et al (2006) An ARC/mediator subunit required for SREBP control of cholesterol and lipid homeostasis. Nature 442:700–704
Zhang J, Bakheet R, Parhar RS, Huang CH, Hussain MM, Pan X et al (2011) Regulation of fat storage and reproduction by Kruppel-like transcription factor KLF3 and fat-associated genes in Caenorhabditis elegans. J Mol Biol 411:537–553

Chapter 15
Remodeling of Membrane Phospholipids by Bacterial Desaturases

Silvia Graciela Altabe, María Cecilia Mansilla, and Diego de Mendoza

Abbreviations

a-BCFAs	Anteiso-branched-chain fatty acids
ACP	Acyl carrier protein
BCFA	Branched-chain fatty acids
BKDH	Branched-chain keto acid dehydrogenase
CoA	Coenzyme A
Des	Desaturase
DHp	Dimerization and histidine phosphotransfer
Fer	Ferredoxin
Fld	Flavodoxin
MUFA	Monounsaturated fatty acid
PUFA	Polyunsaturated fatty acid
SFA	Saturated fatty acid
TM	Transmembrane
TMS	Transmembrane segment
UFA	Unsaturated fatty acid

S.G. Altabe • M.C. Mansilla • D. de Mendoza (✉)
Facultad de Ciencias Bioquímicas y Farmacéuticas, Departamento de Microbiología, Instituto de Biología Molecular y Celular de Rosario (IBR), Consejo Nacional de Investigaciones Científicas y Técnicas, Universidad Nacional de Rosario, Ocampo y Esmeralda, Predio CONICET, Rosario S2000FHQ, Argentina
e-mail: altabe@ibr-conicet.gov.ar; mansilla@ibr-conicet.gov.ar; demendoza@ibr-conicet.gov.ar

J.M. Ntambi (ed.), *Stearoyl-CoA Desaturase Genes in Lipid Metabolism*, 209
DOI 10.1007/978-1-4614-7969-7_15, © Springer Science+Business Media New York 2013

Unsaturated Fatty Acids

The physical properties of a biological membrane are highly dependent on the lipid composition, which shows tremendous variation between different membranes and dependence on conditions (Dowhan 1997). This variation is responsible for the plasticity of biomembranes, which plays a role in cellular adaptation to the environment and allows for cellular processes such as energy storage, signaling, transport, and membrane fusion and fission (Dowhan 1997; Aguilar and de Mendoza 2006; Lee 2011). Membrane fluidity is essential for maintaining the membrane barrier and for optimal functioning of membrane proteins (de Mendoza and Cronan 1983; Kaiser et al. 2011). The fluidity of the membrane lipids is in a large extent determined by the ratio of saturated (SFAs) vs. unsaturated fatty acids (UFAs) (Los and Murata 2004). The presence of *cis*-double bonds in membrane lipid acyl chains causes disturbances in the acyl chain packing, which results in much poorer packing of the acyl chains and therefore in a lower gel-to-liquid crystalline phase transition temperature of the membrane. Thus, UFAs are key molecules in the regulation of cellular membrane fluidity (Cronan and Gelmann 1973). In addition to their structural role, UFAs have recently been recognized as signaling molecules involved in several essential cellular processes, such as cell differentiation and DNA replication (for recent reviews see Heird and Lapillonne 2005; Mansilla and de Mendoza 2005).

Furthermore, the UFAs are important in human nutrition (in edible oils) and may be used as reagents of chemical industry (Napier 2007). Fatty acids determine the properties and nutritional value of edible fats. For example, vegetable oils rich in UFAs generally are considered better in cardiovascular health than fats rich in SFAs. Furthermore, the functional groups related to the double bonds such as hydroxy and epoxy groups confer important functional properties to inedible oils, used in the chemical industry (e.g., castor oil). Currently the strategy of choice to produce specific UFAs for the chemical or food industry is the manipulation of the expression of genes encoding desaturases in plants. However, the exploitation of the rich variety of UFAs produced by microorganisms could have a favorable impact in terms of flexibility and speed of production of unusual UFAs (Bonamore et al. 2006).

Unsaturated Fatty Acids Synthesis in Bacteria

Konrad Bloch, a Nobel Laureate chemist, and coworkers determined that UFAs are synthesized in two different ways; one occurs only in aerobic organisms and requires molecular oxygen, while the other pathway is used in anaerobic conditions (Scheuerbrandt et al. 1961; Scheuerbrandt and Bloch 1962).

The anaerobic pathway for UFA synthesis was established in *Escherichia coli*, where it was shown that only two unique biochemical reactions are required to specifically produce UFAs in this bacterium (Cronan et al. 1969; Bloch 1971). Introduction of the double bond involves a dehydration of the 3-hydroxydecanoyl intermediate of fatty acid synthesis to give a *cis*-3 double bond which would be conserved through subsequent cycles of addition of two carbon atoms to give the membrane lipid UFA moieties (Cronan 2006) (Fig. 15.1a). The double bond is introduced into the growing acyl chain by FabA, an enzyme capable of both the dehydration of β-hydroxydecanoyl-acyl carrier protein (ACP) to *trans*-2-decenoyl-ACP, and the isomerization of *trans*-2 to *cis*-3-decenoyl-ACP. This isomerization is critical to switch the nascent acyl chain into the UFA synthetic pathway, but it is not sufficient. The β-ketoacyl-ACP synthase I (FabB) enzyme is required to elongate the *cis*-3-decenoyl-ACP produced by FabA and its activity is the primary factor in determining cellular UFA content. At this step, the proportion of *cis*-UFA is determined by competition between the use of *cis*-3-decenoyl-ACP by FabB and the use of *trans*-2-decenoyl-ACP by FabI (Zhang and Rock 2008) (Fig. 15.1a). These two proteins, which are involved in type II fatty acid biosynthesis pathway, direct the formation of the two major UFAs that are found in bacteria: 16:1Δ9 and 18:1Δ11 (*cis-vaccenate*). In addition, there is a biochemical mechanism that allows the elongation cycle to respond to the environmental temperature. This mechanism was carefully studied in *E. coli*, and β-ketoacyl-ACP synthase II (FabF) was identified as a key player in the alteration of membrane composition in response to temperature changes (Cronan and Rock 1996; de Mendoza and Cronan 1983). Interestingly, 16:1-ACP is a poor substrate for FabB, and its elongation to 18:1-ACP is controlled by the activity of FabF (Garwin and Cronan 1980; Garwin et al. 1980a, b). It was demonstrated that thermal regulation of membrane fluidity in *E. coli* was due to an intrinsic property of the activity of FabF: the reactivity of FabF toward the palmitoleyl-ACP substrate increases at low temperatures, leading to increased amounts of *cis*-vaccenic acid in membrane phospholipids (Garwin and Cronan 1980). At low temperatures, *E. coli* FabF promotes the production of 18:1Δ11, whereas at higher temperatures the activity of FabF is decreased compared with FabB, the other condensing enzyme, and the formation of 18:1Δ11 is reduced (Garwin et al. 1980a; de Mendoza and Cronan 1983).

Nevertheless, the UFA pathway described in *E. coli* is not widely distributed in anaerobic bacteria that produce UFAs. Genomic analyses indicate that only the alpha and gamma proteobacteria encode the proteins of this pathway (Campbell and Cronan 2001; Heath et al. 2001). However, there are a number of anaerobic bacteria that produce UFAs and do not have a *fabA* homologues in their genomes.

A variation of the FabAB pathway was discovered in *Streptococcus pneumoniae*, where a new enzyme in type II fatty acid synthetase was described (Marrakchi et al. 2002) (Fig. 15.1b). This enzyme, called FabM, carries out the isomerization of *trans*-2-decenoyl-ACP to *cis*-3-decenoyl-ACP, but is not capable of catalyzing the dehydration of a β-hydroxy intermediate. FabM, which lacks homology to FabA, is a member of the hydratase/isomerase superfamily and is present only in streptococcus. *fabM* mutants of *S. pneumoniae* and *Streptococcus mutans* are auxotrophs for UFAs, indicating that FabM is an essential enzyme in streptococci (Altabe et al. 2007). *S. pneumoniae* also possesses a dehydratase that catalyzes the dehydration reaction, FabZ; thus this bacteria uses a combination of FabZ and FabM to introduce the double bond into the growing acyl chain. This pathway utilizes another enzyme, FabK, the enoyl

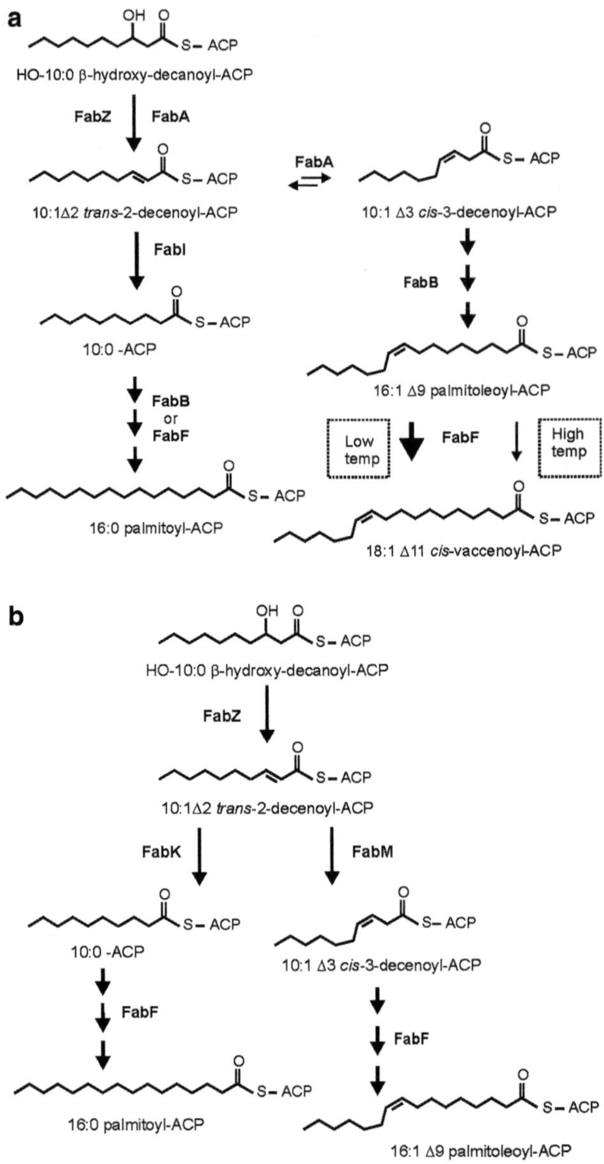

Fig. 15.1 Anaerobic pathways of UFA biosynthesis in *E. coli* and *S. pneumoniae*. (**a**) *E. coli* contains two dehydratases, FabZ and FabA. FabA is a bifunctional enzyme that catalyzes both the removal of water to generate *trans*-2-decenoyl-ACP and the isomerization of this intermediate to *cis*-3-decenoyl-ACP, the key step in UFA production. SFA biosynthesis proceeds by the action of FabI on the *trans*-2 intermediate followed by further elongation cycles initiated by either FabB or FabF condensing enzymes. UFAs require FabB to efficiently utilize the *cis*-3-C10:1-ACP intermediate and initiate the elongation cycles that form 16:1 Δ9 palmitoleoyl-ACP. FabF participates in the elongation of this substrate to 18:1 Δ11 *cis*-vaccenoyl-ACP. The reactivity of this enzyme toward 16:1 Δ9 palmitoleoyl-ACP is increased after a temperature downshift. (**b**) In *S. pneumoniae* there is only a single dehydratase, FabZ, which forms the *trans*-2-decenoyl-ACP intermediate. SFAs are formed by the action of FabK, an enoyl-ACP reductase, followed by further elongation cycles initiated by FabF. UFAs are formed from the isomerization of the *trans*-2 intermediate by FabM, a *trans*-2-*cis*-3-decenoyl-ACP isomerase, followed by further elongation cycles initiated by FabF

reductase that competes with the *cis–trans* isomerase FabM and pulls intermediates to SFA formation. Thus, the branch point in the biosynthesis of UFAs in *S. pneumoniae* occurs following the formation of *trans-2-decenoyl*-ACP, in contrast to *E. coli* where the branch point takes place after the formation of β-hydroxydecanoyl-ACP (Fig. 15.1). The elongation of both saturated and unsaturated acyl-ACPs is carried out by FabF, the only condensing enzyme present in *S. pneumoniae*. The proportion of UFA is controlled by FabT, a transcriptional repressor that regulates the expression of *fabK*, but not *fabM* (Lu and Rock 2006).

Enterococcus faecalis encodes two FabZ homologues and two FabF homologues (FabF is closely related to FabB). It was demonstrated that one of these FabZ homologues, called FabN, has dehydratase/isomerase activity analogous to FabA, whereas the other protein possesses only dehydratase activity (Wang and Cronan 2004). A similar picture was seen for the FabF homologues, one of them, FabO, replaces FabB to perform the synthesis of UFAs, whereas the other functioned only as a FabF enzyme (Wang and Cronan 2004).

On the other hand *Neisseria gonorrhoeae*, a Gram-negative facultative anaerobe, does not encode homologous to any known mechanism of UFA synthesis, either aerobic or anaerobic. Instead, this bacterium possesses a new enzyme, UfaA, that belongs to the 2-nitropropane dioxygenase-like superfamily, which is involved in the synthesis of UFAs when grown anaerobically (Isabella and Clark 2011). Although the mechanism implicated in anaerobic gonococcal UFA synthesis seems to be a distinct form than that of the classical anaerobic UFA synthetic pathway, biochemical understanding of the UfaA reaction mechanism is still fragmentary. Homologues to this enzyme were found in many facultative and obligate anaerobes that produce UFAs but lack *fabA*, suggesting that UfaA is part of a widespread pathway and probably the answer for those organisms that have presented a mystery in their mechanism of UFA synthesis for several years (Isabella and Clark 2011). The aerobic mechanism involved in UFAs synthesis is not yet identified in this bacterium.

De novo fatty acid biosynthesis is finely adjusted by environmental signals to produce acyl chains with the properties that are required to optimize membrane function. However, bacteria are often subjected to abrupt changes, for example in temperature or organic-solvent concentration, that require the immediate modification of existing membrane phospholipid acyl chains to optimize growth under the new conditions. Some prokaryotes such as cyanobacteria, bacilli, mycobacteria, and pseudomonads use a different pathway, which requires molecular oxygen and reduced equivalents, to introduce a *cis* double bond into preexisting fatty acids (Mansilla and de Mendoza 2005; Phetsuksiri et al. 2003; Zhu et al. 2006). This pathway, which is mediated by desaturase enzymes, will be described in the subsequent sections.

Biochemical Characteristics of Desaturation Systems

A desaturase is a special type of oxygenase that can remove two hydrogens from a hydrocarbon chain, especially from a fatty acyl chain, catalyzing the formation of a double bond in the substrate. Unlike normal oxygenases which directly transfer molecular oxygen to a substrate, a desaturase uses activated molecular oxygen to abstract hydrogens from the substrate creating a carbon/carbon double bond in a fatty acid and a water molecule (Shanklin and Cahoon 1998; Los and Murata 1998; Buist 2004). All these enzymes utilize reducing equivalents obtained from an electron transport chain and are able to introduce the double bond in a chemo-, regio-, and stereoselective manner. Desaturases represent a complex group of enzymes that could be classified in different ways, depending on the solubility of the enzyme, the electron donor, subcellular localization, the regioselectivity of the reaction, and the kind of esterification of the fatty acid used as substrate (Sperling et al. 2003).

According to its solubility these enzymes can be classified into two non-evolutively-related groups: the soluble ACP desaturases, found primarily in the plastids of higher plants (e.g., the castor Δ9 stearoyl-ACP desaturase) (Shanklin and Somerville 1991) and a larger group of membrane-bound desaturases found in a wide range of taxa, typified by the yeast Ole1 Δ9-desaturase (Ole1p) (Stukey et al. 1990) and the *Bacillus subtilis* Δ5-desaturase (Δ5-Des) (Aguilar et al. 1998). The acyl-ACP desaturases catalyze the introduction of *cis* double bonds into fatty acid chains bound to ACP (Shanklin and Cahoon 1998), generally at position 9 relative to the carboxyl end of the fatty acids (Δ9). The analysis of the crystal structure of Δ9 stearoyl-ACP desaturase shows that it has an active di-iron center, and that the amino acid residues involved in attaching the di-iron complex forming two characteristic motifs D/EXXH (Lindqvist et al. 1996).

Membrane desaturases are more widely distributed in nature. These proteins belong to a large family of functionally diverse enzymes which include alkane hydroxylases, xylene monooxygenases, carotene ketolases, and sterol and sphingolipid hydroxylases (Shanklin and Cahoon 1998). Based on the differences in substrate specificity, these enzymes were subdivided into two groups: acyl-CoA desaturase and acyl-lipid desaturases (Shanklin and Cahoon 1998). These integral membrane proteins have a different consensus for the active site with a tripartite motif of eight histidines ($HX_{(3-4)}H$, $HX_{(2-3)}HH$, $HX_{(2-3)}HH$), found in conserved positions of the amino acid sequence. It has been observed by site-directed mutagenesis studies that substitution of any of the conserved histidine residues with another amino acid leads to loss of enzyme activity. This is probably due to the inability to bind iron at the active site of the mutant enzymes (Los and Murata 1998).

The complex process of introducing a double bond into fatty acids requires iron cofactors, molecular oxygen, and two reducing equivalents for catalysis (Shanklin et al. 1994; Sperling et al. 2003). The latter are supplied from NAD(P)H by two different electron transport systems, functionally equivalent, that are specific to the subcellular compartment rather than to the class of the desaturase. In the case of desaturases of plant endoplasmic reticulum and the acyl-CoA or acyl-lipid desaturases of animal and fungi, the donor is cytochrome *b5*, either in the form of a fused domain or in free form. For the soluble acyl-ACP desaturases and the integral membrane acyl-lipid desaturases from plastids and cyanobacteria, electrons are delivered by ferredoxins (Fer) which are ubiquitous soluble iron-sulfur proteins involved in a variety of redox reactions (Sperling et al. 2003).

Three classes of regioselectivity have been observed for fatty acid desaturases that would reflect differences in the position of the active site relative to the features of the substrate binding pocket associated with substrate recognition: the Δx desaturases introduce a double bond x carbons from the carboxyl end; $\omega - x$ desaturases dehydrogenate x carbons from the methyl terminus; while $\nu + x$ desaturases use a preexisting double bond as a reference point and dehydrogenate x carbons from the nearest olefinic carbon (Yadav et al. 1993; Hitz et al. 1994; Uttaro 2006). The Δ9 stearoyl-CoA desaturase locates double bonds counting from the carboxyl terminus of the molecule in a manner similar to that employed by the soluble acyl-ACP desaturases. Acyl-lipid desaturases ω6 (Δ12) and ω3 (Δ15) require substrates with a preexisting double bond, most frequently ω6 desaturase prefers Δ9 UFAs, while ω3 desaturases prefer dienoic fatty acids Δ9, 12 (Sperling et al. 2003). "Front-end" desaturases introduce the double bond between one preexisting and the carboxyl (front) end of the fatty acid. These enzymes are involved in polyunsaturated fatty acid (PUFA) biosynthesis and all have *citb5* domain fused to the N-terminus (Sayanova et al. 1999; Uttaro 2006) which operates as electron donor during desaturation. The only exceptions among this kind of desaturases are the Δ6 desaturases of cyanobacteria, which use, at least in vitro, Fer as electron donor (Los and Murata 1998).

Bacterial Desaturases

Many bacterial genomes present ORFs with significant similarity to desaturases (Sperling et al. 2003). In this section we will summarize the current knowledge about those desaturases whose function has been experimentally demonstrated.

Desaturases from the Genus Bacillus

Bloch and coworkers were the first to describe the existence of an oxidative pathway for the biosynthesis of long-chain UFAs by microorganisms, thus initiating the study of bacterial desaturases (Scheuerbrandt and Bloch 1962; Fulco et al. 1964). For a number of years Fulco performed detailed in vivo studies with *B. megaterium* demonstrating that the desaturase activity in this bacterium was induced by low growth temperatures (Fulco 1974). Nevertheless, at that time mechanistic insights in desaturation reactions of *Bacillus* could not be established because the studied organisms were not amenable to genetic analysis. This problem was solved when this phenomenon was examined in *B. subtilis*, which is an excellent model because of its general experimental tractability. Genetic and biochemical experiments successfully established that *B. subtilis* contains a single acyl-lipid desaturase, encoded by the *des* gene (Grau and de Mendoza 1993; Aguilar et al. 1998), and that this enzyme catalyzes the introduction of a *cis*-double bond at the Δ5 position of a wide range of SFAs (Altabe et al. 2003). Thus, this protein was named as Δ5-Des.

Δ5-Des exhibits relatively low sequence identity (typically about 23 %) to membrane desaturases from cyanobacteria and plants (Aguilar et al. 1998). It contains the three His clusters found in the known membrane desaturases with the appropriate spacing which might be involved in non-heme iron binding, constituting the active site of the enzyme (Aguilar et al. 1998). An additional, fourth His region not found in other known desaturases is located downstream from the third His cluster of Δ5-Des, but it is not involved in catalysis (Díaz et al. 2002). However, deletion of the C-terminal part of Δ5-Des, downstream of the third His cluster, eliminates enzyme activity, indicating that this region is essential either for desaturase activity or for appropriate protein folding (Díaz et al. 2002). The amino acid sequence of Δ5-Des contains many hydrophobic stretches, consistent with the presence of several membrane-spanning domains. The hydropathy profile of the primary sequence is similar to those of desaturases from cyanobacteria and to almost all known membrane-associated plant desaturases. However, these hydropathy profiles showed some differences compared to AlkB that has provided the only experimental model for this superfamily of enzymes (van Beilen et al. 1992). Therefore, the topology of Δ5-Des was experimentally analyzed in *E. coli* using a set of fusions of N-terminal fragments of the protein to the reporter alkaline phosphatase and surface Lys biotinylation experiments (Díaz et al. 2002). These data, combined with site-directed mutagenesis of His residues conserved in most desaturases, support a model for Δ5-Des as a polytopic membrane protein with six transmembrane segments (TMSs) and one membrane-associated domain, which might be involved in keeping the catalytic site on the cytoplasmic side of the membrane, close to the substrates (Díaz et al. 2002). The proposed topology of Δ5Des, together with the topological features of the mouse stearoyl-CoA desaturase that contains four TMSs (Man et al. 2006), provides two paradigmatics models for the topology of an acyl-lipid desaturases and an acyl-CoA desaturase, respectively, based on experimental grounds.

The complex process of introducing a double bond into fatty acids requires molecular oxygen and two reducing equivalents for catalysis (Shanklin and Cahoon 1998; Sperling et al. 2003). To study this important question in *B. subtilis*, a genetic approach was undertaken to identify the electron donors required to provide the $2e^-$ required for Δ5-Des activity. The *B. subtilis* genome encodes three proteins that can act as potential electron donors of Δ5-Des: Fer and two flavodoxins (Flds), YkuN and YkuP, which belong to the *ykuNOP* operon (Lawson et al. 2004). Thus, mutagenesis of the corresponding genes was performed and the ability of these proteins to support *B. subtilis* desaturase activity was tested in vivo. The result of these experiments demonstrated that Fer, YkuN, and YkuP can act as redox partners for Δ5-Des and suggest that Fer/Flds proteins could function physiologically in the biosynthesis of UFAs in *B. subtilis* (Chazarreta-Cifre et al. 2011). Although Flds have extensively been described as partners in a number of redox processes, this was the first report describing that flavoproteins can transfer electrons to a lipid desaturase and, in conjunction with Fer, support the desaturation of *B. subtilis* lipids.

The *B. subtilis des* gene is essential for cold-shock survival when this bacterium is grown in glucose-minimal medium (Weber et al. 2001). However, this cold-sensitive phenotype is suppressed by addition of the amino acid isoleucine to the growth media. This protective effect is exclusively based on the function of this amino acid in the biosynthesis of anteiso-branched-chain fatty acids (a-BCFA), which have biophysical properties similar to UFAs (Martin et al. 2009b). Thus, Δ5-Des provides a back up mechanism to survive at low temperatures in the absence of an isoleucine-dependent switch for the branching of fatty acids.

Cyanobacterial Desaturases

Cyanobacteria are a versatile group of bacteria capable of carrying out a plant-like oxygenic photosynthesis. Cyanobacteria comprise over 1,600 species with various morphologies and can survive in very hostile environments (Scanlan 2001). It has been well documented that the content of monounsaturated fatty acids (MUFAs) and PUFAs in membrane lipids of cyanobacteria plays essential roles in maintaining the proper function of biological membranes especially under conditions of temperature stress, and energy limitation, growth and photosynthesis (Jeamton et al. 2008; Murata and Wada 1995; Nishida and Murata 1996; Tasaka et al. 1996; Sakamoto and Murata 2002). All known cyanobacterial desaturases are acyl-lipid desaturases (Des) and have rigorous specificity with respect to the position in fatty acid at which double bond is introduced and to the *sn* position of the glycerol to which fatty acids are esterified. Des proteins and their genes have been extensively described in *Synechocystis* sp. PCC 6803 being *desA* the first gene for an acyl-lipid desaturase to be cloned (Wada et al. 1990). Subsequently, it was determined that this bacterium have four specific acyl-lipid desaturases, designated DesA, DesB, DesC, and DesD, that catalyze desaturation at the Δ12, $\omega 3$, Δ9, and Δ6 position of C18 acids on the *sn*-1 position, respectively (Wada et al. 1990; Sakamoto et al. 1994a, b, c; Reddy et al. 1993).

The order in which desaturases operates is very strictly determined: the first double bond is introduced by the Δ9 desaturase into stearic acid, and the Δ12 and Δ6 desaturases introduce a double bond into mono or diunsaturated fatty acids that already have a double bond in Δ9 position (Murata et al. 1992). The $\omega 3$ desaturase introduces a double bond into fatty acids that have a double bond at the Δ12 position (Murata et al. 1992). Unlike *Synechocystis*, *Synechococcus* has only a single Δ9 desaturase and synthesizes MUFAs exclusively.

The importance of the *des* genes with respect to cold adaptation has been clearly demonstrated by the observation that cyanobacterial mutants defective in these genes are cold-sensitive and grow slower than the wild-type cells (Los and Murata 1999). Disruption of *desA and desD* genes in *Synechocystis* sp. PCC 6803 results in remarkable changes of membrane fatty acids that has deleterious effect on the growth and severe photoinhibition of photosynthesis at low temperatures (Tasaka et al. 1996; Los and Murata 1999).

Pseudomonas aeruginosa *Desaturases*

Pseudomonas aeruginosa is an opportunistic human Gram-negative pathogen that causes nosocomial and life-threatening infections in injured, burned, and immunocompromised patients (Van Delden and Iglewski 1998). The membranes of *P. aeruginosa* contain high quantities of straight-chain UFAs, being *cis*-vaccenate the major species found in the wild-type cells (Zhu et al. 2006). In a recent study it was demonstrated that *P. aeruginosa* has three different pathways to synthesize UFAs, depending on the oxygen availability. In anaerobic conditions, the UFAs synthesis take place exclusively by the actions of the FabAB proteins of the bacterial type II biosynthetic pathway (Fig. 15.1a). In aerobic conditions, besides the FabAB mediated pathway, *P. aeruginosa* employs two oxygen-dependent desaturases, DesA and DesB, which introduce a *cis* double bond into preexisting fatty acids (Zhu et al. 2006). Both desaturases have different substrate specificities and supplement the anaerobic mechanism for UFA synthesis. DesA introduces a double bond at the Δ9 position of fatty acyl chains attached to the *sn*-2 position of existing phospholipids. DesB is also a Δ9 desaturase but introduces double bonds into acyl-CoAs formed from exogenous SFAs, mainly stearate and palmitate, rendering 18:1Δ9 and 16:1Δ9, respectively (Zhu et al. 2006). The *desB* gene is located in an operon with *desC*, a gene encoding an oxidoreductase, which is predicted to function in the electron transport coupled with the DesB desaturation reaction (Zhu et al. 2006). Recently it was demonstrated that DesB is a factor required for full virulence in *P. aeruginosa*, since *desB* mutants are severely deficient in the production of proteolytic enzymes, pyocianin and rhamnolipids, and they show impaired swarming and twitching motilities and reduced virulence in the *Caenorhabditis elegans* infection model (Schweizer and Choi 2011). A *desA* and *desB* double mutant grows well under aerobic conditions, indicating that FabAB is the main pathway to synthesize UFAs (Hoang and Schweizer 1997).

Mycobacterial Desaturases

Mycobacterium tuberculosis is a human pathogen that causes millions of deaths every year. All mycobacteria have a lipid-rich cell envelope containing mycolic acids (Watanabe et al. 2002). Those lipids are complex hydroxylated BCFAs with elevated carbon number (60–90). They may also contain diverse functional groups such as methoxy, keto, epoxy ester groups, cyclopropane rings, and UFAs. Mycolic acids play an important role in reduced cell wall permeability, virulence, and acid fastness characteristic of *M. tuberculosis* (Yuan et al. 1995; Dubnau et al. 2000). Several front-line drugs used for treating tuberculosis inhibit mycolic acid synthesis. Enzymes needed for biosynthesis of mycolic acids are promising drug targets for new anti-TB agents (Tomioka et al. 2008). The mechanisms used to synthesize mycolic acids are not clear at present. Long-chain UFAs containing double bonds in specific positions are required as building blocks, suggesting that desaturation is probably an essential step in the biosynthesis of mycolic acids (Yuan et al. 1995; Dubnau et al. 2000).

The genome sequence of *M. tuberculosis* and related pathogens has revealed the presence of three genes with similarity to known fatty acid desaturases (Cole et al. 1998). The three putative fatty acid desaturases in the *M. tuberculosis* H37Rv genome, *desA1*, *desA2*, and *desA3*, were cloned and expressed in *Mycobacterium bovis* BCG. DesA3 is a Δ9 membrane-bound desaturase responsible for the synthesis of oleic acid (Phetsuksiri et al. 2003). This enzyme appears to be functionally analogous to mammalian stearoyl-CoA desaturase (Ntambi 1995), which has a number of important physiological roles in mammalian cells (Ntambi 1995; Ntambi et al. 2002). Importantly, DesA3 is one out of the 200 genes required for pathogen survival inside the granuloma enclosing a dormant tuberculosis infection (Sassetti and Rubin 2003). It was demonstrated to be essential for the viability of *M. tuberculosis* and was identified as a target of the anti-tuberculosis drug Isoxyl (Phetsuksiri et al. 2003). DesA1 and DesA2 are annotated as homologues of the soluble acyl-ACP desaturases (Cole et al. 1998; Dyer et al. 2005). Nevertheless the role of these enzymes is unknown and further studies of the properties of these putative desaturases are clearly needed.

Modulation of Membrane Fluidity

When bacteria, which are poikilothermic organisms, are exposed to temperatures below those of their normal conditions, the lipids of their membrane become rigidified, leading to a suboptimal functioning of cellular activities (Phadtare 2004; Mansilla and de Mendoza 2005; Al-Fageeh and Smales 2006). These organisms can acclimate to such new conditions decreasing the transition temperature of the membrane; this is the temperature where membrane lipid bilayers undergo a reversible change of state from a fluid (disordered) to a nonfluid (ordered) array of the fatty acyl chains. In most bacteria, the role of introducing acyl chain disorder is fulfilled by UFAs. As phospholipids containing UFAs have a much lower transition temperature than those lipids made of SFAs (Cronan and Gelmann 1973), the desaturation of the acyl chains of membrane phospholipids results in an increase in fluidity, and membrane lipids returns to their original state, or close to it, with restoration of normal cell activity at the lower temperature. This regulatory mechanism system, called thermal control of fatty acid synthesis, seems to be a universally conserved adaptation response allowing cells to maintain the appropriate fluidity of membrane lipids regardless of the ambient temperature. In this section we will focus on the mechanisms of regulation of bacterial desaturases, developed to adapt the membranes to the new temperatures.

The Des Pathway of Bacillus subtilis

B. subtilis cells respond to a decrease in ambient growth temperature by using a dual strategy to restore membrane fluidity. In parallel with a long-term response that leads to an increase in the proportion of saturated a-BCFAs, *B. subtilis* transiently induces the expression of Δ5-Des that desaturates the acyl chain of preexisting membrane phospholipids (Aguilar et al. 1998, 1999, 2001; Altabe et al. 2003). This short-term membrane adaptation requires a canonical two-component regulatory system comprising the histidine kinase DesK and the response regulator DesR (Aguilar et al. 2001). This signal transduction mechanism, termed the Des pathway, enhances the expression of the *des* gene coding for Δ5-Des in response to a decrease in growth temperature (Mansilla et al. 2008). The *B. subtilis* DesK protein is a polytopic protein containing five TMSs that define the sensor domain (Cybulski et al. 2010) and a long cytoplasmic C-terminal tail, which harbors the kinase domain, DesKC. Genetic, physiological, and biochemical evidence proved that DesK is a bifunctional enzyme with both kinase and phosphatase activities that could assume different signaling states in response to changes in membrane fluidity (Aguilar et al. 2001; Albanesi et al. 2004, 2009; Martin et al. 2009a). A kinase-dominant state of DesK predominates upon an increase in the proportion of ordered membrane lipids (Fig. 15.2a). As a consequence of this stimulus perception, DesK undergoes autophosphorylation in the His188 and, subsequently, the phosphoryl group is transferred to the cytoplasmic response regulator DesR. Autophosphorylated DesKC transfers the phosphoryl group to the effector protein DesR that becomes phosphorylated in the predicted Asp 54 residue (DesR-P) (Albanesi et al. 2004). Phosphorylation of the regulatory domain of DesR promotes, in a cooperative fashion, the hierarchical occupation of two adjacent, non-identical, DesR-P DNA-binding sites, so that there is a shift in the equilibrium toward the tetrameric active form of the response regulator (Cybulski et al. 2004). This results in the recruitment of RNA polymerase to the *des* promoter and the activation of *des* transcription, as demonstrated by in vitro experiments (Cybulski et al. 2004) (Fig. 15.2a). Transcriptional activation of *des* is followed by the expression of Δ5-Des and desaturation of the acyl chain of membrane phospholipids (Fig. 15.2b). These newly synthesized UFAs decrease the phase transition temperature of the phospholipids, favoring the phosphatase activity of DesK, resulting in hydrolysis of DesR-P. The unphosphorylated regulator is unable to bind to P*des* and, as a consequence, *des* transcription is turned off (Fig. 15.2b). Using a functional in vitro reconstituted system it was demonstrated that the temperature stimuli comes directly from the membrane lipid bilayer, with no other proteins involved in the sensing or signaling mechanism (Martin et al. 2009a). The opposite activities of DesK, autokinase, and DesR-P phosphatase would be reciprocally regulated by changes in growth temperature that, in turn, adjust the fluidity of membrane phospholipids.

A recent set of crystal structures of DesKC, the catalytic core of DesK, has illuminated the molecular view of how the subdomains of this protein can interact to assemble the three active sites that promote the necessary chemistry for the DesK

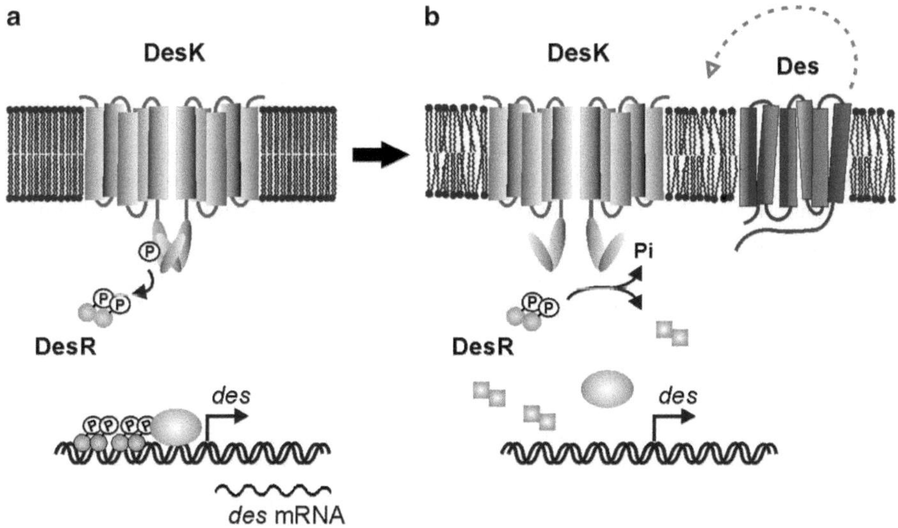

Fig. 15.2 Simplified model of regulation of UFAs synthesis in *B. subtilis*. (**a**) DesK could assume different signaling states in response to changes in membrane fluidity. An increase in the order of the acyl chains of membrane lipids promotes a kinase-dominant state of DesK, which autophosphorylates and transfers the phosphate group to DesR. DesK-mediated phosphorylation of DesR enables interaction of DesR-P dimers with the *des* promoter and RNA polymerase, resulting in transcriptional activation of *des*. (**b**) Δ5-Des is synthesized and desaturates the acyl chains of membrane phospholipids. These newly synthesized UFAs cause a decrease in the order of membrane lipids favoring a phosphatase-dominant state of DesK, leading to dephosphorylation of DesR and thus turning off *des* transcription

regulation cycle and how these interactions might be regulated in response to the cold stimuli (Albanesi et al. 2009). Structural and biochemical approaches showed that DesK is cold-activated through specific interhelical rearrangements in its central four helix bundle domain, known as DHp (for dimerization and histidine phosphotransfer). In a fluid membrane, the transmembrane (TM) domain would stabilize a connecting coiled-coil and the catalytic core into a rigid conformation with the ATP-binding domains attached to the DHp domain. This conformation inhibits autokinase activity and the DHp surface is competent to interact with DesR, resulting in a phosphatase signaling state. Upon cold signal reception, the ensuing structural reorganization would release the ATP-binding domains for histidine phosphorylation (kinase state). Phosphorylation of DesKC induces an as yet different, asymmetric conformation capable of interacting with DesR (phosphotranferase state) (Albanesi et al. 2009). These structural changes of the catalytic domain are promoted by the sensor domain of DesK. One or more of the five TMSs in DesK could undergo a conformational change, in the form of helix rotations and asymmetric helical bends, induced by a modification in the physical state of the membrane lipid bilayer. This information is transmitted to the cytoplasmic domain by the membrane-connecting two-helical coiled-coil, ultimately controlling the alternation between output autokinase and phosphatase activities. Direct support for the functional importance of the TM helices comes from studies of *B. subtilis* strains producing the soluble DesKC domain. In these strains, the *des* gene is constitutively expressed and its transcription is affected neither by growth temperature nor by the presence of UFAs (Albanesi et al. 2004, 2009; Martin et al. 2009a). Moreover, when DesKC is anchored to the membrane by only one TMS, *des* expression is not induced by cold-shock (Cybulski et al. 2010). Therefore, the TMSs of DesK directly participate in signal transduction recognition and play an essential role in the end-product feedback regulation of *des* transcription.

B. subtilis is also capable of anaerobic growth by nitrate or nitrite respiration. In anaerobiosis *des* is strongly upregulated after a temperature downshift, and this transcriptional induction is dependent on the DesK/DesR system (Beranová et al. 2010). However, there is no fatty acid desaturation after *des* induction, under anaerobic conditions, despite the presence of high levels of Δ5Des. This is due to the strict oxygen requirement of Δ5Des to introduce the double bond into fatty acids. The subsequent transfer of cells exposed to cold temperatures from anaerobic to aerobic conditions rapidly restores the downregulation of *des* transcription. Cold adaptation in *B. subtilis* under anaerobic growth is therefore mediated exclusively by the increased synthesis of a-BCFAs, reaching similar degrees of membrane fluidization that in aerobic conditions (Beranová et al. 2010).

Evidence that membrane fluidity, rather than growth temperature, controls transcription of the *des* gene was obtained by experiments in which the proportion of a-BCFAs of *B. subtilis* membranes was varied controlling the provision of branched-chain acyl-CoA molecules, used as primers (Cybulski et al. 2002). Those primers are obtained by deamination of the amino acids valine, leucine, and isoleucine, and oxidative decarboxylation of the corresponding branched-chain alpha-keto acids, a reaction catalyzed by branched-chain keto acid dehydrogenase (BKDH) (Kaneda 1977). The a-BCFAs, which are synthesized using ketoacids derived from isoleucine as primers (Kaneda 1991), are essential to decrease the transition temperature of *B. subtilis* membrane phospholipids to maintain the appropriate fluidity. Limiting the supply of isoleucine dramatically reduces the amount of a-BCFAs of plasma membrane lipids (Klein et al. 1999), resulting in ordered membrane lipids. Growth of cells in the absence of isoleucine results in activation of *des* transcription under isothermal conditions, using a DesK/DesR-dependent mechanism (Cybulski et al. 2002). Moreover in LipA mutants, which are unable to synthesize BCFA precursors, because of BKDH lack of function in the absence of lipoic acid synthesis (Martin et al. 2009b), transcription of *des* gene reaches to levels fourfold higher than those observed in wild-type strains growing in an isoleucine-free medium at 37 °C. These physiological and transcriptional data give further support to the hypothesis that DesK senses membrane fluidity, since the membrane lipids of *lipA*-deficient cells contain a much higher proportion of high-melting-point SFAs than those of cells expressing LipA (Martin et al. 2009b). Thus, an increase in the order of membrane lipids due to a lower content of membrane isoleucine-derived fatty acids at constant temperature or a decrease in temperature at constant lipid composition can be sensed by DesK, leading to induction of UFAs synthesis.

Two Inducible Aerobics Pathways of UFAs Synthesis in **Pseudomonas aeruginosa**

P. aeruginosa is able to regulate its aerobic biosynthesis of UFAs in response to changes in the growth conditions. It was suggested that DesA provides a way to quickly adjust the fluidity of the membrane following changes in the growth conditions given that this desaturase introduces double bonds into preexisting membrane lipids. However, rather than being regulated by temperature, like in *B. subtilis*, *P. aeruginosa desA* expression is upregulated when oxygen is limited, which is reminiscent of the induction of OLE1 expression by Mga2p under hypoxic conditions (Jiang et al. 2001; Vasconcelles et al. 2001). However, the mechanism of regulation of *desA* by oxygen availability is not known.

The second aerobic pathway, the DesCB system, is regulated by DesT, a transcriptional repressor that is able to sense the changes in the composition of cellular acyl-CoA pool and modify its binding to DNA accordingly (Zhang et al. 2007). A unique property of DesT is its ability to differentially bind to the *desT–desCB* intergenic region in response to the presence of different acyl-CoA structures (Zhang et al. 2007). DesT binds tightly to DNA in the presence of UFA-CoA ligands. However, the transcriptional repressor is released from its binding site by SFA-CoAs, thus allowing DesB expression. Structural analyses of the DesT/acyl-CoA complexes revealed that the acyl chain shape directly influences the packing of hydrophobic core residues within the DesT ligand-binding domain. These changes are propagated to the paired DNA-binding domains through conformational changes that modulate DesT-DNA affinity (Miller et al. 2010). Thus, under normal growth conditions, DesT exists in equilibrium between the free and DNA-bound forms, and a basal level of *desCB* transcription is maintained. The presence of SFAs in the growth environment leads to an increase in the intracellular saturated acyl-CoA levels. The binding of saturated long-chain acyl-CoAs, such as palmitoyl-CoA, to DesT releases the protein from the *desCB* promoter and derepresses *desCB* transcription. As a result, saturated acyl CoAs are converted to Δ9 UFA-CoAs and incorporated into membrane phospholipids (Zhang et al. 2007). It was proposed that when the intracellular levels of UFA-CoAs augment, either by DesB desaturation or by uptake of free fatty acids that are then converted to UFA-CoAs by the acy-CoA synthase, they could be available to bind DesT. This would increase the affinity of DesT from its binding site followed by transcriptional repression of the *desBC* operon.

It was previously noted that the *fabAB* operon is repressed by exogenous UFAs and that DesT was able to bind to a palindromic sequence located on the promoter region of this operon, suggesting a coordinate expression of anaerobic (*fabAB*) and aerobic (*desCB*) pathways for UFA synthesis (Subramanian et al. 2010). However it was demonstrated that DesT repression seems to play only a modest role in *fabAB* expression, which seems to be controlled by a complex regulatory network composed of a combination of directly acting factors and indirectly acting transcriptional or translational regulators (Schweizer and Choi 2012).

It was suggested that regulation of *desB* expression by DesT could be important when *P. aeruginosa* grows in an environment rich in fatty acids. Thus, the DesT-DesB system could provide a mechanism that allows *Pseudomonas* cells to substitute energy-intensive synthesized fatty acids for environmental fatty acids and at the same time maintain membrane homeostasis (Cronan 2006).

A Multi-Stress Sensor in Cyanobacteria

In cyanobacterial cells, fatty acid desaturation is one of the crucial steps in the acclimation processes during cold-shock. Low temperatures increase the desaturation of fatty acids of membrane lipids in cyanobacterial cells, as a result of upregulation of the expression of the desaturase genes (Los and Murata 1999). Three out of four genes (*desA*, *desB*, and *desD*) coding for desaturases in *Synechocystis* are cold-inducible (Los et al. 1993, 1997).

The systematical inactivation of each of the 43 putative histidine kinases genes (*hik*) in *Synechocystis* led to the identification of Hik33, involved in the sensing of low temperatures (Suzuki et al. 2000). Subsequently, it was demonstrated that the response regulator RpaB along with Hik33 are involved in low temperature signal transduction (Kappell and van Waasbergen 2007). Hik33 has two membrane-spanning domains, a HAMP linker, a leucine zipper, and a PAS domain located at its amino terminus, while the carboxy-terminus contains a highly conserved histidine kinase catalytic domain (Sakamoto and Murata 2002). Hik33 is involved in the cold-inducible expression of *desB* gene, which encodes the $\omega 3$ desaturase of *Synechocystis* (Suzuki et al. 2000). Several lines of evidence support the hypothesis that the transcriptional regulation of *desB* is mediated by Hik33, which in turn is activated by an increase in the order of lipids. It was proposed that the TM domain of Hik33 acts as a sensor of the state of membrane fluidity. If this is the case, the order of membrane lipids could mediate structural rearrangements in the TM domain that would be transmitted to the HAMP domain, which in turn would adjust the signaling state of the Hik33 kinase catalytic domain. Interestingly, the cold induction of *desB* is light-dependent (Mironov et al. 2012). This additional checkpoint in the signal transduction pathway might be related to the function of the PAS domain contained in Hik33.

The cold-regulated induction of gene expression in *Synechocystis* seems to be similar to the pathway controlled by DesK/DesR in *B. subtilis*. Thus, the control of membrane fluidity by cold-sensor histidine kinases could be conserved in several bacteria that use acyl-lipid desaturases to remodel membrane lipids. In contrast to the *B. subtilis* DesK/DesR system which regulates almost exclusively the expression of the *des* gene (Beckering et al. 2002), Hik33 is a global cold sensor involved in the expression of several genes that are induced at low temperature (Murata and Suzuki 2006). In addition, the analysis of the transcriptome of *Synechocystis* indicated that Hik33 is involved in the sensing of hyperosmotic stress (Mikami et al. 2002; Paithoonrangsarid et al. 2004) and salt stress (Shoumskaya et al. 2005), and is the main sensor of peroxidative stress in this bacterium (Kanesaki et al. 2007). It is remarkable that Hik33 and its cognate regulator(s) controls the expression of distinct respective sets of genes under different kinds of stresses, and this finding cannot be explained by the currently accepted model of two-component systems. A plausible hypothesis is that a non-yet identified factor(s) could determine the specificity of the sensor for a particular stress.

It was shown that a mutation in *hik33* only partially affects the cold induction of *desB* and has not any effect in the temperature regulation of *desA* and *desB* (Suzuki et al. 2001). These findings suggest that a Hik33-independent mechanism operates to regulate these genes. It has early been proposed that temperature-dependent alterations in DNA supercoiling might be involved in the regulation of the expression of the *B. subtilis des* gene (Grau et al. 1994). This observation was extended to cyanobacteria by showing that upon a decrease in ambient temperature the degree of DNA supercoil increases in the promoter region of the *desB* gene (Los 2004). In concert with this finding, it was reported that novobiocin, an inhibitor of the DNA gyrase, prevented the cold-induced expression of several desaturase genes, including *desB* which was the gene more sensitive to inhibition by the antibiotic (Los 2004). DNA microarray-based analysis of gene expression in the presence of novobiocin showed that expression of almost all cold-inducible genes in *Synechocystis* requires DNA gyrase activity (Prakash et al. 2009). Taken together, a plausible model for the perception and transduction of cold-stress signals by cyanobacterial cells would be as follows: the sensory histidine kinase Hik33 perceives the temperature-induced rigidification in the cytoplasmic membrane and acquires a kinase state. Upon autophosphorylation of the Hik33 dimer, it transfers the phosphoryl group to the response regulator RpaB, which binds to promoter region of the genes to induce their transcription. The Hik33-RpaB two-component system regulates 23 out of 48 cold-inducible genes. The cold-inducible genes that are not regulated by Hik33-RpaB could be controlled by a low-temperature induction of negative DNA supercoil. The DNA gyrase-dependent mechanism seems also to be important to control the cold-induced transcription of genes involved in stabilization of mRNA, translation, cell wall metabolism, and regulation of membrane fluidity, which are not dependent of Hik33-RpaB sensing.

Concluding Remarks and Future Perspectives

Much progress has been made over the last several years in the identification and studies of fatty acid desaturases in bacteria. A particularly fundamental question is the mechanistic details of how the acyl-lipid desaturases introduce the double bond into the fatty acid moieties of membrane phospholipids. However, this mechanism has not yet been elucidated because integral membrane proteins are notoriously difficult to study in vitro. The studies of the *B. subtilis* desaturase through the use of genetic techniques coupled with lipid analytics and topological analyses provided important data to understand the structure–function relationship of acyl-lipid desaturases. Advances in genomics and mass-spectrometry-based shotgun lipidomics promise to facilitate the future biochemical investigations of membrane-associated desaturases present in major human pathogens. Finally, insights into the biochemical genetics of bacterial desaturases will impact other fields since these enzymes are ubiquitous membrane proteins found in all kinds of life.

Acknowledgments This work was supported by a grant from Agencia Nacional de Promoción Científica y Tecnológica (FONCYT). S. G. Altabe, M. C. Mansilla, and D. de Mendoza are Career Investigators from Consejo Nacional de Investigaciones Científicas y Técnicas (CONICET).

References

Aguilar PS, de Mendoza D (2006) Control of fatty acid desaturation: a mechanism conserved from bacteria to humans. Mol Microbiol 62:1507–1514

Aguilar PS, Cronan JE Jr, de Mendoza D (1998) A *Bacillus subtilis* gene induced by cold shock encodes a membrane phospholipid desaturase. J Bacteriol 180:2194–2200

Aguilar PS, Lopez P, de Mendoza D (1999) Transcriptional control of the low-temperature-inducible *des* gene, encoding the Δ5 desaturase of *Bacillus subtilis*. J Bacteriol 181:7028–7033

Aguilar PS, Hernandez-Arriaga AM, Cybulski LE, Erazo AC, de Mendoza D (2001) Molecular basis of thermosensing: a two-component signal transduction thermometer in *Bacillus subtilis*. EMBO J 20:1681–1691

Albanesi D, Mansilla MC, de Mendoza D (2004) The membrane fluidity sensor DesK of *Bacillus subtilis* controls the signal decay of its cognate response regulator. J Bacteriol 186:2655–2663

Albanesi D, Martin M, Trajtenberg F, Mansilla MC, Haouz A, Alzari PM, de Mendoza D, Buschiazzo A (2009) Structural plasticity and catalysis regulation of a thermosensor histidine kinase. Proc Natl Acad Sci U S A 106:16185–16190

Al-Fageeh MB, Smales CM (2006) Control and regulation of the cellular responses to cold shock: the responses in yeast and mammalian systems. Biochem J 397:247–259

Altabe SG, Aguilar P, Caballero GM, de Mendoza D (2003) The *Bacillus subtilis* acyl lipid desaturase is a Δ5 desaturase. J Bacteriol 185:3228–3231

Altabe SG, Lopez P, de Mendoza D (2007) Isolation and characterization of unsaturated fatty acid auxotrophs of Streptococcus pneumoniae and Streptococcus mutans. J Bacteriol 189:8139–8144

Beckering CL, Steil L, Weber MH, Volker U, Marahiel MA (2002) Genomewide transcriptional analysis of the cold shock response in *Bacillus subtilis*. J Bacteriol 184:6395–6402

Beranová J, Mansilla MC, de Mendoza D, Elhottová D, Konopásek I (2010) Differences in cold adaptation of Bacillus subtilis under anaerobic and aerobic conditions. J Bacteriol 192:4164–4171

Bloch K (1971) β-Hydroxythioester dehydrase. In: Boyer PD (ed) The enzymes. Academic, New York, NY

Bonamore A, Macone A, Colotti G, Matarese RM, Bolfi A (2006) The desaturase from *Bacillus subtilis*, a promising tool for the selective olefination of phospholipids. J Biotechnol 52:177–183

Buist PH (2004) Fatty acid desaturases: selecting the dehydrogenation channel. Nat Prod Rep 21:249–262

Campbell JW, Cronan JE Jr (2001) Bacterial fatty acid biosynthesis: targets for antibacterial drug discovery. Annu Rev Microbiol 55:305–332

Chazarreta-Cifre L, Martiarena L, de Mendoza D, Altabe SG (2011) Role of ferredoxin and flavodoxins in *Bacillus subtilis* fatty acid desaturation. J Bacteriol 193:4043–4048

Cole ST, Brosch R, Parkhill J, Garnier T, Churcher C, Harris D, Gordon SV, Eiglmeier K, Gas S et al (1998) Deciphering the biology of *Mycobacterium tuberculosis* from the complete genome sequence. Nature 393:537–544

Cronan JE (2006) A bacterium that has three pathways to regulate membrane lipid fluidity. Mol Microbiol 60:256–259

Cronan JE Jr, Gelmann EP (1973) An estimate of the minimum amount of unsaturated fatty acid required for growth of *Escherichia coli*. J Biol Chem 248:1188–1195

Cronan JE Jr, Rock CO (1996) Biosynthesis of membrane lipids. In: Neidhardt FC, Curtiss R III, Ingraham JL, Lin ECC, Low KB, Magasanik B, Reznikoff WS, Riley M, Schaechter M, Umbarger HE (eds) *Escherichia coli* and *Salmonella*: cellular and molecular biology, 2nd edn. ASM, Washington, DC

Cronan JE Jr, Birge CH, Vagelos PR (1969) Evidence for two genes specifically involved in unsaturated fatty acid biosynthesis in *Escherichia coli*. J Bacteriol 100:601–604

Cybulski L, Albanesi D, Mansilla MC, Altabe S, Aguilar PS, de Mendoza D (2002) Mechanism of membrane fluidity optimization in *Bacillus subtilis*: isothermal control of the acyl-lipid desaturase. Mol Microbiol 45:1379–1388

Cybulski L, del Solar G, Craig PO, Espinosa M, de Mendoza D (2004) *Bacillus subtilis* DesR functions as a phosphorylation-activated switch to control membrane lipid fluidity. J Biol Chem 279:39340–39347

Cybulski LE, Martín M, Mansilla MC, Fernandez A, de Mendoza D (2010) Membrane thickness cue for cold sensing in a bacterium. Curr Biol 20:1539–1544

de Mendoza D, Cronan JE Jr (1983) Thermal regulation of membrane lipid fluidity in bacteria. Trends Biochem Sci 8:49–52

Díaz AR, Mansilla MC, Vila AJ, de Mendoza D (2002) Membrane topology of the acyl-lipid desaturase from *Bacillus subtilis*. J Biol Chem 277:48099–48106

Dowhan W (1997) Molecular basis for membrane phospholipid diversity: why are there so many lipids? Annu Rev Biochem 66:199–232

Dubnau E, Chan J, Raynaud C, Mohan VP, Laneelle MA, Yu K, Quemard A, Smith I, Daffe M (2000) Oxygenated mycolic acids are necessary for virulence of *Mycobacterium tuberculosis* in mice. Mol Microbiol 36:630–637

Dyer DH, Lyle KS, Rayment I, Fox BG (2005) X-ray structure of putative acyl-ACP desaturase DesA2 from *Mycobacterium tuberculosis* H37Rv. Protein Sci 14:1508–1517

Fulco AJ, Levy R, Bloch K (1964) The biosynthesis of Δ9 and Δ5-monounsaturated fatty acids by bacteria. J Biol Chem 239:998–1003

Fulco AJ (1974) Metabolic alterations of fatty acids. Annu Rev Biochem 43:215–241

Garwin JL, Cronan JE Jr (1980) Thermal modulation of fatty acid synthesis in *Escherichia coli* does not involve de novo enzyme synthesis. J Bacteriol 141:1457–1459

Garwin JL, Klages AL, Cronan JE Jr (1980a) Beta-ketoacyl-acyl carrier protein synthase II of *Escherichia coli*. Evidence for function in the thermal regulation of fatty acid synthesis. J Biol Chem 255:3263–3265

Garwin JL, Klages AL, Cronan JE Jr (1980b) Structural, enzymatic, and genetic studies of β-ketoacyl-acyl carrier protein synthases I and II of *Escherichia coli*. J Biol Chem 255:11949–11956

Grau R, de Mendoza D (1993) Regulation of the synthesis of unsaturated fatty acids by growth temperature in *Bacillus subtilis*. Mol Microbiol 8:535–542

Grau R, Gardiol D, Glikin GC, de Mendoza D (1994) DNA supercoiling and thermal regulation of unsaturated fatty acid synthesis in Bacillus subtilis. Mol Microbiol 11:933–941

Heath RJ, White SW, Rock CO (2001) Lipid biosynthesis as a target for antibacterial agents. Prog Lipid Res 40:467–497

Heird WC, Lapillonne A (2005) The role of essential fatty acids in development. Annu Rev Nutr 25:549–571

Hitz WD, Carlson TJ, Booth JR, Kinney AJ, Stecca KL, Yadav NS (1994) Cloning of a higher-plant plastid omega-6 fatty acid desaturase cDNA and its expression in a cyanobacterium. Plant Physiol 105:635–641

Hoang TT, Schweizer HP (1997) Fatty acid biosynthesis in *Pseudomonas aeruginosa*: cloning and characterization of the *fabAB* operon encoding beta-hydroxyacyl-acyl carrier protein dehydratase (FabA) and beta-ketoacyl-acyl carrier protein synthase I (FabB). J Bacteriol 179:5326–5332

Isabella VM, Clark VL (2011) Identification of a conserved protein involved in anaerobic unsaturated fatty acid synthesis in Neiserria gonorrhoeae: implications for facultative and obligate anaerobes that lack FabA. Mol Microbiol 82:489–501

Jeamton W, Mungpakdee S, Sirijuntarut M, Prommeenate P, Cheevadhanarak S, Tanticharoen M, Hongsthong A (2008) A combined stress response analysis of *Spirulina platensis* in terms of global diferentially expressed proteins, and mRNA levels and stability of fatty acid biosynthesis genes. FEMS Microbiol Lett 281:121–131

Jiang Y, Vasconcelles MJ, Wretze S, Light A, Martin CE, Goldberg MA (2001) MGA2 is involved in the low-oxygen response element-dependent hypoxic induction of genes in *Saccharomyces cerevisiae*. Mol Cell Biol 21:6161–6169

Kaiser HJ, Surma MA, Mayer F, Levental I, Grzybek M, Klemm RW, Da Cruz S, Meisinger C, Müller V, Simons K, Lingwood D (2011) Molecular convergence of bacterial and eukaryotic surface order. J Biol Chem 286:40631–40637

Kaneda T (1977) Fatty acids of the genus *Bacillus*: an example of branched chain preference. Bacteriol Rev 41:391–418

Kaneda T (1991) Iso- and anteiso-fatty acids in bacteria: biosynthesis, function, and taxonomic significance. Microbiol Rev 55:288–302

Kanesaki Y, Yamamoto H, Paithoonrangsarid K, Shoumskaya M, Suzuki I, Hayashi H, Murata N (2007) Histidine kinases play important roles in the perception and signal transduction of hydrogen peroxide in the cyanobacterium *Synechocystis* sp. PCC 6803. Plant J 49:313–324

Kappell AD, van Waasbergen LG (2007) The response regulator RpaB binds the high light regulatory 1 sequence upstream of the high-light-inducible hliB gene from the cyanobacterium *Synechocystis* PCC 6803. Arch Microbiol 187:337–342

Klein W, Weber MHW, Marahiel MA (1999) Cold shock response of *Bacillus subtilis*: isoleucine-dependent switch in the fatty acid branching pattern for membrane adaptation to low temperatures. J Bacteriol 181:5341–5534

Lawson RJ, von Wachenfeldt C, Haq I, Perkins J, Munro AW (2004) Expression and characterization of the two flavodoxin proteins of *Bacillus subtilis*, YkuN e YkuP: biophysical properties and interactions with cytochrome P450 BioI. Biochemistry 43:12390–12409

Lee AG (2011) Biological membranes: the importance of molecular detail. Trends Biochem Sci 36:493–500

Lindqvist Y, Huang W, Schneider G, Shanklin J (1996) Crystal structure of delta 9 stearoyl-acyl carrier protein desaturase from castor seed and its relationship to other di-iron proteins. EMBO J 15:4081–4092

Los DA (2004) The effect of low-temperature-induced DNA supercoiling on the expression of the desaturase genes in Synechocystis. Cell Mol Biol (Noisy-le-Grand) 50:605–612

Los DA, Murata N (1998) Structure and expression of fatty acid desaturases. Biochim Biophys Acta 1394:1–15

Los DA, Murata N (1999) Responses to cold shock in cyanobacteria. J Mol Microbiol Biotechnol 2:221–230

Los DA, Murata N (2004) Membrane fluidity and its roles in the perception of environmental signals. Biochim Biophys Acta 1666:142–157

Los DA, Horvath I, Vigh L, Murata N (1993) The temperature-dependent expression of the desaturase gene desA in *Synechocystis* PCC6803. FEBS Lett 318:57–60

Los DA, Ray MK, Murata N (1997) Differences in the control of the temperature-dependent expression of four genes for desaturases in *Synechocystis* sp. PCC 6803. Mol Microbiol 25:1167–1175

Lu YJ, Rock CO (2006) Transcriptional regulation of fatty acid biosynthesis in Streptococcus pneumoniae. Mol Microbiol 59:551–566

Man WC, Miyazaki M, Chu K, Ntambi JM (2006) Membrane topology of mouse stearoyl-CoA desaturase 1. J Biol Chem 281:1251–1260
Mansilla MC, de Mendoza D (2005) The *Bacillus subtilis* desaturase: a model to understand phospholipid modification and temperature sensing. Arch Microbiol 183:229–235
Mansilla MC, Banchio CE, de Mendoza D (2008) Signalling pathways controlling fatty acid desaturation. In: Quinn W (ed) Lipids in health and disease. Springer Science+Business Media B.V., Berlin
Martin M, Albanesi D, Alzari PM, de Mendoza D (2009a) Functional in vitro assembly of the integral membrane bacterial thermosensor DesK. Protein Expr Purif 66:39–45
Martin N, Lombardía E, Altabe SG, de Mendoza D, Mansilla MC (2009b) A lipA (yutB) mutant, encoding lipoic acid synthase, provides insight into the interplay between branched-chain and unsaturated fatty acid biosynthesis in Bacillus subtilis. J Bacteriol 191:7447–7455
Marrakchi H, Choi KH, Rock CO (2002) A new mechanism for anaerobic unsaturated fatty acid formation in *Streptococcus pneumoniae*. J Biol Chem 277:44809–44816
Mikami K, Kanesaki Y, Suzuki I, Murata N (2002) The histidine kinase Hik33 perceives osmotic stress and cold stress in *Synechocystis* sp PCC 6803. Mol Microbiol 46:905–915
Miller DJ, Zhang YM, Subramanian C, Rock CO, White SW (2010) Structural basis for the transcriptional regulation of membrane lipid homeostasis. Nat Struct Mol Biol 17:971–975
Mironov KS, Sidorov RA, Trofimova MS, Bedbenov VS, Tsydendambaev VD, Allakhverdiev SI, Los DA (2012) Light-dependent cold-induced fatty acid unsaturation, changes in membrane fluidity, and alterations in gene expression in *Synechocystis*. Biochim Biophys Acta 1817:1352–1359
Murata N, Suzuki I (2006) Exploitation of genomic sequences in a systematic analysis to access how cyanobacteria sense environmental stress. J Exp Bot 57:235–247
Murata N, Wada H (1995) Acyl-lipid desaturases and their importance in the tolerance and acclimatization to cold of cyanobacteria. Biochem J 308:1–8
Murata N, Wada H, Gombos Z (1992) Modes of fatty-acid desaturation in cyanobacteria. Plant Cell Physiol 33:933–941
Napier JA (2007) The production of unusual fatty acids in transgenic plants. Annu Rev Plant Biol 58:295–319
Nishida I, Murata N (1996) Chilling sensitivity in plants and cyanobacteria: the crucial contribution of membrane lipids. Annu Rev Plant Physiol Plant Mol Biol 47:541–568
Ntambi JM (1995) The regulation of stearoyl-CoA desaturase (SCD). Prog Lipid Res 34:139–150
Ntambi JM, Miyazaki M, Stoehr JP, Lan H, Kendziorski CM, Yandell BS, Song Y, Cohen P, Friedman JM, Attie AD (2002) Loss of stearoyl-CoA desaturase-1 function protects mice against adiposity. Proc Natl Acad Sci 99:11482–11486
Paithoonrangsarid K, Shoumskaya MA, Kanesaki Y, Satoh S, Tabata S, Los DA, Zinchenko VV, Hayashi H, Tanticharoen M, Suzuki I, Murata N (2004) Five histidine kinases perceive osmotic stress and regulate distinct sets of genes in Synechocystis. J Biol Chem 279:53078–53086
Phadtare S (2004) Recent developments in bacterial cold-shock response. Curr Issues Mol Biol 6:125–136
Phetsuksiri B, Jackson M, Scherman H, McNeil M, Besra GS, Baulard AR, Slayden RA, DeBarber AE, Barry CE 3rd, Baird MS, Crick DC, Brennan PJ (2003) Unique mechanism of action of the thiourea drug isoxyl on *Mycobacterium tuberculosis*. J Biol Chem 278:53123–53130
Prakash JS, Sinetova M, Zorina A, Kupriyanova E, Suzuki I, Murata N, Los DA (2009) DNA supercoiling regulates the stress-inducible expression of genes in the cyanobacterium *Synechocystis*. Mol Biosyst 5:1904–1912
Reddy AS, Nuccio ML, Gross LM, Thomas TL (1993) Isolation of a delta 6-desaturase gene from the cyanobacterium *Synechocystis* sp. strain PCC 6803 by gain-of-function expression in *Anabaena* sp. strain PCC 7120. Plant Mol Biol 22:293–300
Sakamoto T, Murata N (2002) Regulation of desaturation of fatty acids and its role in tolerance to cold and salt stress. Curr Opin Microbiol 5:208–210

Sakamoto T, Los DA, Higashi S, Wada H, Nishida I, Ohmori M, Murata N (1994a) Cloning of $\omega3$ desaturase from cyanobacteria and its use in altering the degree of membrane-lipid unsaturation. Plant Mol Biol 26:249–263

Sakamoto T, Wada H, Nishida I, Ohmori M, Murata N (1994b) Delta 9 acyl-lipid desaturases of cyanobacteria. Molecular cloning and substrate specificities in terms of fatty acids, sn-positions, and polar head groups. J Biol Chem 269:25576–25580

Sakamoto T, Wada H, Nishida I, Ohmori M, Murata N (1994c) Identification of conserved domains in the delta 12 desaturases of cyanobacteria. Plant Mol Biol 24:643–650

Sassetti CM, Rubin EJ (2003) Genetic requirements for mycobacterial survival during infection. Proc Natl Acad Sci U S A 100:12989–12994

Sayanova O, Shewry PR, Napier JA (1999) Histidine-41 of the cytochrome b5 domain of the borage delta6 fatty acid desaturase is essential for enzyme activity. Plant Physiol 121:641–646

Scanlan D (2001) Cyanobacteria: ecology, niche adaptation and genomics. Microbiol Today 28:128–130

Scheuerbrandt G, Bloch K (1962) Unsaturated fatty acids in microorganisms. J Biol Chem 237:2064–2068

Scheuerbrandt G, Goldfine H, Baronowsky PE, Bloch K (1961) A novel mechanism for the biosynthesis of unsaturated fatty acids. J Biol Chem 236:PC70–PC71

Schweizer HP, Choi KH (2011) Pseudomonas aeruginosa aerobic fatty acid desaturase DesB is important for virulence factor production. Arch Microbiol 193:227–234

Schweizer HP, Choi KH (2012) Characterization of molecular mechanisms controlling fabAB transcription in Pseudomonas aeruginosa. PLoS One 7:e45646. doi:10.1371/journal.pone.0045646

Shanklin JE, Cahoon EB (1998) Desaturation and related modifications of fatty acids. Annu Rev Plant Physiol Plant Mol Biol 49:611–641

Shanklin J, Somerville C (1991) Stearoyl-acyl-carrier-protein desaturase from higher plants is structurally unrelated to the animal and fungal homologs. Proc Natl Acad Sci U S A 88:2510–2514

Shanklin J, Whittle E, Fox BG (1994) Eight histidine residues are catalytically essential in a membrane-associated iron enzyme, stearoyl-CoA desaturase, and are conserved in alkane hydroxylase and xylene monooxygenase. Biochemistry 33:12787–12794

Shoumskaya MA, Paithoonrangsarid K, Kanesaki Y, Los DA, Zinchenko VV, Tanticharoen M, Suzuki I, Murata N (2005) Identical Hik-Rre systems are involved in perception and transduction of salt signals and hyperosmotic signals but regulate the expression of individual genes to different extents in synechocystis. J Biol Chem 280:21531–21538

Sperling P, Ternes P, Zank TK, Heinz E (2003) The evolution of desaturases. Prostaglandins Leukot Essent Fatty Acids 68:73–95

Stukey JE, McDonough VM, Martin CE (1990) The OLE1 gene of *Saccharomyces cerevisiae* encodes the delta 9 fatty acid desaturase and can be functionally replaced by the rat stearoyl-CoA desaturase gene. J Biol Chem 265:20144–20149

Subramanian C, Rock CO, Zhang YM (2010) DesT coordinates the expression of anaerobic and aerobic pathways for unsaturated fatty acid biosynthesis in Pseudomonas aeruginosa. J Bacteriol 192:280–285

Suzuki I, Los DA, Kanesaki Y, Mikami K, Murata N (2000) The pathway for perception and transduction of low-temperature signals in *Synechocystis*. EMBO J 19:1327–1334

Suzuki I, Kanesaki Y, Mikami K, Kanehisa M, Murata N (2001) Cold-regulated genes under control of the cold sensor Hik33 in Synechocystis. Mol Microbiol 40:235–244

Tasaka Y, Gombos Z, Nishiyama Y, Mohanty P, Ohba T, Ohki K, Murata N (1996) Targeted mutagenesis of acyl-lipid desaturases in *Synechocystis*: evidence for the important roles of polyunsaturated membrane lipids in growth, respiration and photosynthesis. EMBO J 15:6416–6425

Tomioka H, Tatano Y, Yasumoto K, Shimizu T (2008) Recent advances in antituberculous drug development and novel drug targets. Expert Rev Respir Med 2:455–471

Uttaro AD (2006) Biosynthesis of polyunsaturated fatty acids in lower eukaryotes. IUBMB Life 58:563–571

Van Beilen JB, Penning D, Witholt B (1992) Topology of the membrane-bound alkane hydroxylase of *Pseudomonas oleovorans*. J Biol Chem 267:9194–9201

Van Delden C, Iglewski BH (1998) Cell-to-cell signaling and *Pseudomonas aeruginosa* infections. Emerg Infect Dis 4:551–560

Vasconcelles MJ, Jiang Y, McDaid K, Gilooly L, Wretzel S, Porter DL, Martin CE, Goldberg MA (2001) Identification and characterization of a low oxygen response element involved in the hypoxic induction of a family of *Saccharomyces cerevisiae* genes. Implications for the conservation of oxygen sensing in eukaryotes. J Biol Chem 276:14374–14384

Wada H, Gombos Z, Murata N (1990) Enhancement of chilling tolerance of a cyanobacterium by genetic manipulation of fatty acids desaturation. Nature 347:200–203

Wang H, Cronan JE (2004) Functional replacement of the FabA and FabB proteins of *Escherichia coli* fatty acid synthesis by *Enterococcus faecalis* FabZ and FabF homologues. J Biol Chem 279:34489–34495

Watanabe M, Aoyagi Y, Mitome H, Fujita T, Naoka H, Ridell M, Minnikin DE (2002) Location of functional groups in mycobacterial meromycolate chains; the recognition of new structural principles in mycolic acids. Microbiology 148:1881–1902

Weber MHW, Klein W, Muller L, Niess UM, Marahiel M (2001) Role of the *Bacillus subtilis* fatty acid desaturase in membrane adaptation during cold shock. Mol Microbiol 39:1321–1329

Yadav NS, Wierzbicki A, Aegerter M, Caster CS, Perez-Grau L, Kinney AJ, Hitz WD, Booth JR Jr, Schweiger B, Stecca KL, Allen SM, Blackwell M, Reiter RS, Carlson TJ, Russell SH, Feldmann KA, Pierce J, Browse J (1993) Cloning of higher plant omega-3 fatty acid desaturases. Plant Physiol 103:467–476

Yuan Y, Lee RE, Besra GS, Belisle JT, Barry CE 3rd (1995) Identification of a gene involved in the biosynthesis of cyclopropanated mycolic acids in *Mycobacterium tuberculosis*. Proc Natl Acad Sci 92:6630–6634

Zhang YM, Rock CO (2008) Membrane lipid homeostasis in bacteria. Nat Rev Microbiol 6:222–233

Zhang YM, Zhu K, Frank MW, Rock CO (2007) A Pseudomonas aeruginosa transcriptional regulator which senses fatty acid structure. Mol Microbiol 66:622–632

Zhu K, Choi KH, Schweizer HP, Rock CO, Zhang YM (2006) Two aerobic pathways for the formation of unsaturated fatty acids in Pseudomonas aeruginosa. Mol Microbiol 60:260–273

Index

J.M. Ntambi (ed.), *Stearoyl-CoA Desaturase Genes in Lipid Metabolism*,
DOI 10.1007/978-1-4614-7969-7, © Springer Science+Business Media New York 2013

MIX
Papier aus verantwortungsvollen Quellen
Paper from responsible sources
FSC® C105338
www.fsc.org

If you have any concerns about our products,
you can contact us on
ProductSafety@springernature.com

In case Publisher is established outside the EU,
the EU authorized representative is:
Springer Nature Customer Service Center GmbH
Europaplatz 3, 69115 Heidelberg, Germany

Printed by Libri Plureos GmbH
in Hamburg, Germany